Defending Illusions

The Political Economy Forum

Sponsored by the Political Economy Research Center (PERC)
Series Editor: Terry L. Anderson

Land Rights: The 1900s' Property Rights Rebellion
edited by Bruce Yandle

The Political Economy of Customs and Culture: Informal Solutions to the Commons Problem
edited by Terry L. Anderson and Randy T. Simmons

The Political Economy of the American West
edited by Terry L. Anderson and Peter J. Hill

Property Rights and Indian Economies
edited by Terry L. Anderson

Public Lands and Private Rights: The Failure of Scientific Management
by Robert H. Nelson

Taking the Environment Seriously
edited by Roger E. Meiners and Bruce Yandle

Wildlife in the Marketplace
edited by Terry L. Anderson and Peter J. Hill

The Privatization Process: A Worldwide Perspective
edited by Terry L. Anderson and Peter J. Hill

Enviro-Capitalists: Doing Good While Doing Well
by Terry L. Anderson and Donald R. Leal

Environmental Federalism
edited by Terry L. Anderson and Peter J. Hill

Defending Illusions: Federal Protection of Ecosystems
by Allan K. Fitzsimmons

The Market Meets the Environment: Economic Analysis of Environmental Policy
edited by Bruce Yandle

Defending Illusions

Federal Protection of Ecosystems

Allan K. Fitzsimmons

ROWMAN & LITTLEFIELD PUBLISHERS, INC.
Lanham • Boulder • New York • Oxford

ROWMAN & LITTLEFIELD PUBLISHERS, INC.

Published in the United States of America
by Rowman & Littlefield Publishers, Inc.
4720 Boston Way, Lanham, Maryland 20706

12 Hid's Copse Road
Cumnor Hill, Oxford OX2 9JJ, England

British Library Cataloguing in Publication Information Available

Library of Congress Cataloging-in-Publication Data

Fitzsimmons, Allan K.
Defending illusions : federal protection of ecosystems / Allan K. Fitzsimmons.
p. cm.—(The political economy forum)
Includes bibliographical references.
ISBN 0-8476-9421-6 (cloth : alk. paper).—ISBN 0-8476-9422-4 (pbk. : alk. paper)
1. Ecosystem management—Government policy—United States. I. Title. II. Series.
QH76.F58 1999
577′.0973—dc21 99-17318
CIP

Printed in the United States of America

∞™ The paper used in this publication meets the minimum requirements of American National Standard for Information Sciences—Permanence of Paper for Printed Library Materials, ANSI/NISO Z39.48—1992.

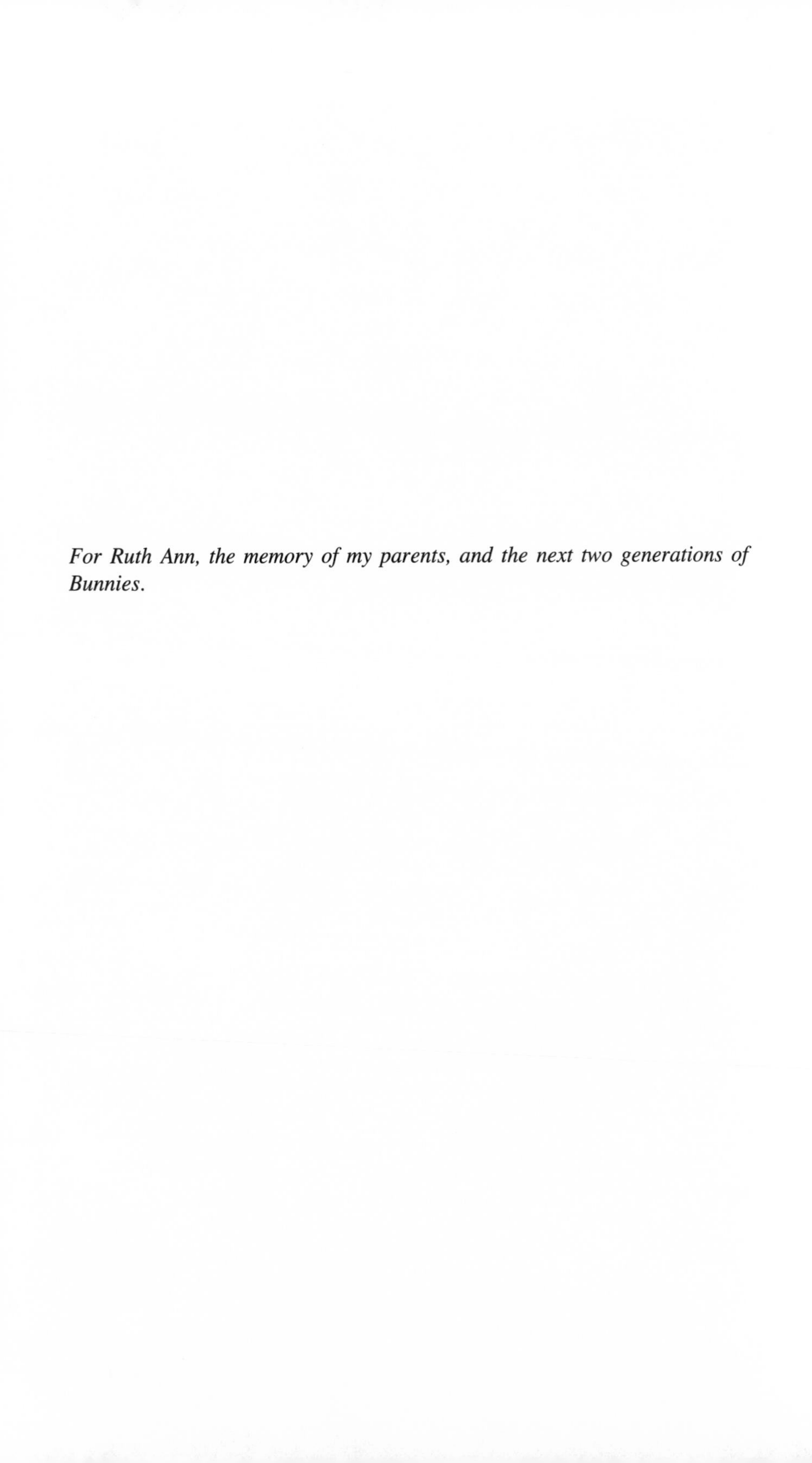

For Ruth Ann, the memory of my parents, and the next two generations of Bunnies.

Contents

Maps and Tables

Maps

Tables

Preface

When Dot and Fitz took their young son to Yosemite Valley, the Giant Forest, Death Valley, and the Grand Canyon in the early 1950s, they had no idea they were planting the seeds for this book. Of course, I did not know it either. But over time one thing led to another. A boy's awe at the beauty of those places fostered a love of geography that turned into degrees and university professorships. My growing interest in public policies governing the land led me to spend nearly a decade working as an advisor to senior decision makers in the U.S. Departments of Interior (DOI) and Energy (DOE). As professor and staffer I witnessed growing efforts to forsake Americans' long-standing balanced approach to natural resource policy in favor of one elevating protection of Mother Nature above all other government policy goals. I became aware that too few people grasp the societal implications of the new policy or have a clear picture of its foundation. I write this book to increase public understanding of ecosystem management—what its champions hail as the new paradigm, which is intended to put aright what they see as Americans' present unethical relationship with the earth.

Many people helped me in my journey leading to this book. Joe Spencer and Tom McKnight at UCLA tirelessly taught me the geographer's craft and the need for diligence in research. Faculty colleagues and my students provided sounding boards and challenged arguments as did my fellow participants in the Keystone Center's National Dialogue on Ecosystem Management. They, and many career federal professionals with whom I was privileged to work, gave me valuable insights and revealed perspectives I had not previously considered or adequately explored. Mary Lou Grier (as deputy director of the National Park Service), Bill Horn (as assistant secretary for fish, wildlife, and parks in DOI), and Linda Stuntz (as deputy undersecretary for policy, planning, and analysis at DOE) gave me opportunities to evaluate natural resource policies and participate in their development.

With regard to the book, Terry Anderson of the Political Economy Research Center (PERC) and Steve Wrinn of Rowman & Littlefield blessed the project based on a simple outline. Antony Sullivan and the Earhart Foundation provided financial support. Staff at the Fenwick Library of George Mason University and the Chinn Park Regional Library of Prince William County aided in tracking down books, articles, and various other research materials. Jim Lacy of the National Wilderness Institute helped with the maps. Rick Stroup of PERC read the manuscript and made valuable suggestions. Our daughter Shannon reviewed all the draft chapters. I found her comments invaluable, especially those that helped me focus and clarify arguments. She retrained her father to write in the active voice, no small feat on her part after I spent decades writing in a passive academic style. Should anyone think this book reasonably well written, much of the credit is hers. Despite all this help, however, I am solely responsible for any errors that readers may find.

Just as authors do not write in a vacuum, they do not live in one. My wife Ruth Ann is a gift from God, and I owe her more than I can recount or repay. Without her, this book would have remained but an idea.

The Plan of This Book

Scholars, government officials, land users, and others do not agree on one definition of ecosystem management. Some observers believe that it is simply the next incremental step in the evolution of traditional environmental and land use policies. Others see it as the new paradigm, as a once-in-a-century opportunity to get our relations with nature right. I begin the introduction by looking at different views of ecosystem management and then compare the new paradigm with traditional approaches to natural resource management in the United States. Beginning with the Reagan presidency, I show that successive administrations increasingly embraced ecosystem management in broad terms but that it was not endorsed in its new paradigm form until the arrival of the Clinton administration. The introduction concludes by showing the reader the vagueness of the new paradigm and how it portends a greatly expanded role for the federal government in resource and land use decision making throughout the nation.

The first three chapters lead the reader to a firm grasp of what the ecosystem concept means when people take it out of the textbook and apply it to the landscape. The public, activists, company executives, journalists, teachers, students, professors, and policymakers all use the term *ecosystem management,* often in some variation of a phrase such as "we are trying to maintain the ecosystem" or "the ecosystem is dying." Such usage betrays their poor understanding of the concept, a point that I illustrate by looking at how journalists often treat ecosystems. Chapter 1 shows what ecosystems are and are not. Scientific definitions and practice reveal ecosystems to be geographic fictions that researchers construct to facilitate some human inquiry. Ecosystems are not objective realities, not living and discrete entities on the landscape. Thus, for example, ecosystems cannot become sick or die, as they do not live in the first place. This is a vital point for Americans to understand because proponents of the new paradigm frequently assert that ecosystems need healing as justification for government intervention on their behalf. In the new paradigm the government will use ecosystems as geographic units to guide the application of government rules and regulations, so I compare the spatial attributes of ecosystems to those of other geographic units of governance, namely, states and counties. This comparison illustrates the land use chaos that Americans would be subjected to if states and counties had the same spatial characteristics as ecosystems.

In chapter 2, I focus on ecosystem mapping because if ecosystems are to become the geographic guide for government polices, then ecosystem maps become indispensable arbiters of where regulators and land managers will permit or forbid activities to occur. The chapter looks at the common characteristics of all maps and then at ecosystem maps in particular. I analyze the variables (e.g., climate, vegetation, and soils) that cartographers use in producing ecosystem maps for their objectivity and then review maps of these phenomena for their precision. I compare four national ecosystem maps that portray quite different ecosystem patterns for the nation.

The Greater Yellowstone Ecosystem (GYE) is the subject of chapter 3. People began to float the idea of the GYE in the 1970s. In the intervening years, scientists, environmental activists, policymakers, and others have attempted to define it, and in chapter 3, I further demonstrate the vague and arbitrary nature of the ecosystem concept by comparing several maps of the GYE and the variables that lie behind them. I offer a glimpse at the new paradigm world by examining the first attempt at large-scale federal ecosystem management, the development of a "Vision for the Greater Yellowstone Ecosystem."

Supporters of ecosystem management portray it as the morally and scientifically correct means for avoiding ecological disaster. The next three chapters help the reader assess these claims. Chapter 4 lays out America's present level of societal commitment to environmental protection. I outline how Americans use land nationally and go on to survey the condition of the environment in the United States, giving particular attention to biological diversity, as its protection is routinely interwoven with calls for ecosystem management by scientists, activists, and others. In chapter 4, I identify and assess the standard champions of ecosystem management use in making their proclamations of portending environmental doom. They derive this standard—deviation from so-called natural conditions—not from the findings of scientists but from a worldview in which, for some, nature is a source of goodness and wisdom and, for others, deserving of worship and veneration as Mother Earth.

Scholars understand that environmentalism is rooted in a religious veneration of nature. In chapter 5, I trace the religious underpinnings of ecosystem management from the work of John Muir through contemporary environmental prophets. The chapter illustrates that biocentrism—the driving force of today's environmental movement—is the melding of nature worship with fuzzy science. I relate the extent to which Congress includes devotion to nature in federal statutes. I end the chapter by looking at the nature of nature. Is nature a beneficent goddess dispensing blessings from the wilderness, as some would have us believe, or is it simply a collection of forces, processes, and biota that can bring death and misery to ourselves and other species as well as offer opportunities for sustenance?

Chapter 6 addresses the status of ecosystem science. Supporters of ecosystem management claim that it is firmly based on science, but what is the state of that science? I answer this question in this chapter. The term *ecosystem* dates from 1935, for example, but researchers have yet to produce a cogent body of theory that provides a foundation for accurate ecological predictions, and scientists do not agree on fundamental issues, such as identifying the core characteristics of ecosystems. Advocates call for the government to protect ecosystem health, integrity, and sustainability, but do these concepts have wide acceptance among scientists, and do they possess substantive and agreed-on meanings in the research community? I analyze claims made by researchers that ecosystems provide services that benefit humans as well as the oft-repeated contention that healthy economies depend on healthy ecosystems.

Chapter 7 explores ecosystem protection in existing laws and treaties. Congress puts the word *ecosystem* into more than 100 federal statutes but in doing so does not give generic authority for ecosystem protection to either government regulators or federal land managers. At the direction of administration officials, however, bureaucrats are pushing forward on many fronts to make ecosystem protection the basis for federal environmental and land management policies. I examine two such efforts closely: One would make ecosystem protection the centerpiece for managing the national forest system nationwide, and a second would impose ecosystem management on 75 million acres of public land in Oregon, Washington, Idaho, Montana, Wyoming, Utah, and Nevada. In addition to domestic programs, new paradigmists seek to use international agreements to protect ecosystems within the United States. Consequently, I explore the significance of the World Heritage Convention, the Man and the Biosphere Programme, and especially the UN Convention on Biological Diversity for ecosystem protection in the United States.

New paradigmists advance several proposals to abet their cause; these are the subjects of chapter 8. The Wildlands Project, for example, would place 50 percent of the contiguous states into wilderness or near wilderness status in order to protect ecosystems. Inspired by the Endangered Species Act, some advocates of the new paradigm propose an endangered ecosystems act while others call for a constitutional amendment to protect ecosystems. I examine arguments made for these initiatives by their proponents in the light of earlier analyses of ecosystem characteristics, the state of ecosystem science, and ecosystem protection as a manifestation of nature-based religion. Concerns expressed by members of Congress and others about these efforts serve as a transition to the last two chapters of the book, which address the societal implications of the new paradigm and point out a way for Americans to continue to enhance the quality of our environment without sacrificing authority to central government or improvements in human well-being.

Chapter 9 points out that, if the revolution succeeds, federal regulators and land managers would be responsible for national land use planning. I explain what such a policy means for the expansion of government regulatory authority over land and resource use nationwide and how it would concentrate power into the hands of the federal bureaucracy and narrow special interests. I make it clear that in a new paradigm world, Americans would surrender individual freedoms in order to protect Mother Earth. I reveal how new paradigmist legal scholars are already constructing arguments to use federal protection of ecosystems as a way to get around the Takings Clause of the Fifth Amendment to the Constitution. If they can persuade the judiciary to accept their analysis, then federal regulators would be free to engage in national land use planning unfettered by concerns that judges would require government to pay for what it took from landowners. An analysis of Section 404 of the Clean Water Act provides a glimpse of the impact of new paradigm thinking on human well-being.

The overwhelming weight of the evidence demonstrates that the quality of the American environment is good and getting better. As with all human endeavors, however, there is always room for improvement. Chapter 10 suggests how we can become better stewards without sacrificing continuing improvement in the human condition. I argue that for policies to be politically sustainable, the chief beneficiaries must bear the greatest burden of the costs. It is unfair to shift the cost of providing public benefits to a relatively few unlucky landowners. It is even more unjust to expect private landowners to shoulder the costs of providing benefits to free-riding new paradigmists. Effective and enduring policies are those that provide for balance among agreed-on societal objectives and that harness the strengths of American society: democratic processes, individual freedom, decentralized and market-driven decision making, and protection of private property rights. To illustrate these points, I present a market-based approach to preserving the nation's wetlands base.

In this book I contend that in our efforts to become worthier keepers of the land, knowledge and understanding must guide us rather than the nature-worshiping mysticism of biocentrists or the false claims of environmental upheaval that thrive only in the dark of ignorance.

Introduction

Government and the New Relationship with Nature

> Nothing better characterizes the Administration's new approach to natural resource management than the emphasis . . . on managing whole ecosystems. This Administration is seeking new ways . . . to manage whole ecosystems, rather than small disparate pieces of natural habitat.
>
> —President Bill Clinton and Vice President Al Gore, *Budget of the United States for Fiscal Year 1995; "Dear Friend" letter of April 14, 1994*

A revolution is upon us. Its backers want to make protection of ecosystems the number one goal of federal environmental and land management policies. They would replace traditional ideas of using land and natural resources to enhance human well-being with new thinking that places protection of nature at the center of our relationship with the environment. This is the new paradigm intended by its champions to completely restructure how Americans see and interact with nature. Advocates of this new paradigm tell us that we must adopt their views in order to prevent long-proclaimed environmental breakdown. For some, the revolution is a sin offering to their offended Earth Goddess. Should the revolution succeed, and it is well on its way to doing so, the role of the federal government and unelected special interests in our lives would expand in ways never before seen in American history while opportunities to improve the human condition and environmental quality would be forfeited.

Americans favor protecting the environment.[1] The public does not support careless or wasteful interaction with our lands and waters or the living things found there. Over the past several decades, we have enacted thousands of laws, ordinances, and regulations at all levels of government to protect and enhance environmental quality so that today "we have the most comprehensive environmental policy, statutory and regulatory framework of any country in the world."[2] This commitment has yielded substantial improvements in the condition of the environment in the United States. By nearly any measure, environmental quality in the United States is good and getting better (see chapter 4). Environmentalists and other advocates of the new paradigm deserve credit for helping bring about these achievements, but they were accomplished by using our traditional approach to land and natural resource policy, which blends conservation, multiple use, and preservation with a concern for human health.

Despite proven success, new paradigm adherents proclaim that "put simply, the old paradigms are no longer scientifically or politically valid!"[3] The old ways, they tell us, lead "to collapse of life and living as we know it," whereas the new path of "ecocentrism, environmental ethics, and ecosystem management" leads to the just revealed nirvana of sustainability.[4] Those who do not subscribe to the new ethical relationship with nature are addicted to an unhealthy and unharmonious relationship with Mother Earth, according to Vice President Al Gore, who likens the recalcitrant to drug addicts in a state of denial about their addiction.[5]

This book is about ecosystem management—the new paradigm that calls for government to manage and protect ecosystems as real objects on the landscape that possess characteristics such as health, integrity, and sustainability—and what it can mean for Americans of this and future generations. I propose alternatives that enable us to become better stewards of the environment without relinquishing more power to the government or unaccountable special interests and without sacrificing improvements in human well-being and individual liberty.

What Is Ecosystem Management?

Experts concur that scientists, land managers, land users, and others do not agree on one definition of ecosystem management.[6] The Congressional Research Service reports that there is "little agreement" about the goals of ecosystem management.[7] Jack Ward Thomas, the Clinton administration's first chief of the U.S. Forest Service, remarks that "I promise you I can do anything you want to do by saying it is ecosystem management. . . . But right now it's incredibly nebulous."[8] About the term *ecosystem management,* Micheal Bean of the Environmental Defense Fund suggests that "rarely has a term of public discourse gone so directly from obscurity to meaninglessness without any intervening period of coherence."[9] Researchers identify a wide variety of activities that some call *ecosystem management.*[10] Federal agencies are engaged in a 145-million-acre ecosystem management project involving thousands of landowners in a region extending from the Cascade Mountains in Washington and Oregon to the Rocky Mountains in Montana and Wyoming (see chapter 7).[11] In contrast, local government calls an effort by a single landowner in suburban Seattle covering forty acres in the Bear Creek watershed "ecosystem management."[12]

From a policy perspective, some people see ecosystem management as the next logical step in the evolution of traditional approaches to natural resources policy and land management. Still others see ecosystem management more as a process for decision making.[13] The interpretation of ecosystem management that I analyze here sees it as the scientific and moral foundation of a new relationship with Mother Earth. However, before delving into the new paradigm, I outline Americans' traditional approach to natural resources management and touch on ecosystem management as a process.

The traditional paradigm is people centered. With it, policymakers emphasize land use management intended to produce outputs useful to humans: food, fiber, timber, minerals, energy, and other tangible substances as well as intangibles, such as outdoor recreation ranging from motorboating to backpacking. They look to protect human health. Moreover, they also recognize the intrinsic value of other species and special places and the need for sound stewardship in order to pass on a quality environment to future generations. Those who view ecosystem management as an incremental change in traditional resource policy often see it as a better way to do these kinds of things while taking into consideration advances in scientific understanding of the world around us as well as changes in societal attitudes. In December 1995, federal agencies held an ecological stewardship workshop in Tucson, Arizona. Some participants argued that ecosystem management allows them to consider new issues, such as ecosystem health, integrity, and sustainability, as well as factors, such as biological diversity. However, in contrast with the new paradigm agenda, participants did not necessarily consider these items important as goals in their own right; rather, they provide a better way to reach traditional ends, such as the production of forest products, watershed protection, and the provision of outdoor recreation opportunities. The well-being of people remains at the center of things. Robert Lackey of the Environmental Protection Agency offers a definition of ecosystem management that captures this view: "the application of ecological and social information, options, and constraints to achieve desired social benefits within a defined geographic area over a specified period."[14]

The traditional American approach to natural resource management and environmental policy evolved over the nineteenth century and is a rich admixture of ideas.[15] It includes conservation, preservation, and protection of public health. Gifford Pinchot, the first chief of the Forest Service, espoused conservation and wise use of natural resources, as did President Teddy Roosevelt and others around the turn of the twentieth century. Pinchot credits WJ McGee with coining the catch phrase that came to describe conservation, that is, "the greatest good for the greatest number for the longest time."[16] The idea includes consideration of intergenerational obligation and preservation but is fundamentally utilitarian and about the wise use of resources to benefit present generations. Pinchot sought to make clear the meaning of conservation in his book *The Fight for Conservation*.[17] Here Pinchot emphatically notes that while conservation includes "provision for the future . . . it means also and first of all the recognition of the right of the present generation to the fullest necessary use of all resources with which this country is so abundantly blessed." The first principle of conservation, according to Pinchot, is development: the use of natural resources to enhance the human condition. The second principle is the prevention of waste that human or natural actions cause. His time was a period of enormous carelessness in resource use. For example, the thoughtless destruction brought about both by ill-considered timber cutting practices and by forest fires repelled conservationists. The third principle is that the benefits of natural resources, whether they result from development or preservation, should flow to the people rather than merely be generated "for the profit of a few." Here he is concerned about the giving of public lands and natural resources to monopolists.

Another giant of that time was John Muir, founder of the Sierra Club. Muir was a staunch advocate of preservation: the setting aside of areas to be free of forestry, grazing, and other developmental pursuits. He led the effort to establish Yosemite National Park. Muir revered nature and saw such reservations as temples, but the government established them, in Muir's view, as places were people should come and partake of nature's gifts. In 1901, he wrote that

> the tendency nowadays to wander in wildernesses is delightful to see . . . tired, nerve-shaken, overcivilized people are beginning to find out that going to the mountains is going home; that wildness is a necessity . . . [because it allows them to] enrich their own little ongoings with Nature . . . [by] washing off sins and cobwebs of the devil's spinning . . . [through] getting in touch with the nerves of Mother Earth.[18]

In his view, wilderness visitors were to go "jumping from rock to rock, feeling the life of them; learning the songs of them."[19]

The basic tenets of federal land management and wildlife management policy find their roots in the ideas of Pinchot and Muir. On the one hand we have sustained yield and multiple use, and on the other we have preservation. Congress directs the Forest Service and the Bureau of Land Management to manage much of the public land in such a fashion as to produce a continuous flow—sustained yield—of renewable resources such as timber, forage, and game. The public can use their commonly owned lands in many ways (e.g., timber production, energy and mineral resource extraction, generation of hydroelectricity, and outdoor recreation of many kinds). Land managers accommodating multiple uses know that it requires making trade-offs. Camping and oil production are not compatible uses of the same piece of ground, although they can coexist nicely in the same neighborhood. Critics of multiple use claim that noncommodity uses such as recreation get short shrift under multiple-use management, while others say that sustained yield, at least for some kinds of renewable resources such as fisheries or at maximum levels, is not really possible because managers use outmoded understandings of ecosystems in that effort.[20] Those who view ecosystem management as an evolutionary change in traditional policies believe that it would help address these concerns.

Muir's legacy is equally evident. At the beginning of 1997, our national park system contained 374 units covering nearly 81 million acres, and over 103 million acres of public lands were set aside as wilderness. According the Wilderness Act, these areas are to be places "where the earth and its community of life are untrammeled by man, where man himself is a visitor and does not remain."[21] The "wilderness" classification prohibits economic development as well as the building of roads and other trappings of modern society. The influence of Muir and others is also evident in statutes and policies that protect nature and its components for their intrinsic value regardless of who may own the land, as with the Endangered Species Act or the federal protection given to wetlands through Section 404 of the Clean Water Act.[22]

Americans' concerns for public health provide another component of the traditional approach to natural resource and environmental policy. This, too, dates back to the turn of the twentieth century. In 1899, Congress passed the Rivers and Harbors Act (also known as the Refuse Act) to prevent the dumping of garbage into Lake Michigan around the city of Chicago and thus began efforts by the federal government to prevent dumping into navigable waterways.[23] It enacted other statutes tied to human health during the next several decades. Passing health-related environmental legislation became a growth industry for Congress after the middle of the twentieth century when public concerns over air and water quality, toxic and hazardous substances, and similar matters grew significantly.[24]

Some people think of ecosystem management as a decision-making process. The Keystone Center, a neutral, nonprofit education and public policy organization that seeks to find common ground on controversial public policy issues, made one effort to sort out the meaning of ecosystem management. At the request of the Clinton administration, the center convened its National Policy Dialogue on Ecosystem Management. It brought together members of the administration, federal agency employees and other public sector personnel, environmentalists, farmers, cattlemen, foresters, academics, and others (including myself). We visited dozens of projects labeled "ecosystem management" throughout the country and met extensively with those involved. In October 1996, after eighteen months of often intense debates, the dialogue produced a final report that defines ecosystem management as "a collaborative process that strives to reconcile the promotion of economic opportunities and livable communities with the conservation of ecological integrity and biological diversity."[25]

The Clinton administration also periodically seeks to portray ecosystem management as a process, usually as a means of deflecting criticism. The White House Interagency Ecosystem Management Task Force, for example, writes that "the ecosystem approach is a process, not a mapping convention," when it addressed the concerns of "skeptics."[26] During a congressional hearing on ecosystem management, one administration witness told the committee that "in general, ecosystem management tries to create a process where you have enough people at the table so that you make a societal decision about your value choices," while another proclaims, "Let's not lose sight of the fact that ecosystem management is a process."[27] However, the process that the administration has in mind is to implement only those goals that it already embraces without the benefit of widespread public debate. The White House task force concludes that "the goal of the ecosystem approach is to restore and maintain the health, sustainability, and biological diversity of ecosystems while supporting sustainable economies and communities."[28] What they are proposing is to elevate protection of nature above all else because ecosystem health, sustainability, and diversity are arbitrarily judged in terms of deviation from what is natural, that is, in terms of what the landscape might look like had Europeans not occupied North America (see chapter 4). There is a strong flavor of nature worship (see chapter 5) here that advocates of this view of ecosystem management often cloak in secular definitions, such as those of Edward Grumbine of the Sierra Institute. "Ecosystem management," he says, "integrates scientific knowledge of ecological relationships within a complex sociopolitical and values framework toward the general goal of protecting native ecosystem integrity over the long term."[29] It is not difficult for one to break the new paradigmists' code: "native ecosystem integrity" translates to "Mother Nature."

Ecosystem Management as the New Paradigm

"'Ecosystem management,' represents a fundamental change in approaches to the management of lands, oceans, natural resources, and the human activities affecting them . . . [it] . . . is much more than the latest buzzword; it represents a fundamental paradigm shift," says Jane Lubchenco, Distinguished Professor of Zoology at Oregon State University, telling us that it is "a concept whose time has come."[30] But what is a paradigm?

Thomas Kuhn, in his classic book *The Structure of Scientific Revolutions,* gives us much of our current understanding of scientific paradigms and paradigm shifts.[31] Kuhn uses the term in two ways.[32] In its broadest usage, a paradigm encompasses all the "beliefs, values, techniques, and so on" of a community of scientists. In its more restricted use, "it denotes the concrete puzzle-solutions which, employed as models or examples, can replace explicit rules as a basis for the solution of the remaining puzzles of normal science." The new paradigm view of ecosystem management supposes changes within each of these meanings. In Kuhn's widest sense, new paradigmists are indeed changing their beliefs, values, and techniques. They are forsaking emphasis on the roles played by theory, empirical evidence, and verification—cornerstones of our traditional understanding of scientific method—in favor of high-sounding but subjective, insubstantial, and unverifiable concepts, such as the notions of ecosystem integrity, health, and sustainability (see chapter 6). The change in beliefs allows new paradigmists to advocate a public policy of federal protection of ecosystems whose success or failure cannot be judged by objective and verifiable criteria or through the use of independent and replicable testing. The Forest Service provides a practical example of this shift away from the tenets of science to a value system based on opinion and subjectivity (see chapter 7).

The Forest Service seeks to shift to the new paradigm in managing the nation's national forest system. They want to make the maintenance and restoration of ecosystem "sustainability" the agency's prime goal. As clear evidence of an effort based on a subjective rather than objective value system, the Forest Service specifically rejects the establishment of "concrete standards regarding ecosystem sustainability or diversity," does not say what either a degraded or a restored ecosystem might be, and, amazingly, does not even define the term *ecosystem.*[33] It jettisons theory, verification, and objectivity in favor of dogma.

The new paradigm is eco- or biocentric. Nature, not man or God, is at its heart. Philosophy-professor-turned-journalist Alston Chase neatly sums up biocentrism in his book *In a Dark Wood,* in which he aptly describes it is as a blend of weak, or even pseudo, science and nature worship (see chapter 5).[34] Nature-worshiping environmental activists discovered the emerging discipline of ecology in the 1960s. Chase observes that ecology told the activists that nature was "organized into interconnected parts called ecosystems . . . [and] that conditions were good so long as ecosystems kept all their parts and remained in balance."[35] Activists then reason that "if every thing is connected to everything else . . . then all living things are of equal worth, and the health of the whole—the ecosystem—takes precedence over the needs and interests of individuals," including people. They can argue that protecting ecosystem health is the highest good and that to do otherwise is immoral. From a public relations and advocacy perspective, this is perfect: science and the moral high ground seemingly rolled into one cause, thus enabling biocentrists to paint their opponents as scientifically ignorant and morally bankrupt.

New paradigmists want the presumed needs of nature to constrain and take precedence over human actions (see chapters 4 and 7). One supporter writes that "the goal of ecological integrity places the protection of ecosystem patterns and processes before the satisfaction of human needs."[36] In advocating the new paradigm, the University of Vermont's Carl Reidel and Jean Richardson proclaim that "such revered principles as multiple use, sustained yield and even conservation"[37] should be replaced by a new land ethic that first and foremost requires the safeguarding of "ecological sustainability, natural diversity, and productivity of the landscape."[38] Believers in the new paradigm support public policy proposals such as requiring federal officials to manage all federal land (30 percent of the country) for the principal purpose of keeping ecosystems in an unimpaired condition, establishing an endangered ecosystem program for the country as a whole that is analogous to the endangered species program, and placing 50 percent of the nation into wilderness or near-wilderness status (see chapter 8).[39] Such proposals emanate from a worldview that is radically different from those Americans who value human well-being more than that of animals, plants, and inanimate objects. The new paradigm is without doubt revolutionary.

Officials in the Reagan and Bush administrations were the first to entertain the idea of ecosystem management, but chiefly as the next iteration of traditional natural resource policy that keeps people at the center of public policy. At the federal level, the new paradigm version of ecosystem management had to await the arrival of the Clinton administration, in which Vice President Al Gore, Secretary of the Interior Bruce Babbitt, and others actively embrace it.

Ecosystem Management in the Reagan and Bush Administrations

Ecosystem management made a tenuous appearance during the Reagan administration, but it did not reach the White House level. In December 1988, for example, the director of the National Park Service issued a new version of *Management Policies,* the agency's chief policy document, covering how all the units of the national park system should be managed. Chapter 4 of that document, "Natural Resource Management," calls for trying "to maintain all the components and processes of naturally evolving park ecosystems."[40] This policy applies natural zones within parks, that is, areas already designated as portions of park landscapes where the dominant management goal is the conservation of natural resources and processes. Thus, administrators confine ecosystem protection to particular areas within particular parks rather than make it a principal goal of park management.

Officials in the Bush administration took much greater note of ecosystem management and even lapsed into new paradigmist enviro-babble. For example, the president writes, in an annual report of the President's Council on Environmental Quality, that "the environment is composed of a seamless web of relationships between living organisms and the air, water, and land that surrounds them."[41] In a speech to members of Ducks Unlimited, he remarks that "our natural heritage must be recovered and restored. . . . We can and should become nature's advocate."[42]

President Bush nonetheless held to traditional approaches to environmental stewardship despite his rhetorical lapses. He did not seek to expand the role of government in the name of environmental protection or to elevate environmental protection over the achievement of other legitimate societal goals. His failure to do so, coupled with attempts to cut back on the intrusiveness of federal environmental regulations, reaped him a whirlwind of criticism from many new paradigm advocates. A particular affront was his rejection of the UN Convention on Biological Diversity, which calls for worldwide protection of ecosystems and the subordination of human activity to that cause (see chapter 7). This precipitated strident criticism of Bush at the 1992 Earth Summit in Rio de Janeiro, where a Sierra Club official said that Bush was presiding over the "environmental destruction presidency."[43]

Mike Deland, the chairman and only member of the President's Council on Environmental Quality, found the new paradigm view of ecosystem management attractive. He includes a lengthy chapter, "Linking Ecosystems and Biodiversity" in his *21st Annual Report of the Council on Environmental Quality* and recites part of the new paradigm mantra.[44] "The longterm wellbeing of ecosystems can be ignored only at the nation's great peril . . . [because] a healthy environment makes wealth possible. . . . And ecosystems are . . . the key to ecological health."[45] According to Deland and the report, ecosystems are "living forms" that Americans must preserve as part of the nation's biological diversity. Biological diversity has "intrinsic worth" and is the basis for ecological health. Therefore, in Deland's view, the United States needs a national goal to protect biological diversity, and we must "ensure that human actions remain compatible with ecological health." Deland's work did not result in radical policy changes because other members of the administration endorsed the traditional approach to natural resource policy and rejected the new paradigm.

In June 1992, the Departments of the Interior and Agriculture issued a report titled "America's Biodiversity Strategy: Actions to Conserve Species and Habitats."[46] In their letter transmitting the report, Interior Secretary Manual Lujan and Agriculture Secretary Edward Madigan acknowledge the value of biological diversity and "stable ecosystems" but argue that the report "shows that over the past 100 years, a framework has evolved to help effectively manage and conserve this Nation's biological resources for the use and enjoyment of present and future generations."[47] More important, they felt that any national goals regarding the maintenance or restoration of biological diversity or ecosystems must be "practicable;" in "consonance with the nation's other social and economic goals, including robust economic growth;" and remain "within a legal and institutional system that respects private property rights, state-Federal relationships and stimulates technological progress and sustainable economic growth." This is not the new paradigm.

Perhaps Dale Robertson, chief of the Forest Service, offered the only major policy directive on ecosystem management put out during the Bush presidency. On June 4, 1992, he produced a directive titled "Ecosystem Management in the National Forests and Grasslands."[48] It is a mixture of the old and the new paradigms. He announces that it is time to employ ecosystem management as the means to achieve multiple-use management in the belief that it demonstrates "a higher sensitivity to all of the environmental values" of the national forests. Here he reflects the evolutionary view of ecosystem management—that it is a new tool to do an expanded version of the old job.

The accompanying guidelines, however, are a study in the arcane. Robertson instructs managers to focus on both present and desired future landscape conditions while thinking in terms of multiple geographic and temporal scales. Robertson tells the managers to balance goals for the land with goals for people. Goals for the land are said to be beauty; soil fertility and stability; quality and flow of its waters; clarity of its air; diversity of plants, animals, and communities; and "the interconnectedness and character of habitats and landscapes that provide for the health and resilience of ecological systems and processes." Robertson does not indicate how he arrived at these goals or suggest what they might mean. People goals are less numerous: prosperity, diversity, and "the health and vitality of the people who depend on the land for their livelihoods, outdoor recreation opportunities, and inspirational experiences." In the end Robertson admonishes managers to "use good judgment" in implementing the guidelines, a most difficult assignment given the imprecision of his guidance.

Robertson's ecosystem management initiative is long on style and short on substance, but administration officials recognized its "green" value. They cited it as an accomplishment in a document that was intended to highlight the administration's environmental achievements prior to the 1992 Earth Summit.[49] Overall, however, there was little real action to develop or implement a policy of federal management and protection of ecosystems during the Bush years. All that changed dramatically with the arrival of President Clinton, Vice President Gore, and their senior environmental and natural resource team.

Ecosystem Management in the Clinton Administration

New paradigm oratory began immediately. During his confirmation hearing in January 1993, Department of the Interior (DOI) secretary-designate Babbitt told Congress that "the western environment . . . appears to be a vast and complex but fragile web." He said that "when we start extinguishing links in the ecological web . . . we take enormous risks and ultimately threaten our ability to live in harmony" with that environment.[50] At the behest of President Clinton, he quickly turned confirmation hearing talk into action by asking Congress to establish a new bureau within the DOI: the National Biological Survey (NBS).[51] The NBS, he said, "is my top priority as Secretary of the Interior."[52] In presenting his case for the new agency to Congress, Babbitt said that "we are all learning that ecosystem management is the most effective and efficient natural resource management strategy . . . [and] the NBS will be the biological underpinning that allows Interior to manage on an ecosystem basis."[53] When Congress refused to pass legislation creating the NBS, the secretary acted unilaterally, establishing the organization administratively on September 29, 1993.[54] He gave it an initial budget of over $160 million and more than 1,700 employees, largely taken from other agencies within the DOI, such as the Fish and Wildlife Service and the National Park Service.

No one can question Babbitt's commitment to the new paradigm. He had undergone a spiritual awakening in which he came to believe "that the vast landscape was somehow sacred and holy."[55] His remarks at the 1994 Earth Day celebration at the White House reflect this mysticism. He comments that "someone once said when a butterfly flaps its wings in the Amazon, it may create a snowstorm in Chicago." He goes on to suggest that this sort of profound thinking requires us to "move across the entire landscape" to protect ecosystems.[56] Babbitt's early work is barely a hint of what was to come.

The administration set the stage for the new paradigm and its implementation throughout the federal bureaucracy with the publication of Vice President Gore's *Reinventing Environmental Management* in September 1993.[57] Gore declares that "maintaining healthy ecosystems sustains their productivity and is vital to ensuring a high quality of life for future generations of Americans" while lamenting that development diminishes the "health and value of our ecosystems." The administration suggests that the economic use of land and resources is little more than a necessary evil.

Protection of ecosystems became the driving force of environmental policy with the report's recommendation that "agencies should interpret their existing mandates as broadly as possible to implement the ecosystem management policy and process." While theoretically only a suggestion, no cabinet secretary or agency head was going to ignore it. It was, after all, a major priority of the vice president. Gore writes in his book *Earth in the Balance: Ecology and the Human Spirit* that we must "use every policy and program, every law and institution, every treaty and alliance, every tactic and strategy, every plan and course of action—to use, in short, every means to halt the destruction of the environment and to preserve and nurture our ecological systems."[58] Thus, Gore wants to start a crusade to save the planet. He seeks a radical conversion of society, believing it imperative that Americans adopt the new paradigm and all its trappings. "Minor shifts in policy, marginal adjustments in ongoing programs, moderate improvements in laws and regulations," he says, "are forms of appeasement" intended to hide from the public the necessity of the "sacrifice, struggle, and [the] wrenching transformation of society" required to put our relationship with the Earth Mother aright.[59] This is strong language that leaves little room for misinterpretation. President Clinton hailed Gore's "passion, his commitment, his vision, and his sheer knowledge of environmental and natural heritage issues" during the signing ceremony establishing the controversial Grand Staircase–Escalante National Monument just prior to the 1996 presidential election, when he singled out *Earth in the Balance* for special praise (see chapter 10).[60] Yet economist Robert Hahn, in his review of the book for the *Yale Law Journal,* finds Gore's evidence for environmental meltdown "vastly overstated" and is critical of the "religious fervor" with which the vice president "derides people who do not share his perspective, most notably mainstream economists, scientists who beg to differ with his conclusions, and people who would be hurt by his proposed policies."[61]

The White House made sure that no one missed Gore's message or the president's endorsement of it. It established the White House Interagency Ecosystem Management Task Force, chaired by Katie McGinty, a Gore protégée who also headed the President's Council on Environmental Quality. Its purpose is to ensure that the new policy is driven into the federal bureaucracy. The task force consists of assistant-secretary-level representatives from the Departments of Agriculture, Army, Commerce, Defense, Energy, Housing, Interior, Justice, Labor, State, and Transportation as well as people from the Environmental Protection Agency and the president's Office of Management and Budget and Office of Science and Technology Policy. It published a two-volume report that lays out the goals of ecosystem management and makes thirty-one recommendations for their achievement.[62] Recommendations include having agencies develop "regional ecosystem plans," revising agency budgets to "reflect long-term needs of ecosystems," taking the necessary steps to "ensure that an ecosystem approach can be effectively applied across administrative boundaries," and being sure to "obtain the human resources needed to cope with the changing requirements of ecosystems" as well as to "infuse new ideas into traditional organizations." On December 15, 1995, the cabinet departments represented on the task force entered into a "Memorandum of Understanding to Foster the Ecosystem Approach," which was intended to begin the implementation of the recommendations. The administration believes that ecosystems are living entities that have both "long-term needs" and "changing requirements." It sees the proper role for government as providing for those needs and requirements, and the White House insists that agencies reconfigure their budgets, planning, and allocation of personnel to do so. This is classic new paradigm thinking.

The administration quickly established goals for ecosystem management. The task force writes that

> the ecosystem approach is a method for sustaining or restoring natural systems and their functions and values . . . [and] . . . the goal of the ecosystem approach is to restore and sustain the health, productivity, and biological diversity of ecosystems and the overall quality of life through a natural resource management approach that is fully integrated with social and economic goals. This is essential to maintain the air we breath, the water we drink, the food we eat, and to sustain natural resources for future generations.[63]

The DOI's Bureau of Land Management, which oversees over 260 million acres of public land in twenty-nine states, views ecosystem management this way:

> The primary goal of ecosystem management is to develop and implement management that conserves, restores, and maintains the ecological integrity, productivity, and biological diversity of public lands . . . since the production of all goods and services is dependent on ecosystem health, BLM's overriding objective will be to maintain naturally diverse and sustainable ecological systems.[64]

The late Mollie Beattie, then director of the Fish and Wildlife Service (FWS) in the DOI, wrote that the agency's goal for ecosystem management is "to contribute to the effective conservation of natural biological diversity through perpetuation of dynamic, healthy ecosystems."[65]

The nature-worshiping elements of the new paradigm are present in each of these goal statements. The task force calls for "sustaining or restoring natural systems," the BLM wants to "maintain naturally diverse . . . systems," and the FWS is driven to conserve "natural biological diversity." Nature is to be protected for nature's sake. Since the United States is mainly a Christian nation whose people reject nature worship, this aspect of the new paradigm is masked by appeals to its human benefits and through muddying the waters with assertions that human concerns will be factored into the policy. The task force claims that ecosystem management is "essential to maintain the air we breath, the water we drink, the food we eat, and to sustain natural resources for future generations" while calling for it to be "fully integrated with social and economic goals." "Fully integrated" does not mean that policymakers will treat social and economic goals as equivalent with nature protection. Indeed, the claim that ecosystem management is "essential" if we are to breathe, drink, and eat makes the subordination of social and economic goals inevitable. Thus, moving to the new paradigm is deemed necessary for our survival, so the protection of ecosystems must become the government's top priority, and the achievement of other goals must yield when there are conflicts.

It is an enticing argument, but one that does not stand even cursory inspection by informed analysts who know that if we depended on natural systems for our food, then most Americans would starve to death. Every major foodstuff that our farmers produce—grains, fruits, vegetables, and animals—is an exotic species and was never part of a natural ecosystem in the United States or is in any way connected with our natural (i.e., undisturbed by Europeans and their descendants) biological diversity.

The administration made it clear to career federal employees that it was playing hardball on the issue. Within weeks of President Clinton's second inauguration, it forced two career deputy chiefs of the Forest Service into retirement. According to press accounts, many in the Forest Service had angered the administration's "environmental standard-bearers." Former associate chief George Leonard observes that "longtime career employees are committed to carrying out the programs that have been funded by Congress. I'm not sure this administration wants to carry out those programs."[66] Congress authorizes the traditional approach to natural resource management, not the new paradigm (see chapter 7). Leonard goes on to say that "the Forest Service has been led by career civil servants . . . I had 20 years in Washington, and there wasn't a single instance where a secretary directed a civil-service firing. . . . But this Administration has interjected itself into the heart of the agency." They were especially angry when the agency, in response to a request by Congress, told the members that the administration's new paradigm plans to cancel existing logging contracts could cost $1.6 billion.

Interestingly, as I show in chapter 7, legal scholars agree that no existing federal law makes protecting ecosystems the principal mission of any federal agency. Nonetheless, the administration continues to drive its new paradigm agenda into federal agencies. In its 1997 *Strategic Plan,* the Environmental Protection Agency asserts that its mission includes safeguarding ecosystems and proclaims its intent to "protect the environmental integrity of ecosystems" and restore "life in damaged ecosystems."[67] In unveiling agency plans for the next century at a ceremony in March 1998, Forest Service Chief Mike Dombeck says that "our first priority is to maintain and restore the health of our ecosystems and watersheds."[68] Over at the FWS—and after three years of confusion and resistance by some within the ranks regarding directives to make ecosystem management the primary goal of the agency—the agency directorate tells employees in February 1998 that "ecosystem conservation is the job of the U.S. Fish and Wildlife Service; it is the 'normal work' of all Service employees, to which all of our individual and collective efforts must contribute. . . . The Directorate agrees that those of us who will not support the direction of the Service must be prepared to step aside."[69] Meanwhile, Gore presses on and calls on federal agencies to issue a report card on the nation's ecosystems by the year 2000.[70]

Implications of the New Paradigm

Fog enshrouds the new paradigmists' interpretation of ecosystem management.[71] This is true even though they write and speak constantly about the need for science and the development of specific, measurable goals. Indeed, it is difficult for researchers to find an article or report on the subject that does not cite the importance of identifying quantifiable measures. Yet even staunch advocates of the new approach acknowledge that producing such measures has been difficult. As Lubchenco so eloquently understates, the goal of sustaining ecosystems "is difficult to translate into specific objectives" in practice.[72] Other new paradigm scholars concede that it is "lacking in clear definition" and offers "little of the discipline and direction once provided" by the old paradigm.[73] Scholars conclude that virtually all the key terms associated with the new paradigm are controversial within the scientific community (see chapter 6).[74] The new paradigm must remain vague and indeterminate because its advocates weave it out of insubstantial notions that defy independent verification or objective evaluation.

In contrast, Americans' traditional approach to nature resource and environmental policy provides for the establishment of specific and measurable goals. Under the old paradigm, we can objectively judge success; under the new we cannot. For example, we can take a species off the endangered species list as soon as a predetermined number of breeding pairs has been identified in the wild. This kind of certainty does not exist in the new paradigm world. Since ecosystems are neither real nor living, we do not know when they might be endangered or threatened or subsequently returned to so-called good health. As I point out in chapter 1, we do not even know where they are. In the new paradigm world, policymakers can never demonstrate policy success because they can never demonstrate that goals have been reached. Thus, all the government plans and regulations put into place to achieve new paradigm goals will remain with us in perpetuity; or, worse, since no one can show that the first batch of regulations result in success, new paradigmists will use that to justify additional federal controls and so on ad nauseum. It is important to remember that with ecosystem management, it is people, not ecosystems, that the government will manage.

The new paradigm, if fully implemented, would greatly intrude on the property rights of all Americans. It not only lays the foundation for new and never-ending federal regulations but also drastically expands their geographic reach. Under the traditional paradigm, only private lands with specific characteristics fall under federal regulations. For example, they must be a wetland or the home of an endangered species. Moreover, the federal government confines its interest in such land to that particular characteristic. Ecosystems, however, are everywhere. Each square foot of the country can be said to be an ecosystem, so that every bit of land and water in the nation could be subjected to a federal regulatory interest under the new paradigm without regard to its condition or characteristics.

If federal protection of ecosystems is beginning to sound like national land use planning wherein nature protection takes precedence over improvement in human well-being, that is clearly what it portends. In 1997, Secretary Babbitt told an audience at Boise State University that we must find "the missing link in nature's self-regulating cycle of life" in order to meet the goal of "[restoring] our forests to presettlement equilibrium." Importantly, we must "structure economic use in a way that helps us reach that goal."[75]

Judging from how journalists, politicians, environmentalists, and many others employ the term, there is a good deal of ignorance about the ecosystem concept abroad in our nation. You might want to test your own knowledge with the following question: Which state has more ecosystems—California, Kansas, Florida, or Rhode Island? In chapter 1 you will learn the answer.

Notes

1. Everett Ladd and Karlyn Bowman, *Attitudes toward the Environment: Twenty-Five Years after Earth Day* (Washington, D.C.: American Enterprise Institute, 1995).

2. Richard Haeuber, "Setting the Environmental Policy Agenda: The Case of Ecosystem Management," *Natural Resources Journal* 36, no. 1 (winter 1996): 1.

3. Carl Reidel and Jean Richardson, "Strategic Environmental Leadership in a Time of Change," Inaugural Donlon Lecture (Syracuse: State University of New York, College of Environmental Science and Forestry, spring 1995), 11.

4. Roderick Nash, "Historical and Philosophical Considerations of Ecosystem Management," in *Ecosystem Management: Status and Potential,* report of a workshop convened by the Congressional Research Service, March 24–25, Senate Committee on Environment and Public Works, 103rd Cong., 2nd sess., December 1994, S. Prt. 103–98, 31.

5. Al Gore, *Earth in the Balance: Ecology and the Human Spirit* (New York: Penguin, 1993), 220–26.

6. Haeuber, "Setting the Environmental Policy Agenda, 3–4; Christopher A. Frissell and David Bayles, "Ecosystem Management and the Conservation of Biodiversity and Ecological Integrity," *Water Resources Bulletin* 32, no. 2 (April 1996): 229–40; Mark E. Jensen, Patrick Bourgeron, Richard Everitt, and Iris Goodman, "Ecosystem Management: A Landscape Ecology Perspective," *Water Resources Bulletin* 32, no. 2 (April 1996): 203–16; Norman L. Christensen et al., "Report of the Ecological Society of America Committee on the Scientific Basis for Ecosystem Management," *Ecological Applications* 6, no. 3 (1996): 665–91; Fred B. Samson and Fritz L. Knopf, "Preface," in *Ecosystem Management: Selected Readings,* ed. Fred B. Samson and Fritz L. Knopf (New York: Springer, 1996); and Margaret Moote, Sabrina Burke, Hanna J. Cortner, and Mary G. Wallace, *Principles of Ecosystem Management* (Tucson, Ariz.: Water Resources Research Center, University of Arizona, January 1994).

7. Roderick Nash, "Historical and Philosophical Considerations of Ecosystem Management," in *Ecosystem Management: Status and Potential,* report of a workshop convened by the Congressional Research Service, March 24–25, Senate Committee on Environment and Public Works, 103rd Cong., 2d sess., December 1994, S. Prt. 103–98, 3.

8. Jack Ward Thomas, "Ecosystem Management," speech delivered to U.S. Forest Service's public affairs personnel, Washington, D.C., April 11, 1993, transcribed from video by Patty Burel of the U.S. Forest Service (transcription available from the author).

9. Micheal J. Bean, "A Policy Perspective on Biodiversity Protection and Ecosystem Management," in *The Ecological Basis for Conservation,* ed. S. T. A. Pickett, R. S. Ostfeld, M. Shachak, and G. E. Likens (New York: Chapman & Hall, 1997), 23.

10. Stephen Yaffee et al., *Ecosystem Management in the United States* (Washington, D.C.: Island Press, 1996); The Keystone Center, *The Keystone National Policy Dialogue on Ecosystem Management—Final Report* (Keystone, Colo.: The Keystone Center, 1996).

11. U.S. Department of Agriculture, Forest Service, *A Framework for Ecosystem Management in the Interior Columbia Basin and Portions of the Klamath and Great Basins,* Pacific Northwest Research Station General Technical Report No. PNW-GTR-374, ed. Richard W. Haynes, Russell T. Graham, and Thomas M. Quigley, June 1996.

12. The forty-acre parcel is owned by Wendy Walsh and was visited by members of the Keystone Center's National Dialogue on Ecosystem Management in June 1995.

13. Robert Lackey, "Ecosystem Management: Implications for Fisheries Management," *Renewable Resources Journal* 13, no. 4 (winter 1995–96): 11–13, succinctly reviews the evolutionary and revolutionary views of ecosystem management. Additional views of ecosystem management may be found in David W. Crumpacker, "Prospects for Sustainability of Biodiversity Based on Conservation Biology and US Forest Service Approaches to Ecosystem Management," *Landscape and Urban Planning* 40 (1998): 47–71; Jack Ward Thomas, "Forest Service Perspective on Ecosystem Management," *Ecological Applications* 6, no. 3 (1996): 703–5; David R. Montgomery, Gordon E. Grant, and Kathleen Sullivan, "Watershed Analysis as a Framework for Implementing Ecosystem Management," *Water Resources Bulletin* 31, no. 3 (June 1995): 1–18; Thomas R. Stanley Jr., "Ecosystem Management and the Arrogance of Humanism," *Conservation Biology* 9, no. 2 (April 1995): 255–62; David N. Bengston, "Changing Forest Values and Ecosystem Management," *Society and Natural Resources* 7 (1994): 515–33; Christopher A. Wood, "Ecosystem Management: Achieving the New

Thomas Lovejoy (who also accepted an appointment with Secretary Babbitt) and Harvard University's E. O. Wilson, both longtime advocates of the new paradigm.

45. Council on Environmental Quality, *Environmental Quality: 21st Annual Report* (Washington, D.C.: U.S. Government Printing Office, 1991), 135.

46. U.S. Department of the Interior, U.S. Department of Agriculture, *America's Biodiversity Strategy: Actions to Conserve Species and Habitats* (Washington, D.C.: DOI Office of Program Analysis and USDA Natural Resources and Environment, June 1992).

47. Edward Madigan and Manual Lujan, "Dear Friend," letter, June 18, 1992.

48. F. Dale Robertson, "Ecosystem Management of the National Forests and Grasslands," U.S. Forest Service memorandum to regional foresters and station directors, 1330–1, Washington office, June 4, 1992. The memo came after a three-year courtship with ecosystem management under the New Perspective program. See Kessler et al., "New Perspectives for Sustainable Natural Resources Management."

49. "US Actions for a Better Environment: A Sustained Commitment." The document is somewhat unique in that it bears the seal of the president of the United States on its cover and was put together under direction of the White House's Office of Management and Budget, but it contains no specific information regarding the place or date of publication or the publisher.

50. Bruce Babbitt, testimony before Congress on his nomination as secretary of the interior, Senate Committee on Energy and Natural Resources, 103rd Cong., 1st sess., January 19 and 21, 1993, S. Hrg. 103–3, 35, 59.

51. Bill Clinton, "Remarks by the President in Earth Day Speech," press release, Office of the Press Secretary, The White House, April 21, 1993.

52. Bruce Babbitt, testimony on H.R. 1845, "The National Biological Survey Act of 1993," before Congress, House Subcommittee of Technology, Environment and Aviation, and the Subcommittee on Investigations and Oversight of the Committee on Science, Space, and Technology, 103rd Cong., 1st sess., September 14, 1993, Committee Print 56, 12.

53. Babbitt, Hearing on H.R. 1845, 13.

54. Bruce Babbitt, U.S. Department of the Interior Secretarial Order No. 3173, dated September 29, 1993. The 104th Congress ceased funding the National Biological Survey as a separate agency with Interior and instead folded it into the long-established U.S. Geological Survey.

55. Bruce Babbitt, "Between the Flood and the Rainbow: Our Covenant: To Protect the Whole of Creation," article from the secretary, downloaded from the Department of the Interior's Web page at http://www.doi.gov, December 16, 1995; also cited by Alston Chase in "Prophets for the Temple of Green," *Washington Times,* January 26, 1996, A16, as being from a speech titled "Our Covenant: To Protect the Whole of Creation," given in Boston on November 11, 1995, to a joint meeting of the National Religious Partnership for the Environment and the American Association for the Advancement of Science.

56. Bruce Babbitt, Press briefing, Office of the Press Secretary, The White House, April 21,1994.

57. Al Gore, *Creating a Government That Works Better and Costs Less: Reinventing Environmental Management* (Washington, D.C.: Office of the Vice President, September 1993).

58. Gore, *Earth in the Balance,* 274.

59. Gore, *Earth in the Balance,* 274.

60. Bill Clinton, "Remarks by the President in Making Environmental Announcement Outside the El Tovar Lodge in Grand Canyon National Park," press release, Office of the Press Secretary, The White House, September 18, 1996.

61. Robert W. Hahn, "Toward a New Environmental Paradigm," *Yale Law Journal* 102, no. 7 (May 1993): 1760.

62. White House Interagency Ecosystem Management Task Force, *The Ecosystem Approach. Vol. I: Overview,* and *Vol. II: Implementation Issues* (Washington, D.C.: The White House, November 1995).

63. White House Interagency Ecosystem Management Task Force, *The Ecosystem Approach. Vol. I: Overview,* 3.

64. Bureau of Land Management, U.S. Department of the Interior, "Ecosystem Management from Concept to Commitment" Washington, D.C., n.d., 2.

65. Mollie Beattie, "An Ecosystem Approach to Fish and Wildlife Conservation," *Ecological Applications* 6, no. 3 (August 1996): 696.

66. George Leonard, as quoted in Valerie Richardson, "Deputy Forest Service Chiefs Forced to Retire," *Washington Times,* February 8, 1997, A3.

67. Environmental Protection Agency, *EPA Strategic Plan,* EPA/190-R-97–002 (Washington, D.C.: Environmental Protection Agency, September 1997).

68. Mike Dombeck, "A Gradual Unfolding of a National Purpose: A Natural Resource Agenda for the 21st Century," speech delivered to Forest Service employees in Washington, D.C., March 2, 1998, downloaded from the Forest Service's Web site at http://www.fs.fed.us/news/agenda/sp30298.htm, July 10, 1998.

69. Significant difficulties with implementing ecosystem management with the FWS are documented in James E. Christensen, K. Jeffrey Danter, Debra L. Griest, Gary W. Mullins, and Emmalou Norland, *U.S. Fish and Wildlife Service Ecosystem Approach to Fish and Wildlife Conservation: An Assessment* (Columbus: Ecological Communications Lab, School of Natural Resources, Ohio State University, January 1998). Fish and Wildlife Service, U.S. Department of the Interior, "Directorate Decision on the U.S. Fish and Wildlife Service Approach to Ecosystem Conservation: An Assessment by Ohio State University," February 1998. The directorate posted the decision, addressed "To All Employees," on the FWS's Web site, downloaded from http://www.fws.ecoreport, March 24, 1998.

70. Ed Rykeil, "Ecosystem Science for the Twenty-First Century," *BioScience* 47, no. 10 (November 1997): 705–7. At the request of the White House Office of Science and Technology, the H. John Heinz III Center for Science, Economic, and the Environment undertook the project and expects to have some results available in 1999; see its Web site at http://www.heinzctr.org.

71. Hereafter, the phrase "ecosystem management" should be assumed to mean the new paradigm view of ecosystem management, unless otherwise noted.

72. Lubchenco, "The Scientific Basis of Ecosystem Management," 36.

73. Reidel and Richardson, "The Scientific Basis of Ecosystem Management," 11.

74. Cortner and Moote, "Setting the Political Agenda."

75. Bruce Babbitt, "A Coordinated Campaign: Fight Fire with Fire," remarks at Boise State University, February 11, 1997, downloaded from "The Secretary's Alcove" at the Department of the Interior's Web site at http://www.doi.gov, April 28, 1997.

Chapter 1

Everyone Knows What an Ecosystem Is . . . Or Do They?

> A dung pile or whale carcass are ecosystems as much as a watershed or a lake.
>
> —Ad Hoc Committee on Ecosystem Management,
> Ecological Society of America

What is an ecosystem? What are their geographic characteristics? Do they provide a rational way to partition the landscape for government purposes? Sir Arthur Tansley, a British plant ecologist, coined the term in 1935. Environmentalists, politicians, journalists, teachers, bureaucrats, land users, and many others all use the term, so one might assume that there is a general understanding of the ecosystem concept outside scientific circles. Unfortunately, that is not the case. Ecologist Gene Likens observes that "although the term, ecosystem, has gained widespread acceptance, its popular usage and application frequently abuse the concept that Tansley intended and that is understood by practicing ecosystem scientists."[1]

The ecosystem concept comes to us from science, so it is useful to examine definitions offered by members of that fraternity, who researchers closely identify with the concept. Tansley does not offer a precise definition; rather, he portrays ecosystems as physical systems encompassing living and nonliving things plus all the interactions between all such components.[2] In 1966, George Van Dyne, a noted American systems ecologist, defines an ecosystem as "a functional unit consisting of organisms (including man) and environmental variables of a specific area."[3] Eugene Odum, a driving force in ecosystem ecology, writes that an ecosystem "is any unit that includes all the organisms that function together in a given area interacting with the physical environment so that a flow of energy leads to clearly defined biotic structures and cycling of materials between living and non living parts."[4] Likens deems ecosystems to be "a spatially explicit unit of the Earth that includes all of the organisms, along with all components of the abiotic community within its boundaries."[5] The Ecological Society of America's Ad Hoc Committee on Ecosystem Management adopted his definition.[6] Harvard entomologist Edward O. Wilson, in his book the *Diversity of Life,* regards an ecosystem as "the organisms living in a particular environment, such as a lake or forest (or, at an increasing scale, an ocean or the whole planet), and the physical part of the environment within its boundaries."[7] Several points emerge from these definitions that those outside the scientific community do not widely understand but that are of vital significance if policymakers are to consider ecosystems as the new geographic basis for land use management and the application of federal law and regulation.

For advocates of the new paradigm, these definitions seem to hold a good deal of promise. They point to ecosystems as bounded space, a necessary condition for applying government rules. Van Dyne refers to "a specific area." Odum looks to "a given area." Likens and the Ecological Society of America see ecosystems as "a spatially explicit unit of the Earth." Wilson views ecosystems as having "boundaries." The point that many nonscientists do not understand well, however, is that researchers arbitrarily determine the bounded space referred to by the authors of these definitions. Ecosystems are but mental constructs, a convenient geographic shorthand that analysts use to describe some particular slice of the landscape rather than examples of objective reality. As Robert Ricklefs of the University of Pennsylvania writes, "Ecological systems . . . have no boundaries in space or time—they are not discrete, identifiable units like organisms."[8]

Scientists do not discover ecosystems on the landscape through diligent fieldwork or the application of ecological theories; instead, researchers determine them arbitrarily to suit their own purposes. According to James Agee and Darryll Johnson of the University of Washington, an ecosystem is "any part of the universe chosen as an area of interest, with the line around that area being the ecosystem boundary."[9] Evidence of the arbitrary geographic nature of ecosystems abounds. For nearly three decades a common trait among scholarly books presenting collections of ecosystem studies is the lack of any geographic consistency among the ecosystems serving as units of study.[10] These works fully reflect geographer Bruce Hannon's observation that "the delimitation of the [eco]system is strictly up to the observer; i.e., the system boundaries and the list of internal elements may be chosen at will."[11]

This geographic free-for-all exists because there are no theoretical or methodological requirements imposed by ecology or any other branch of science regarding the size, shape, or location of any portion of the landscape that is to bear the ecosystem label. There is, of course, nothing at all wrong with a researcher fixing the geographic limits of the area that he or she will study by using whatever spatial parameters are appropriate to the research enterprise. As a research tool the ecosystem concept may aid scientists gain a better understanding of the workings of the world. The problem arises when policymakers and others seek to transfer this tool from the world of science to that of governance and public policy. Researchers concocting ecosystems as objects of study is one thing—a valid endeavor—but pretending that they are discrete and real entities on the landscape whose protection should become the centerpiece of federal environmental and natural resource rules and regulation is quite another.

Consider ecosystem boundaries. Like the old Frank Sinatra hit song, researchers decide on boundaries "my way" because there are no rules that everyone agrees on for their identification. There are no generally accepted procedures regarding, among other things: 1) which spatial variables researchers should consider in formulating ecosystem boundaries, 2) how many variables these same researchers must use when establishing boundaries, 3) how the scientists should meld the individual distributions of variables that they use to depict an ecosystem into a single ecosystem boundary, 4) the scale at which researchers should identify ecosystems, or 5) how to account for constant changes in the distribution of living things (to say nothing of the fact that the distributions of individual species change at different rates).

Robert Bailey, a geographer with the U.S. Forest Service, outlines three broad approaches to determining ecosystem boundaries: gestalt, map overlay, and controlling factors.[12] The gestalt method is essentially a touchy-feely approach to the determination of ecosystems that Bailey dismisses as "nothing more than 'place-name regions'" that others cannot verify or confirm. The Sierra Club uses this approach in preparing its ecosystem map found in chapter 2. Overlaying maps of different variables then merging the separate patterns into a single set of ecosystems is a traditional method and the one that James Omernik adopts in preparing a national ecosystem map for the Environmental Protection Agency (see chapter 2). Bailey notes that this method can confront the analyst with a "staggering complexity of boundary patterns." As a way to simplify things, he advocates using a third approach, employing different variables to construct ecosystems, depending on the geographic scale of interest. He calls this the "controlling factors approach" and uses it to produce national ecosystem maps for the Forest Service.

The scientific community acknowledges the vague geographic and temporal characteristics of ecosystems and has done so from the beginning. In his original article describing the concept, Tansley concludes that "the [eco]systems we isolate mentally are not only included as parts of larger ones, but also overlap, interlock, and interact with one another."[13] Ecologist Simon Levin observes that "what we call an ecosystem . . . is really just an arbitrary subdivision of a continuous gradation of local species assemblages."[14] They are, after all, "merely . . . localized, transient experiment[s] in species interaction"[15] that "ecological scientists . . . [treat] as temporary conveniences to be discarded and replaced at will."[16] Environmental historian Donald Worster concludes that "we must conceive of ecosystems then, not as permanent entities engraved on the face of the earth but as shifting patterns in the endless flux, always new always different."[17]

Some scientists seek to introduce a bit of order into this chaos by thinking of ecosystems as nested hierarchies.[18] Bailey's controlled factors approach to mapping ecosystems represents this view and is at the bottom of former Forest Service Chief Dale Robertson's admonition to Forest Service managers regarding moving forward with ecosystem management of the national forests. Robertson calls on Forest Service managers to "integrate thinking and actions at multiple spatial and temporal scales . . . at least one scale larger and one scale smaller than the scale you are working at and at least for several decades into the future; more and longer if possible."[19] Nested hierarchies may be a convenient way to visualize the spatial ordering of ecosystems, but it remains an intellectual exercise since the subjects of the nesting—ecosystems—have no agreed-on spatial attributes that may serve as an organizing principle with which to determine the fundamental geographic characteristics reflected in the nesting approach.

Geographic scale is a permanent Achilles' heel for advocates of the new paradigm. There is no inherently right or wrong scale regarding ecosystems. In his book *Ecosystem Geography,* Bailey identifies three scales of ecosystem inquiry: microecosystems occupying a few acres, mesoecosystems being on the order of 250 acres, and macroecosystems reaching 25,000 acres.[20] Meanwhile, the Forest Service adopted a "National Hierarchical Framework of Ecological Units" that envisions an eight-level nesting of ecosystems ranging from less than 10 acres to over 600 million acres.[21] In practice we find the federal government involved in the analysis of 144 million acres in developing "A Framework for Ecosystem Management in the Interior Columbia Basin" and parts of adjacent drainages, in the 19-million-acre Greater Yellowstone Ecosystem (see chapter 3), and in California's less-than-one-acre vernal pools.[22] Scholars arbitrarily select an ecosystem's scale on the basis of the question to be investigated rather than through the application of ecological truths. A question asked at one scale may well be answered quite differently when asked at another. For example, the assessment of species richness—the number of species per unit area—is highly dependent on the size of the area researchers choose to investigate.[23]

One can liken the actuality of the ecosystem idea to your kitchen table (representing some portion of the landscape) covered with layer upon layer of jigsaw puzzles, each having a different number of pieces. On the bottom you may have a picture of a duck cut into five parts for assembly by a three-year-old and on the top a 5,000-piece solid-blue mind-boggler intended to challenge the most dedicated puzzle wizard. In between are dozens of other puzzles, each with its own designer and its own purpose. The individual pieces represent individual ecosystems. Now imagine changing the size and shape of all the pieces in all the layers at different rates while simultaneously adding some new pieces and removing others. To make matters even more realistic, also imagine that you cannot really see the edges of most of the pieces. That is the spatial reality of the ecosystem concept.

Are Ecosystems Alive?

The use of language likening ecosystems to living organisms has introduced an enormous amount of confusion into the discussion of ecosystem management. Scientists (who should know better) and nonscientists alike routinely treat ecosystems as living things, as with references to ecosystem health. How many times have you heard or read about an ecosystem somewhere that humans have degraded and is in need of restoration as though someone had just discovered a sick child and sought to heal her? How often has someone recited the dogma that economic health is dependent on ecosystem health? I explore the idea of ecosystem health in greater detail in chapter 6; for now the muddleheadedness surrounding this notion is captured by one scientist's view that "an ecological system is healthy and free from 'distress syndrome' if it is stable and sustainable—that is, if it is active and maintains its autonomy over time and is resilient to stress."[24] Some say that "fairness to the future of our culture and our species" demands that we protect ecosystem health and that we must do so to "stop the deplorable destruction of natural processes and the dynamic creativity they embody."[25]

The rhetoric of ecosystem health confuses landscape change with disease. When someone asserts that humans have degraded an ecosystem or that one is in poor health, what they really mean is that they preferred some portion of the landscape the way it used to be and want to reestablish earlier conditions to the greatest extent possible. There may be legitimate reasons for changing how people manage or use land, but claiming that we need to restore an ill ecosystem to a condition of health is not one of them. Nonetheless, it can be an effective rallying cry for political action. Who, after all, is going to argue in favor of disease?

It is easy for people to lapse into equating ecosystems with living things, especially in nonscientific discourse. It is also incorrect to do so. Frank Golley, longtime head of the University of Georgia's Institute of Ecology, observes that the claim "that ecosystems are self-regulating superorganisms . . . [intent on] . . . maintaining stability" justifies criticism of the ecosystem concept.[26] Because biota are key components of ecosystems, it is tempting to transfer characteristics of these living parts to the ecosystem as a whole. A tree is alive, and so are the birds, squirrels, insects, and other species that may inhabit it for all or part of their individual lives. But the tree-bird-squirrel-insect ecosystem does not form a separate organismal entity in its own right. Each species pursues strategies to ensure its own survival. In case of fire, the birds willingly abandon the tree to its fate. Consider the example of aquatic ecosystems. A stream may contain all manner of different species, each of which strives to maintain its own kind. Each individual organism is alive. The stream itself is not. Rain falls. Responding to gravity and surface conditions, some water molecules run over the land as sheet flow and eventually gather into rivulets and into ever larger types of channelized flows. This is not an indication of life; rather, it is an example of the functioning of physical laws. The tributary that they form is simply a collection of water molecules moving downhill in the same way that the water from your kitchen sink flows down a sewer. Few people would claim that a running sewer is a living thing in its own right, trying to maintain its health, integrity, or sustainability. Why, then, should anyone seriously propose that an aquatic ecosystem has independent life?

The Ecological Society of America tells us that a dung pile is an ecosystem. Does anyone consider their dog's droppings to be a living thing? How would the idea that "an ecological system is healthy and free from 'distress syndrome' if it is stable and sustainable—that is, if it is active and maintains its autonomy over time and is resilient to stress" apply in this case? Likens defines ecosystems as small drainage basins in his notable work at the Forest Service's Hubbard Brook Experimental Forest in New Hampshire. Van Dyne, on the other hand, spent much of his professional career studying grassland ecosystems. Are Likens's watersheds and Van Dyne's grasslands alive as independent entities? Clearly not, and neither are old growth forests in the Pacific Northwest, Florida's Everglades, or any other bit of the landscape someone categorizes as an ecosystem. Yet advocates of the new paradigm adhere to the mistaken belief that ecosystems are superorganisms when they call for the government to protect ecosystem health.[27]

Ecosystems are mental constructs—heuristic devices, not living things. They do not sprout from seeds, hatch from eggs, or emerge from wombs. They do not procreate or breathe. They do not eat, drink, or make merry. They do not think, strategize, or seek to maintain their existence. Humans—including governments—cannot heal them because ecosystems do not have the capacity to hurt. Ecosystems cannot die because they were never alive.

An Ecosystem Quiz

A little quiz on ecosystems can illustrate the practical result of these ecosystem realities.

Question 1. In the contiguous forty-eight states, there are approximately how many ecosystems?

a. less than 100
b. 100 to 999
c. 1,000 to 9,999
d. 10,000 to 99,999
e. more than 100,000

The answer is e. Actually, there are an indefinitely large number of ecosystems in the United States. That is because anyone can use whatever variables they choose in defining an ecosystem. As we have seen, an ecosystem can be a dung pile, a watershed, a forest, or the entire planet. It can be a leaf or a tree or a river basin or a spring or a patch of meadow or your backyard.

Question 2. True or False. An ordinary homesite anywhere in the United States can be thought of as existing within at least 100 different ecosystems at the same time.

The answer is True. It can be in as many different ecosystems as there are individuals out there using different approaches, variables, scales, and so on to define ecosystems. A home can simultaneously be in multiple ecosystems defined solely on the basis of watersheds simply by changing scale. For example, it can be in the drainage of the East Fork of Tiny Creek, which lies within the drainage of Tiny Creek, which lies within the drainage of Not-Quite-so-Tiny Creek, and so on. The homesite is also located within multiple ecosystems that scientists define using different combinations of geographic variables associated with plants, animals, and physical features of the environment.

Question 3. True or False. Nature establishes ecosystem boundaries.

The answer is False. People, not nature, delimit ecosystem boundaries because ecosystems are mental constructs, and so are their boundaries. Humans, selecting components of the physical environment or using subjective classifications of climate, vegetation, and the like, create ecosystem boundaries. I address this issue in much greater detail in chapter 2.

Question 4. True or False. The boundaries of ecosystems are always clearly visible on the landscape.

The answer is False. In many, if not most, cases the ecosystem boundary that appears on an ecosystem map cannot be seen on the land. That is because the boundaries are composites. Even the comparatively simple idea of basing ecosystem boundaries on a single variable like watersheds does not necessarily produce visible boundaries because drainage networks are poorly defined for about one-third of the contiguous states.[28]

Question 5. Scientists consider about what percentage of the possible information concerning the living things within an area when determining the boundaries of an ecosystem of, say, at least 1,000 acres in size?

a. at least 75 percent
b. 50–75 percent
c. 25–50 percent
d. 1–25 percent
e. less than 1 percent

The answer is e. Some ecosystem boundaries make no use of information concerning living things at all, for example, those based on watersheds. When scientists factor living things into the determination of boundaries, they use only the distributions of less than a dozen out of perhaps thousands of species present in the area. Scientists do not normally consider nondistributional information about the species (e.g., population size, population dynamics, and species health) when constructing boundaries. More on this in chapter 2.

Question 6. True or False. Multiple ecosystem boundaries can crisscross a given tract of land.

The answer is True. Because those determining ecosystem boundaries can define them as they please, researchers can festoon a specific area with many different ecosystem boundaries.

Many of us obtain knowledge from the news. If you did not answer all six questions correctly, perhaps you are having difficulty overcoming the poor quality of media reporting about ecosystems.

Ecosystems and the Press

Ecosystem ignorance is widespread, but it is no wonder, as the information that people receive is often incorrect or incomplete. In the classroom, texts and teaching too often mix advocacy with science at all levels of instruction so that many observers increasingly criticize the quality of environmental education.[29] Larry Gigliotte of Cornell University writes that it "has produced ecologically concerned citizens . . . armed with ecological myths."[30] The college text *Environmental Geology,* for example, strays far from science and into the realm of the occult by telling its readers that an ethical approach to the environment "affirms the right of all resources, including plants, animals, and earth materials, to continued existence . . . in a natural state."[31] Most of us are no longer in school, however, and we ordinarily receive our information on environmental matters from the press, where a lack of understanding regarding ecosystems is particularly notable.[32] The reality that ecosystems are "merely . . . localized, transient experiment[s] in species interaction" that "ecological scientists . . . [treat] . . . as temporary conveniences to be discarded and replaced at will" has not yet penetrated the collective mind of journalists. Instead, newspapers around the country print numerous stories each year based on the erroneous assumption the ecosystems are real entities on the landscape rather than geographic fabrications.

Journalists typically accept statements about the decline of ecosystem health without understanding that they are not living things. When Jasper Carlton, director of the Biodiversity Legal Foundation, proclaims that "we are wiping out a major ecosystem," the *Christian Science Monitor* dutifully prints the claim.[33] The *Chicago Tribune*'s environmental writer, Peter Kimball, writes "that fragmentation of ecosystems is just a long-term way of killing them."[34] The *Washington Post*'s lead for its story on the release of the United Nations book *Global Biodiversity Assessment* reads, "Thanks to the relentless predations of just one species, Homo sapiens, the global threat to biodiversity—the total variety of organisms and ecosystems that exist on Earth—is now 'unprecedented in human history.'"[35] This sort of breathless rhetoric may help sell newspapers, but it is more akin to the world of Chicken Little than to the world of science. These journalists fail to realize that neither people nor nature can kill an ecosystem because ecosystems are not living things and that there will always be an infinite number of ecosystems on the earth regardless of human actions.

The press, and many others, mistakenly label change in the distribution of species as the death of a living thing. The ecosystem that Carlton rails about is, in fact, a disjoint collection of spots that prairie dog colonies occupy in the short-grass plains extending from western Texas to Montana. If every prairie dog died tomorrow, there would be changes in the distribution of other species, but there would not be the body of a dead ecosystem lying prone on the plains available for viewing by tourists or ecologists. Indeed, the acreage that prairie dogs occupy is far less than in times past, yet much vegetation and many animals and insects now call the now-prairie-dog-less portions of the plains states home. Humans should not view the differences in species mix in the absence of prairie dogs as the death of an ecosystem; it is simply a change in the landscape, which individual observers may like or dislike.

In December 1995, Defenders of Wildlife, a Washington-based environmental organization, issued a report titled "Endangered Ecosystems: A Status Report on America's Vanishing Habitat and Wildlife."[36] Several newspapers carried stories on the monograph, and the Knight-Ridder/Tribune News Service put out a wire story. Even though the Knight-Ridder story identifies coauthor Reed Noss as a "combat biologist" and says that the report "had a frankly political purpose," the press dutifully relays to its readers the specter of dying ecosystems that the Defenders' report conjures up.[37] The *Atlanta Journal and Atlanta Constitution* headlines its story "Georgia Cited among Top States Where Mother Nature Is at Risk—Wake-up Call: A 'Report Card on . . . Ecological Health' Ranks the State the Third Most Imperiled in the Nation."[38] This newspaper and the *Austin American-Statesman* quote coauthor Robert Peters as saying human activity has placed the nation in peril of "the imminent loss of hundreds and perhaps thousands of natural ecosystems" and that there is "an overall tide of biotic impoverishment" in the United States.[39] Only one journalist seems to suspect that something may be amiss. Writing in the *St. Petersburg Times,* David Olinger notes that "the Defenders of Wildlife report stretched normal definitions of a biological community" by lumping together under a single ecosystem umbrella portions of the Everglades, pine forests on the Florida Keys, coral reefs, and the Florida Bay.[40] Regrettably, Olinger does not pursue his apparent unease about the report.

It is unfortunate that reporters fail to examine the report's "stretches," inconsistencies, and assumptions. For example, the ecological yardstick that Peters and Noss use to determine whether an ecosystem is endangered is the extent to which it now differs from what they imagine the landscape would look like if Europeans had never landed in North America, that is, its deviation from what an uninhibited Nature would have done to the land if left alone since 1492. Reporters simply accept this standard even though it stems entirely from Noss's and Peters's personal preferences rather than having any objective or scientific basis (see chapter 4). No one seems to find it odd that the authors change geographic scale when it is convenient to do so in order to paint a more desperate picture of ecosystem health. Peters says that the "Atlantic white cedar is a highly endangered ecosystem," yet it occupies only few hundred acres in New Jersey's Great Cedar Swamp.[41] At the opposite end of the geographic scale, the report lists the much larger, but wondrously vague, "South Florida Landscape" as the nation's most endangered ecosystem.[42] If journalists had read the report or asked some questions before filing their stories, they would have uncovered other serious shortcomings.

The report plays fast and loose with the ecosystem concept as it is normally understood by scientists. The authors essentially concede this point when they admit to adopting a definition of ecosystems that includes "what ecologists call plant communities or associations."[43] The scientific community has distinguished between ecosystems and plant communities for over half a century. Peters and Noss also note that "many of our ecosystem types are defined in part by adjectives indicating condition, though these are not part of traditional plant community classifications." They tell us that their "use of the term 'ecosystem' should not be confused with another common use, which is to signify a geographic area functioning as an ecological unit." But this is the essence of the scientific community's understanding of ecosystems! In other words, the authors freely abandon whatever scientific agreement there is regarding the definition of ecosystems yet claim that their work is grounded in science. Peters invokes the mantle of science at the press conference announcing the report, and the Defenders' press release quotes Peters as saying, "This is a science-based report card on our nation's ecological health."[44]

The authors trumpet their inventive approach to defining ecosystems as allowing ecosystems to be "measured and mapped." Curiously, reporters do not seem to notice that Peters and Noss do not produce a single ecosystem map even though the report refers to hundreds of ecosystems. It is not as though the authors do not like maps. The report contains thirty outline maps of the United States, but not a single ecosystem boundary appears in its pages. Maps claiming to show "endangered ecosystems" do no such thing. They just color in whole states. For example, the map of the "Ancient Ponderosa Pine Forest" ecosystem shades all of fourteen midwestern and western states where a ponderosa pine forest is thought by the authors to have been present somewhere in the state at some time in the past. This technique is highly misleading because it grossly exaggerates the amount of land where a ponderosa pine forest ever existed. For example, the report includes all of Nevada in this ecosystem, yet forests of all kinds presently cover only 10 percent of the state.[45] The press does not seem to notice these sorts of distortions. They are not struck that the report depicts the entire state of California on eight separate maps showing completely different "endangered ecosystems," meaning that the entire state was simultaneously covered by eight different ecological communities—an obvious impossibility. An inquisitive reporter might also have asked why the authors label the categories of risk to ecosystems as "moderate," "high," and "extreme" in the map legends instead of more common designations like low, medium,

that *arbitrary* means "based on or subject to one's opinion, judgment, prejudice, etc." These describe very nicely what occurs when anyone attempts to locate an ecosystem. No political or scientific body has established rules to demarcate ecosystems. We are left with only the opinions, judgments, and prejudices of those formulating the boundaries. Using ecosystems as a geographic basis for government authority, therefore, represents a radical departure from our national experience with establishing the spatial boundaries guiding the application of government power.

How do ecosystems compare with states and counties as geographic units (table 1.1)? An area of obvious contrast is the number of divisions. There are precisely fifty states and slightly over 3,000 counties or their equivalents.[52]

TABLE 1.1
Characteristics of Counties and Ecosystems as Geographic Units

Characteristic	Counties	Ecosystems
Number of units in the U.S.	3,043	Unlimited
Size range (1,000s square miles)	.026–20	<.00001–3,100
Overlap one another on the landscape?	No	Yes
Possess agreed-on ways to define and locate boundaries?	Yes	No
Possess a single set of agreed-on boundaries for the U.S.?	Yes	No
Boundary locations accurately known?	Yes	No
Boundary locations change over time?	No	Yes
Size and shape routinely change over time?	No	Yes
Boundaries arbitrarily determined?	Yes	Yes
Originally conceived as a means of organizing the landscape to guide government actions?	Yes	No

Note: It is possible to niggle over the exact number of counties; Louisiana, for example, labels its chief substate jurisdictions "parishes." The largest ecosystem in the United States can be thought of as the 3.1 million square miles of the contiguous forty-eight states. Substate jurisdictions in Alaska overseen by native corporations can reach 87,000 square miles.

Everyone knows and agrees to them. In contrast, there are an infinite number of ecosystems that anyone may identify within the territory of the United States and there is no agreement on either their number or their location. On the matter of size, ecosystems can range from a leaf a few inches across to the 3.1 million square miles that comprise the contiguous forty-eight states. States and counties exhibit a considerable range in size but nothing to approach the variability that ecosystems exhibit; Alaska, after all, is only 425 times the size of Rhode Island.

States and counties form single, contiguous sets of geographic units without overlap or gaps; ecosystems do not. Using ecosystems to determine jurisdictions will inevitably result in overlapping units because that is an inherent spatial characteristic of ecosystems. We can already see this at work in an embryonic way with the territory that the Applegate Partnership embraces. The partnership is a community-based effort begun in 1992 "to encourage and facilitate the use of natural resources principles that promote ecosystem health and diversity."[53] New paradigmists often use the partnership as a promising example of ecosystem management. Geographically, it is based on the approximately 500,000-acre watershed of the Applegate River in southern Oregon. But several other, quite different ecosystems encompass this same acreage. For example, researchers define an ecosystem based on the range of the northern spotted owl to include the Applegate watershed. In addition, the watershed is subsumed in larger ecosystems depicted on maps prepared by the Forest Service, the Environmental Protection Agency, the Fish and Wildlife Service, and the Sierra Club (I examine these maps in the next chapter). If the government were to establish separate management goals for each one of the several ecosystems that include the Applegate watershed, which set of goals would control land use in the area?

To illustrate the confusion that one must associate with using ecosystems as geographic guides to the application of government authority, suppose that counties possessed the same spatial characteristics as ecosystems. Counties would overlap on the landscape. Each county would have its own zoning requirements and land use plan, its own tax structure, and its own set of administrative offices, bureaucracies, and courts. For the sake of simplicity, I will assume that a segment of the landscape is included within only three counties. Those wishing to make use of land in this segment would have to satisfy the requirements of three different county governments. What happens if one county emphasizes industrial development, another emphasizes the preservation of farmland, and the third sees the future in terms of commercial and residential expansion? Land use gridlock. Suppose that our three counties agreed on how to use at least a portion of their mutual territory. Yet, remembering that ecosystem boundaries shift and that there are an unlimited number of ecosystems in the United States, imagine that a fourth and a fifth county now assert jurisdiction over this zone of consensus or that one of the three original counties altered its boundaries and relinquished its claims. Negotiations begin anew. This is an irrational means of organizing the landscape for government purposes.

One way out of the spatial dilemma caused by the inherent geographic characteristics of ecosystems is for Congress or state governments to legislate ecosystem boundaries in the same manner that governments fix state and county boundaries.[54] Legislatively fossilizing a pattern of ecosystems would be arbitrary and contrary to the understanding among scholars that ecosystems are constantly changing in time and space. Thus, while governments can codify state and county boundaries in law, there is no rational way to do so with ecosystem boundaries without contravening the ecosystem concept itself. Efforts that policymakers would make to preserve a government-imposed pattern of ecosystems would be doomed to failure, and societal energies devoted to such quixotic adventures would be wasted in a pursuit of the impossible.

A second approach to dealing with the infirmities of the ecosystem concept is to pretend that they do not exist. In the spring of 1994, the Sierra Club launched its Critical Ecoregions Program. It announced in its magazine, *Sierra,* that "the 21 Ecoregions we have identified will send those who need strict definitions into a tizzy. Nature has messy boundaries, and systems blend into each other—and so do our ecoregions."[55] Readers have good reason to be tizzified because a scant sixteen months later *Sierra* carries another article asserting the United States has 261 types of ecosystems.[56] The author praises all manner of ecosystem preservation initiatives regardless of scale and blissfully overlooks the multitude of problems that preserving overlapping and constantly shifting ecosystems create. Nothing matters beyond adoption of the new paradigm.

The White House Interagency Ecosystem Management Task Force takes yet another tack in dealing with difficulties inherent in the use of ecosystems in public policy.[57] It acknowledges that problems exist and then deny that they are important. The task force notes that "geographic boundaries appropriate for addressing one issue may not work for another" but concludes that ecosystem management "does not rely on prior definitions of precise, scientifically valid delineations of ecosystems that apply to all situations."[58] Ecosystems, it says, "need to be characterized and studied at scales appropriate to the issues at hand." It defends this idiosyncratic approach by claiming that ecosystem management is a "process not a mapping convention." In other words, it does not matter where we draw the lines, how many we draw, or how we may draw them, as long as we are protecting ecosystems. For citizens who expect that their government will be neither arbitrary nor capricious, this is not a welcome position.

People constantly abuse and misuse the ecosystem concept. This is especially evident in its use by activists and others attempting to recast government environmental and land use policies so as to save, protect, or restore ecosystems in the mistaken belief that they are real entities on the landscape. New paradigm advocates must constantly try to hide, ignore, or disguise the geographic characteristics of ecosystems, lest the public and decision makers become fully aware of them and reject the new paradigm out of hand.

Policies, rules, regulations, and plans directing how government should manage land and guide the use of natural resources require the drawing of lines on maps as a prerequisite to the application of government authority. The next chapter addresses the matter of ecosystem maps.

Notes

1. Gene Likens, *The Ecosystem Approach: Its Use and Abuse* (Oldendorf/Luhe, Germany: Ecology Institute, 1992), 16.
2. A. G. Tansley, "The Use and Abuse of Vegetational Concepts and Terms," *Ecology* 16, no. 3 (1935): 284–307.
3. George M. Van Dyne, "Ecosystems, Systems Ecology, and Systems Ecologists," in *Complex Ecology: The Part-Whole Relation in Ecosystems,* ed. Bernard C. Pattern and Sven Jorgensen (Englewood Cliffs, N.J.: Prentice Hall, 1995), 1–27.
4. Eugene Odum, *Basic Ecology* (Philadelphia: Saunders College, 1983), 13.
5. Likens, *The Ecosystem Approach,* 9.
6. Norman L. Christensen et al., "Report of the Ecological Society," Committee on the Scientific Basis for Ecosystem Management, *Ecological Applications* 6, no. 3 (1966): 665–91.
7. E. O. Wilson, *Diversity of Life* (Cambridge, Mass.: Harvard University Press, 1992), 396.
8. Robert Ricklefs, "Structure in Ecology," *Science* 236 (April 10, 1987): 206.
9. James K. Agee and Darryll R. Johnson, eds., *Ecosystem Management for Parks and Wilderness* (Seattle: University of Washington Press, 1988), 233.
10. Fred B. Samson and Fritz L. Knopf, eds., *Ecosystem Management: Selected Readings* (New York: Springer, 1996); Stephen Yaffee et al., *Ecosystem Management in the United States* (Washington, D.C.: Island Press, 1996); and George M. Van Dyne, ed., *The Ecosystem Concept in Natural Resource Management* (New York: Academic Press, 1969).
11. Bruce Hannon, "Accounting in Ecological Systems," in *Ecological Economics: The Science of Management of Sustainability,* ed. Robert Costanza (New York: Columbia University Press, 1991), 238.
12. Robert G. Bailey, *Ecosystem Geography* (New York: Springer, 1996). Bailey includes two other approaches, multivariate clustering and digital image processing, but devotes very little attention to them. Neither of these methods has resulted in widely discussed ecosystem maps at a national scale.
13. Tansley, "The Use and Abuse of Vegetational Concepts and Terms," 300.

14. Simon Levin, "The Problem of Pattern and Scale in Ecology," *Ecology* 73, no. 6 (December 1992): 1960.

15. R. V. O'Neill, "Get Ready to Rewrite Your Lecture Notes," *Ecology* 77, no. 2 (March 1996): 660.

16. Bryan G. Norton, "A New Paradigm for Environmental Management," in *Ecosystem Health: New Goals for Environmental Management,* ed. Robert Costanza, Bryan G. Norton, and Benjamin D. Haskell (Washington, D.C.: Island Press, 1992), 35.

17. Donald Worster, *Nature's Economy—a History of Ecological Ideas,* 2nd ed. (Cambridge: Cambridge University Press, 1994), 412.

18. R. V. O'Neill et al., *A Hierarchical Concept of Ecosystems* (Princeton, N.J.: Princeton University Press, 1986).

19. F. Dale Robertson, "Ecosystem Management of the National Forests and Grasslands," U.S. Forest Service memorandum to regional foresters and station directors, 1330–1, Attachment 1, Washington office, June 4, 1992.

20. Bailey, *Ecosystem Geography.*

21. U.S. Forest Service, *National Hierarchical Framework of Ecological Units* (Washington, D.C.: ECOMAP, October 29, 1993).

22. Richard W. Haynes, Russell T. Graham, and Thomas M. Quigley, eds., *A Framework for Ecosystem Management in the Interior Columbia Basin and Portions of the Klamath and Great Basins,* Pacific Northwest Research Station General Technical Report No. PNW-GTR-374, U.S. Forest Service, U.S. Department of Agriculture, Seattle, Washington, June 1996.

23. See generally Michael J. Conroy and Barry Noon, "Mapping Species Richness for Conservation of Biological Diversity: Conceptual and Methodological Issues," *Ecological Applications* 6, no. 3 (1996): 763–73; Monica G. Turner, Robert H. Gardner, and Robert V. O'Neill, "Ecological Dynamics at Broad Scales," *The Science and Biodiversity Policy Supplement to BioScience* (June 1995): S29–37; David Stoms, "Scale Dependence of Species Richness Maps," *Professional Geographer* 46, no. 3 (August 1994): 346–58; and Ling Bain and Stephen J. Walsh, "Scale Dependencies of Vegetation and Topography in a Mountainous Environment of Montana," *Professional Geographer* 45, no. 1 (February 1993): 1–11.

24. Benjamin D. Haskell, Bryan D. Norton, and Robert Costanza, "Introduction: What Is Ecosystem Health and Why Should We Worry about It?" in *Ecosystem Health: New Goals for Environmental Management,* ed. Benjamin D. Haskell, Bryan D. Norton, and Robert Costanza (Washington, D.C.: Island Press, 1992), 9.

25. Haskell, Norton, and Costanza, eds., *Ecosystem Health,* 5.

26. Frank Golley, *A History of the Ecosystem Concept in Ecology* (New Haven, Conn.: Yale University Press, 1993), 204.

27. White House Interagency Ecosystem Management Task Force, *The Ecosystem Approach: Healthy Ecosystems and Sustainable Economies. Vol. I: Overview* (Washington, D.C.: The White House, June 1995), 3.

28. James M. Omernik and Robert G. Bailey, "Distinguishing between Watersheds and Ecoregions," *Journal of the American Water Resources Association* (formerly *Water Resources Bulletin*) 33, no. 5 (October 1997): 935–49.

29. Independent Commission on Environmental Education, *Are We Building Envi-*

ronmental Literacy? (Washington, D.C.: ICEE, 1997); Michael Sanera and Jane S. Shaw, *Facts Not Fear* (Washington, D.C.: Regnery, 1996). These works evaluate K-12 grade level materials.

30. Quoted in Independent Commission on Environmental Education, *Are We Building Environmental Literacy?*, 8.

31. Edward A. Keller, *Environmental Geology,* 7th ed. (Upper Saddle River, N.J.: Prentice Hall, 1996), 5.

32. It is possible to find sound reporting on ecological issues. For example, Alan Cutler, "What Is a Species?" *Washington Post,* August 9, 1995, H1.

33. Julian Lloyd, "The Prairie-Dog Divide: To Reduce or Protect," *Christian Science Monitor,* May 8, 1996, 3.

34. Peter Kimball, "Shawnee Forest Fight Cuts Both Ways," *Chicago Tribune,* September 23, 1996.

35. Curt Suplee, "Earth's Biotic Wealth Faces Unprecedented Threat," *Washington Post,* November 20, 1995, A3.

36. Reed F. Noss and Robert L. Peters, *Endangered Ecosystems: A Status Report on America's Vanishing Habitat and Wildlife* (Washington, D.C.: Defenders of Wildlife, 1995).

37. Heather Dewar, "Biologists Say Florida, California, Southeast Face Greatest Risk of Losing Their Wild Lands," Knight-Ridder/Tribune News Service, December 20, 1995.

38. *Atlanta Journal and Atlanta Constitution,* "Georgia Cited among Top States Where Mother Nature Is at Risk, Wake-Up Call: A 'Report Card . . . on Ecological Health' Ranks the State the Third Most Imperiled in the Nation," December 21, 1995, A20.

39. Jeff Nesmith, "Texas 'Ecosystems' Ranked among Endangered in Report," *Austin American-Statesman,* December 21, 1995, A1.

40. David Olinger, "Florida Tops List of Endangered Habitats," *St. Petersburg Times,* December 21, 1995, B1, B6.

41. Richard Degner, "Five of 21 'Endangered Ecosystems' Found in N.J.," *Press* (Atlantic City), February 10, 1996.

42. Noss and Peters, *Endangered Ecosystems,* 10.

43. Noss and Peters, *Endangered Ecosystems,* 4.

44. Defenders of Wildlife, "Many Ecosystems Nationwide Near Breaking Point," press release, December 20, 1995.

45. Arthur B. Daugherty, *Major Uses of Land in the United States, 1992,* Agriculture Economic Report No. 723, Economic Research Service, U.S. Department of Agriculture, Washington, D.C., 1995.

46. Bill Dietrich, "Natural Balance," *Seattle Times,* March 30, 1995, A13; Lynn Van Matre, "Fox River Groups Join to Help Ecosystem Partnership Qualify for State Program to Safeguard Resources," *Chicago Tribune,* September 25, 1996.

47. Tom Knudsen, "Sierra's Problem: Too Many People," *Sacramento Bee,* June 11, 1996, B1.

48. Bruce Hannon, in Costanza, ed., *Ecological Economics.*

49. For an excellent discussion of early boundary issues between territories and

states, see Ralph H. Brown, *The Historical Geography of the United States* (New York: Harcourt, Brace & World, 1948).

50. Local difficulties can arise when a river or stream used as a political boundary alters course, but such changes are of little consequence in the overall scheme of things.

51. Christensen et al., "Report of the Ecological Society," 670.

52. According to the National Association of Counties, there were 3,043 counties or their equivalents in 1995.

53. Applegate Partnership, "Applegate Partnership, Practicing Trust Them Is Us," Grants Pass, Oregon, n.d. The organization is composed of local people representing industry, environmental groups, land users, and landowners. Federal lands constitute 70 percent of the watershed, but representatives of the agencies involved do not directly participate.

54. John W. Wuichert, "Toward an Ecosystem Management Policy Grounded in Hierarchy Theory," *Ecosystem Health* 1, no. 3 (September 1995): 161–69.

55. Jane Elder, "At Home on the Planet," *Sierra,* March/April 1994, 55.

56. Dave Foreman, "Missing Links," *Sierra,* September/October 1995, 52–57, 96–98.

57. White House Interagency Ecosystem Management Task Force, *The Ecosystem Approach.*

58. White House Interagency Ecosystem Management Task Force, *The Ecosystem Approach,* 26.

Chapter 2

Mapping Ecosystems

A single map is but one of an indefinitely large number of maps that might be produced for the same situation or from the same data.

—Mark Monmonier, *How to Lie with Maps,* 2

The General Accounting Office concludes that ecosystem maps are "a prerequisite to planning for, budgeting, authorizing, and appropriating funds for, and ultimately managing activities on the basis of ecological units."[1] If policymakers are to use ecosystems as a geographic guide for the application of government authority, then the ecosystem map becomes of paramount importance. This chapter examines the characteristics of maps in general and ecosystem maps and their components in particular.

Print advertisers often use the phrase "As Seen on TV" to convey legitimacy to whatever product their ad is touting, as though television is a magical medium that presents only balanced, verifiable, and unprejudiced information. Similarly, many people seem to believe that if something is on a map, then it must be so. The widespread view that maps are an objective depiction of reality is naive and incorrect. Geographer Mark Monmonier explains that "maps, like speeches and paintings, are authored collections of information and also are subject to distortions arising from ignorance, greed, ideological blindness, or malice."[2] Maps are the end products of many subjective decisions. Cartographers must determine what data will be gathered, how it will be obtained, what information will eventually appear on the map, and (of equal importance) what will be left behind. What projection will be used? What scale? What colors, symbols, and words will be employed? Even the most scrupulously prepared maps "incorporate assumptions and conventions of the society and the individuals who create them."[3] Nonetheless, maps are of enormous value and an indispensable means of conveying knowledge about location, the attributes of places, and the distribution of spatial phenomena.

Maps can serve many needs. They can inform, serve as tools of analysis, and be instruments of propaganda. A highway map seeks to inform a user about the location and general condition of roadways with reference to destinations such as cities, airports, parks, or ski resorts. When the international community wants to analyze the feasibility of getting foodstuffs to refugees, they turn to maps showing transportation routes and the condition of those routes, the location and size of refugee camps, sources of water, and a host of other critical items. I show in chapter 1 how the Defenders of Wildlife use maps as propaganda tools by greatly exaggerating the geographic extent of so-called endangered ecosystems.

All maps have certain common attributes.[4] They are a subjective—not objective—means to portray spatial data. Subjectivity enters into the development of a map at all stages of map preparation. For maps that intend to show something about what is actually on the landscape, cartographers must make decisions about which landscape features they will show and which they will ignore. Cartographic axioms do not guide their judgments; rather, their actions are influenced by the purpose of the map, the availability and quality of data, project budget and time constraints, and the assumptions, biases, and predilections of those making the decisions. Even a map of a comparatively small area, such as Washington, D.C., cannot show every house, tree, and parking lot. The map may show and identify some structures (e.g., museums and major government buildings), but most will be missing. For example, a map showing the Smithsonian's National Air and Space Museum is unlikely to also identify crack houses and massage parlors. Building materials, structure height and condition, the location of potholes, and much else will be absent, reflecting the fact that maps are generalizations and include only a minuscule fraction of the total body of spatial information that could be presented about the area they show.

Subjectivity is manifest in the choice of projections a cartographer uses as a basis for the map. Researchers use maps to portray a three-dimensional world on a two-dimensional sheet of paper or computer screen. They immediately and unavoidably introduce misrepresentations of the real world because no map projection—the systematic method that cartographers use to translate points from three dimensions to two—can faithfully preserve shape, size, distance, and direction as it actually exists on the earth's surface.[5] The larger the area being mapped, the greater the amount of distortion. A map of a city block can reproduce the relative size, shape, and location of individual buildings and lots quite accurately, whereas world maps based on differing projections present dramatically different visual representations of the planet, all of which are inaccurate. The classic Mercator projection, for example, badly distorts size at high latitudes so that Alaska and Brazil appear as having about the same area even though Brazil is five times larger. Projections that accurately reproduce size relationships distort shapes, some much more than others. Some projections seek a middle ground, preserving neither size nor shape entirely. The inescapable fact is that all map projections distort geographic truth so that the selection of a projection depends on which truth the mapmaker wishes to most faithfully represent and which are less vital to the project at hand. Projection choice may also depend on simple aesthetics or on the existence of an agenda that the map is intended to serve.

Mapmakers convey information and create impressions through the use of symbols, color, and words. The cartographer selects these map components to serve whatever purpose he or she desires. For example, using a skull and crossbones to designate Superfund sites in the United States may evoke a different response from readers than would a neutral symbol, such as a circle. Coloration can increase impact. Suppose that the skull and crossbones were red as opposed to green. Size can also affect impact. By increasing the size of the symbol, the map can easily create a false impression about the amount of land actually involved in some activity. On a national map of Superfund sites, a site symbol the size of the word *symbol* in this sentence would present a much different image of the extent of these sites than if the symbol was akin to the dot over the letter *i* in this book. The dot would be far more accurate in terms of relative size but may be so small as to go unnoticed, so a cartographer intent on producing as unbiased a map as possible might settle for a symbol more the size of the letter *o*.

No general discussion of maps would be complete without attention to the question of scale.[6] Scale is the proportional relationship between distance on the map and distance on the surface of the earth. It is often displayed on a map as a ratio, for example, 1:1,000,000, meaning that 1 unit of distance on the map represents 1 million of the same units on the land. Several maps of the United States in *Goode's World Atlas* are done at a scale of 1:28,000,000, which translates to 1 inch on the map representing about 440 miles on the land. A square inch on the map represents approximately 193,600 square miles on the earth's surface. The same atlas contains urban area maps at a scale of 1:300,000, or 1 inch equals 4.7 miles. Here a square inch represents about 22 square miles. The city maps are said to be at a larger scale than are the national maps because their ratio of 1:300,000 is a larger number than is 1:28,000,000 when each are thought of as fractions; that is, 1/300,000 is a bigger number than 1/28,000,000.

Scale is a limiting factor on the amount of information that a map can portray. The smaller the scale, the greater the amount of information that must be left off the map and the greater the amount of data generalization that must take place. At 1:28,000,000, major cities appear simply as small circles on the map. In more densely settled areas, such as the Northeast and southern California, many individual cities do not appear at all, while others show up only as unidentified dots. On urban area maps done at 1:300,000, it is possible to show freeways and some major streets, selected places, and some physical features as well as to portray the boundaries of discrete local political units. On Goode's map of the Los Angeles area, you can find the San Diego Freeway, Dodger Stadium, the Los Angeles River, and the city of Burbank, but you do not find Victory Boulevard, Los Encinos State Historic Park, or the La Brea Tar Pits. You can find universities but not high schools or elementary schools. The choice of map scale, like so many other decisions in the map-making process, depends on the parameters guiding the project. In chapter 6, I point out that the choice of scale can significantly influence the outcome of the analysis of landscape patterns and processes.

Geographic Regions and the Question of Boundaries

Geographers have studied regions for thousands of years. As a field of study, geography dates to ancient Greece as scholars thought to more systematically address age-old questions such as Where is it? and What is it like there?[7] The regional concept remains a key element in geographic scholarship as we seek to gain a better understanding of why and how the landscape is occupied, used, organized, and transformed.

Geographers identify many kinds of regions, but the two most basic types are uniform (or formal) and nodal (or functional).[8] Nodal regions have a core and surrounding zone of influence. New York City is at the center of numerous nodal regions, all of which are geographically different and whose boundaries are loosely known and normally invisible on the landscape. It has international influence with regard to financial decision making, a national reach in terms of television's evening news, and is a regional node for employment in various occupations. Uniform regions display homogeneity regarding the spatial variable or variables that people use to define them. Rarely are they totally homogeneous; rather, the variables of interest simply have a relatively small gradient of change across the region. Often the best one can say about the internal consistency of uniform regions is that with respect to the defining variables, a place within region A is more like another place in region A than any area within adjacent region B. A map of the United States showing population density by county is an example. There is one kind of uniform region that is entirely homogeneous with respect to its defining variable. They are political regions. On a political map of the United States, each state represents an absolutely homogeneous region in that every point within every state is only in that state. There is no gradient of homogeneity within a state, no cores or peripheries. Any point in New York is just as much in New York as any other point within its boundaries.

Geographers define uniform regions in a variety of ways: characteristics of the physical environment, features of human occupancy, or perceptions of space. *Goode's World Atlas* contains several maps of uniform regions in the United States, including groundwater hardness, natural vegetation, physiography, land use, generalized types of farming, and average annual precipitation.[9] Other examples of uniform regions abound, many based on impressions: the Midwest, the South, the rust belt, the sun belt, the corn belt, the Bible belt, Indian country, Cajun country, lake country, hill country, gold country, cattle country, and Tiger country (in reference to a school mascot). Ecosystems are simply another kind of uniform region.

Uniform regions are often spatially fuzzy places with often remarkably imprecise boundaries. Where are the Great Plains? In the *Regional Geography of Anglo America,* the authors describe the difficulties of trying to identify the boundaries of "one of the most universally recognized and precisely bounded regions of Anglo-America."[10] They position the eastern boundary by using several measures. In Nebraska, for example, emphasis is on the change from corn-dominated agricultural landscapes to wheat-dominated landscapes. Field investigation shows that southeastern Nebraska is clearly corn and southwestern Nebraska is clearly wheat, but finding a precise boundary is "an exercise in frustration." The boundary, they say, is best thought of as a "broad transition zone that approaches 300 miles in width."

In the north, they emphasize the shift from a landscape dominated by grass and grain south of the boundary to forest north of it. Observing the landscape reveals an "erratically transitional" zone nearly 100 miles wide. This is because, as with many kinds of vegetational patterns, there is extensive interfingering of vegetative communities. The western boundary is based on the distinction between flat land in the east and slopes in the west. Even here the boundary is a transition zone because of the foothills, hogbacks, and mesas interspersed between Denver, which is clearly on the plains, and the slopes of the Rocky Mountains some twenty-five miles to the west. It is worth pointing out that even though a single region—the Great Plains—is at issue, the authors used different variables at different places to delimit its boundary. On the east the emphasis is on differentiating between two types of exotic species, corn and wheat. To the north the authors base the boundary on distinguishing between exotic grains and natural forests, while to the west the key lies in topography. This is art, not science, and the principal author, Tom McKnight, would be the first to agree.[11]

Are regions real? Are they objective entities? The geographic fraternity has considered these questions for some time, and most reject the idea that regions are actual objects on the landscape. In his 1982 presidential address to the Association of American Geographers (AAG), John Fraser Hart observes that as "preposterous" as it may seem, geographers once invested an "enormous" amount of energy trying to identify real regions.[12] More than twenty years earlier, Richard Hartshorne, a major twentieth-century figure in American geography, laments that an effort was under way to resurrect the idea of regions as real entities. He felt that the notion had already been proven "illusory" in the first half of the nineteenth century and was gratified that most such efforts had "passed into history."[13] Indeed, in 1947 a committee of scholars, convened by the AAG to study the regional concept, ultimately found regions to be a useful intellectual mechanism for arranging and displaying information after emphatically rejecting the idea that regions are objective reality.[14]

Uniform regions that researchers show on maps are artificial. They simply represent one way in which analysts can display phenomena in a spatial context. These regions are frequently representations of subjective classification schemes, as with those related to climate, soils, vegetation, and physiography that are frequently used by cartographers in the construction of ecosystem maps. The vital point here is that maps routinely do not represent boundaries; rather, they create them.[15] This is especially true in the case of ecosystems.

Variables That Cartographers Use to Make Ecosystem Maps: What Do We Know?

Climate is a basic component of the environment. It acts on all living and nonliving terrestrial components of the earth's surface and significantly affects life in the oceans. It sets limits and provides opportunities for biota. It directly sculpts the land surface through weathering and erosion while affecting a variety of physical and chemical processes, from soil formation to the cycling of elements through the environment. In turn, climate can be influenced by features and conditions on the earth's surface, as with mountains blocking airflows and creating dry areas in their "rain shadow" and by snow and ice reflecting incoming solar radiation. Climate is understood to be the "characteristic condition of the atmosphere near the Earth's surface over . . . a particular region."[16] It is the cumulative long-term expression of multiple atmospheric characteristics, such as temperature and precipitation, a "mental construct, composed of statistical abstractions of measured elements . . . and . . . observed weather episodes."[17] As an abstraction, climate does not exist at a given moment in a given place. No one can observe it directly.

Scholars use numerous measures in their study of climate. *Goode's World Atlas,* for example, contains maps of January normal temperatures, July normal temperatures, and normal annual range in temperature as well as temperature-related maps, such as those showing the first killing frost, last killing frost, and average frost-free period. Maps of average annual precipitation, November-to-April precipitation, May-to-October precipitation, and annual variability of precipitation appear. Climatologists, geographers, and atmospheric physicists study barometric pressure, winds, humidity, clouds and cloud cover, dew point, type and intensity of precipitation, evaporation, reflectivity, movement of air masses, and much else in their efforts to gain a better understanding of climate. Indeed, the *Climatic Atlas of the United States* contains some 270 maps of various atmospheric phenomena.[18]

We can say two things with certainty about these maps. They do not portray tangible features of our world, and the location of every boundary is guesswork. This is true even though those who map climatic data work carefully and professionally and have a constantly growing array of techniques available to ensure that the resulting maps are as accurate as possible. Nonetheless, many of the variables that cartographers are mapping are constructs that have no reality on the surface of the earth. Average annual precipitation, normal July temperatures, or annual variability in temperature or precipitation cannot be found on the landscape.

Analysts seek to make sense out of the vast array of climatological data. A long-standing approach has been to construct climate classifications.[19] In doing so, researchers use a small fraction of the available climate variables. These are chosen depending on the purpose of the classification and the factors believed to be most pertinent by its author. Scholars have advanced dozens of different schemes over the years.[20] Robert Bailey of the U.S. Forest Service uses the system devised by Wladimir Köppen in 1931 and revised by Glenn Trewartha in 1968 as the first spatial filter for mapping the ecosystems of the United States.[21] Indeed, the Köppen classification is probably the most widely used in the world today.

Köppen's classification is hierarchical and dependent on hypothetical temperature and precipitation factors. The first level has five categories, which Köppen defines by considerations of mean monthly temperatures. For example, he defines his category of "humid mesothermal climates" on the basis of the coldest month having an average temperature of below 64.4°F but above 26.6°F and having at least one month with an average temperature over 50°F. Mapping the boundaries of Köppen's—or anyone else's—classification always "poses a major problem to classifiers" because the boundaries are as artificial as the classifications.[22] This should not be surprising when you stop to consider that what researchers are mapping are multiple abstractions that they fashion from geographically discontinuous and incomplete data. As a consequence, the thousands of miles of boundaries that appear on the Köppen map of climates for the United States are spatial estimates of a human conceptualization rather than precise geographical statements about a fact of nature.

with natural/seminatural vegetation being native plants plus exotics that have established self-sustaining populations. "Formation" is the finest physiognomic level within the hierarchy. Researchers base it on "ecological groupings of vegetation units with broadly defined environmental and additional physiognomic factors in common." For example, one of the dozens of formations within the herbaceous class is "medium-tall tropical or subtropical grassland with needle-leaved evergreen or mixed trees." The two floristic categories are "alliance" and "community association" that classifiers define as a "physiognomically uniform group of Community Associations sharing one or more . . . diagnostic species" and "a uniform group of vegetation stands that share one or more . . . diagnostic species."

This system is subjective. It incorporates the use of abstractions (mean monthly temperatures), other classifications (leaf morphology), and artificial categorizations (planted/cultivated vs. natural/seminatural vegetation). Subjectivity is apparent throughout. Why is the distinction between a vegetated and nonvegetated area set at 1 percent plant cover instead of 2 percent or 5 percent? There is no ecological significance in differentiating between an area with 25 percent of its vegetative cover dominated by trees with their crowns not touching and an area with 24 percent of its vegetative cover so dominated. Yet under the government's proposal, the first area would be classified and mapped as an area of "open tree cover," but the second would fall into a different category.

Federal reviewers have mixed opinions on the proposed classification. Many managers support standards of some kind but want them to be practical from both scientific and management standpoints. Land managers from Alaska point out that the classification is not well suited to vegetation found in that state. Others note the difficulty of actually obtaining the data called for in the classification for both technical and budgetary reasons. For example, one manager observes that "height of strata beneath the dominant canopy layer, additive canopy closure across all strata, and crown shape are all difficult if not impossible to map with remote sensing techniques presently being used . . . other remote sensing data, such as radar, may prove to be useful to some extent in defining characteristics of the canopy beneath the dominant layer, but this has yet to be demonstrated in an operational capacity." Still others question the temporal usefulness of the data:

> Our experiences on the grasslands of the Northern Great Plains indicates a high level of field verification required to produce a product useful for only a short period of time. Our variable distribution of both precipitation and temperature are tempered by aspect and the many soil types. This often causes yearly as well as seasonal fluctuations in the amount and types of grasses and forbs present false readings as riparian areas. These factors result in yearly variances in interpreting

ground cover, compositions, or production. Annual grasslands in California respond in similar fashion, changing both composition and production in response to the yearly climatic changes.[29]

The proposed federal classification looks at vegetation that is presently on the land. Other views are possible, especially those emphasizing natural vegetation; that is, what the nation's vegetation might look like absent the hand of humans. Potential natural vegetation is considered in major national ecosystem mapping efforts undertaken by the Forest Service and the Environmental Protection Agency examined in the next section. The definitive source for information on potential natural vegetation comes from a classification and map done by Küchler.[30]

For Küchler, potential natural vegetation is "the vegetation that would exist today if man were removed from the scene and if the plant succession after his removal were telescoped into a single moment."[31] The purpose of collapsing time is to eliminate vegetation changes due to climatic fluctuation, an interesting but unrealistic proposition. The definition presupposes that vegetative complexes proceed in a more or less orderly manner down a single discernible path toward a single climax state. Members of the scientific community have seriously questioned this idea for some time so that accurate prediction of the distribution and makeup of future vegetation complexes, even using the definition's implausible assumptions, seems dubious at best.[32] Since there is no way to achieve the "telescoping" of time envisioned in the definition, products resulting from it are entirely hypothetical and not subject to empirical testing or verification.

Küchler considers both physiognomic and floristic factors in his classification. The first level uses physiognomic types—forests, shrubs, and grasslands—with forests subdivided into broad-leaf and needle-leaf classes. He acknowledges multiple different kinds of grassland/forestland areas and several shrub/grassland regions. What Küchler envisions as dominant genera drives his next level, which is floristic. The result is a system with 106 separate categories for the contiguous forty-eight states and an additional seventeen for Alaska and Hawaii. The distribution of the vegetation belonging to many of these categories is discontinuous, even at a national scale, so that several hundred or more separate potential natural vegetation regions appear on his map. Northeastern spruce-fir forests, for example, occur in more than thirty separate places, as do conifer bogs. When looking at Küchler's map of potential natural vegetation with its many hundreds of definitive-looking vegetation regions and their tens of thousands of miles of boundaries, it is important for one to remember that it is all supposition and guesswork.

The preparation of vegetation (as well as all other) maps requires analysts to gather and integrate data to construct a map. A vegetation mapper faces data from multiple sources: field observations, aerial photographs, satellite imagery, and other vegetation maps. These data are of differing quality and completeness, they may have been obtained in different years or seasons and so reflect vegetational responses to unlike temperature and precipitation regimes, and they may have been collected at different scales and for different purposes.[33] No data set can faithfully represent all aspects of vegetation as it exists on the landscape. With field data, for example, the investigator must first decide on what sampling methods he or she will employ, what kind of data he or she will collect (and not collect), and how he or she will obtain and record data. Field investigators can generate data by measurement and estimation. They can record nominal data (presence or absence of a species), ordinal data (plant height categories), or ratio scale data (plant densities), which affect how they can subsequently use the information in their analyses. Data gained from aerial photos and satellite imagery are likewise incomplete, and researchers can use this data only after they have applied numerous interpretive and error correction measures.[34] Essentially every step in the data-handling process requires judgment calls. Researchers leave behind information at every decision point via omission or generalization because it is impossible to identify and record all there is to know about every plant, species, or plant community, regardless of the methods researchers use.

Scholars must integrate multiple data sets to form the boundaries or transition zones that appear on vegetation maps. Resolving the problem of boundaries has bedeviled generations of vegetation mappers. This is because distinctive vegetation boundaries are normally nonexistent, the distributions of individual species constantly shift, and communities of vegetation form and dissolve, and new ones appear in a given area. Cartographers can combine data by visually comparing separate maps of various plant distributions and determining where compromise boundaries are best placed. They can use geographic information systems (GIS). With a GIS, a plethora of spatial data from multiple sources is placed into computerized databases for manipulation. The advent of GIS is a powerful and welcome new tool for use in the analysis of spatial questions of all kinds, including vegetation. It permits much faster processing of much more data than was the case just a decade ago. Nonetheless, data integration requires mental activity rather than number crunching so that while a GIS can aid an analyst in adding, subtracting, and recombining data, their use does not change the fact that the boundaries appearing on vegetation maps at most scales create a false sense of spatial precision as well as the illusion that the landscape is neatly divided into concise patches of discrete and fixed vegetation units.[35]

Scientists often consider soils another vital component of the physical environment when identifying ecosystems. Soils are a key component of the national map of ecosystems done by James Omernik for the Environmental Protection Agency, and they are an essential feature in trying to identify wetland ecosystems. As with climate and vegetation, the mapping of soils means giving spatial expression to an arbitrary classification system.

Humans have been classifying soils for several thousand years. In China, efforts date back some 4,000 years, and Roman scholars were ranking soils with regard to their suitability for plant growth 2,000 years ago.[36] The Russian scientist V. W. Dokuchaev initiated modern thinking about soils in the later quarter of the 1800s when he recognized that soils were much more than a surface covering of weathered rocks. He argued that soil is a substance in its own right that results from complex interactions over time between parent rocks, climate, plants, animals, and topography. Approaching soil from a perspective that it may be seen as something having its own characteristics—rather than simply as a place to grow plants—gave rise to new and diverse efforts at soil classification. In this century numerous formulations appeared in Europe and the United States, often under the aegis of national governments.[37]

In 1975, the U.S. Soil Conservation Service published the principal soil classification system—*Soil Taxonomy*—that scientists use in the United States.[38] They base their classification on soil morphology. That is, they use physical properties of soil, such as structure, texture, parent material, color, horizons, and moisture, to group soils into a six-tiered hierarchy. The most general class is "order," with eleven taxa. Next is "suborder," with fifty-two members, followed by "great groups," containing some 230 taxa. The lowest or finest level is the "series," of which there are more than 12,000 kinds.

Mapping soils as classified by the *Soil Taxonomy* means that cartographers must deal with many of the same kind of spatial uncertainties that we see with regard to climate and vegetation classifications. This is because, as with climate and vegetation, the cartographers are dealing with abstractions. According to S. W. Boul et al., "A taxonomic class, be it a soil order or series, is an abstract concept established to encompass specific ranges of soil properties as a grouping for a taxonomic categorization . . . [and that] . . . these taxa are not 'truths' in and of themselves."[39] Individual soil characteristics, such as color and structure, do not vary uniformly across the landscape. Their individual distributions are imperfectly known, so differing soil types normally grade into one another as opposed to abutting one another along clear lines of demarcation. Thus, in deciding on the placement of soil boundaries on a map, the mapper must make a series of judgments. The results are generalizations and are but one of many possible boundaries.

To summarize, over time maps have been made for many reasons. Arthur Robinson takes a broad view when concluding that maps serve three general func-

Bailey's national maps include four tiers that he calls (in descending order by size) domains, divisions, provinces, and sections (not shown on map 2.1). He uses the highest level of Köppen's climate classification to divide the nation into four domains. He then subdivides these domains into divisions, again using Köppen's ideas of regional climates. Next he uses physiognomic considerations of macrovegetation to partition divisions into provinces and then further subdivides provinces into sections based chiefly on considerations of physiography and floristic characteristics of vegetation. The precise way in which all this is done is not clear since he did not describe how he rolls the various boundaries of the distributions of the individual variables into a single ecosystem boundary. Moreover, for the 1994 map, "Ecoregions and Subregions of the United States," regional teams from the Forest Service create the section-level regions, making it nearly impossible to achieve a consistent weighting of variables across the nation. Recall Monmonier's admonition that "a single map is but one of an indefinitely large number of maps that might be produced . . . from the same data."

What is clear is that changes have occurred between 1976 and 1994. Boundaries at all levels of the hierarchy shift, and new provinces and sections appear. Within the contiguous states, Bailey's 1994 map shows fifty-seven ecosystems at the province level compared to thirty in 1976.[44] The 1994 map by Bailey and his coeditors depicts 164 ecosystems down to the section level in contrast with sixty-one on the 1976 national ecosystem map. Imagine the chaos that would have resulted if we had a federal policy of protecting ecosystems in place in 1976 that policymakers were using to defend the sixty-one section-level ecosystems identified by Bailey at that time, only to have those sixty-one ecosystems be replaced in 1994 by 164 new ecosystems.

James Omernik of the Environmental Protection Agency published an ecosystem map of the forty-eight states in 1987 (map 2.2).[45] Omernik's view is that ecosystems "are reflected in spatially variable combinations of causal factors," which he considers to include "climate, mineral availability (soils and geology), vegetation, and physiography." His map depicts 104 ecosystems.[46] Like Bailey, Omernik factors in climate, potential natural vegetation, soils, and physiography and draws heavily on classifications and maps prepared by earlier researchers.

Omernik takes a fundamentally different path to defining ecosystems than Bailey. Bailey rejects use of a map-overlay approach, but Omernik embraces it.[47] Omernik does not recognize ecosystems as nested hierarchies, so he applies variables simultaneously rather than sequentially. For example, Bailey does not consider potential natural vegetation until the third tier in his hierarchy is reached, but it is factored in at the outset by Omernik. Moreover, while Omernik cites climate as a causal factor in identifying ecosystems, he does not

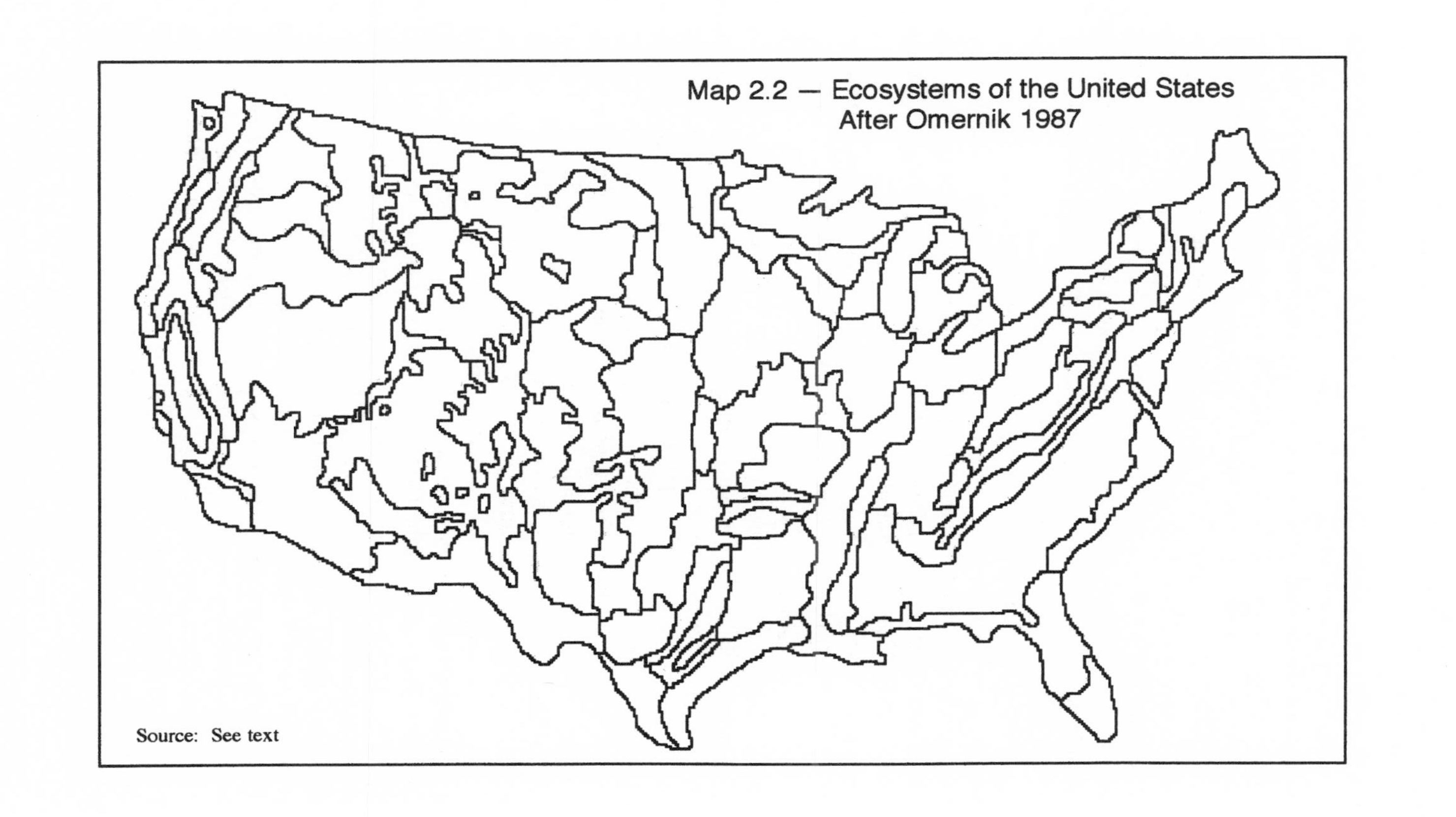
Map 2.2 — Ecosystems of the United States
After Omernik 1987
Source: See text

use Köppen's or any other climate classification in delimiting ecosystems. Instead, he turns to what he calls his "four component maps."

Omernik uses a map-overlay technique for depicting ecosystems. He combines maps of land use, potential natural vegetation, physiography, and soils to "sketch out regions that were relatively homogeneous" for these variables.[48] He recognizes that his source maps vary in accuracy and level of generality and seeks to overcome these inherent deficiencies by consulting other, often larger-scale maps and additional reference sources and then simply making qualitative determinations of where ecosystem boundaries should be placed. There is virtually no possibility that other researchers using exactly the same data considered by Omernik would produce the same map.

In 1995 the U.S. Fish and Wildlife Service (FWS) became the third federal agency to produce an ecosystem map for the nation (map 2.3).[49] Although they consider the work of Bailey and Omernik, they ultimately go in a new direction. Bailey and Omernik use multiple criteria that include both biotic and abiotic components of the environment to craft their notions of the nation's ecosystems. The FWS rejects the multiple-variable route in favor of selecting a single factor on which to base the depiction of ecosystems: watersheds. A 1987 map, "The Hydrologic Units of the United States," prepared by the U.S. Geological Survey (USGS), is the basis of the FWS effort.[50] The USGS identifies 2,150 "cataloging units" in the contiguous states and aggregates these into 352 "accounting units," which are combined into 222 "subregions" for organization into twenty-two "regions." The FWS considers these data and reorganizes them into forty-two ecosystems for the contiguous states.

The FWS ignores living things and interactions among living and nonliving components of the environment that are critical to scientific definitions of ecosystems in crafting their ecosystem map.[51] Its selection of watersheds is even more remarkable because the distribution of most of the nation's biota (as well as abiotic features of the environment, such as climate and soils) are not closely correlated with watersheds. The processes that drive climate, soil formation, and biotic distribution are quite different and by no means find any sort of nexus in watersheds, as a reading of any basic text on physical geography will confirm.[52] Furthermore, as Omernik and Bailey note, surface drainage is poorly defined for about one-third of the contiguous states.[53] Yet so unsettled are the spatial matters surrounding the ecosystem concept that the FWS map includes in the same ecosystem such ecologically dissimilar areas as the dry, barren, and mountainous region of Death Valley; the low-lying well-vegetated rolling scrublands of the southern California coast; and the forested mountains around Monterey Bay south of San Francisco. It is difficult for one to find much science here, but such is the world of delimiting ecosystems.

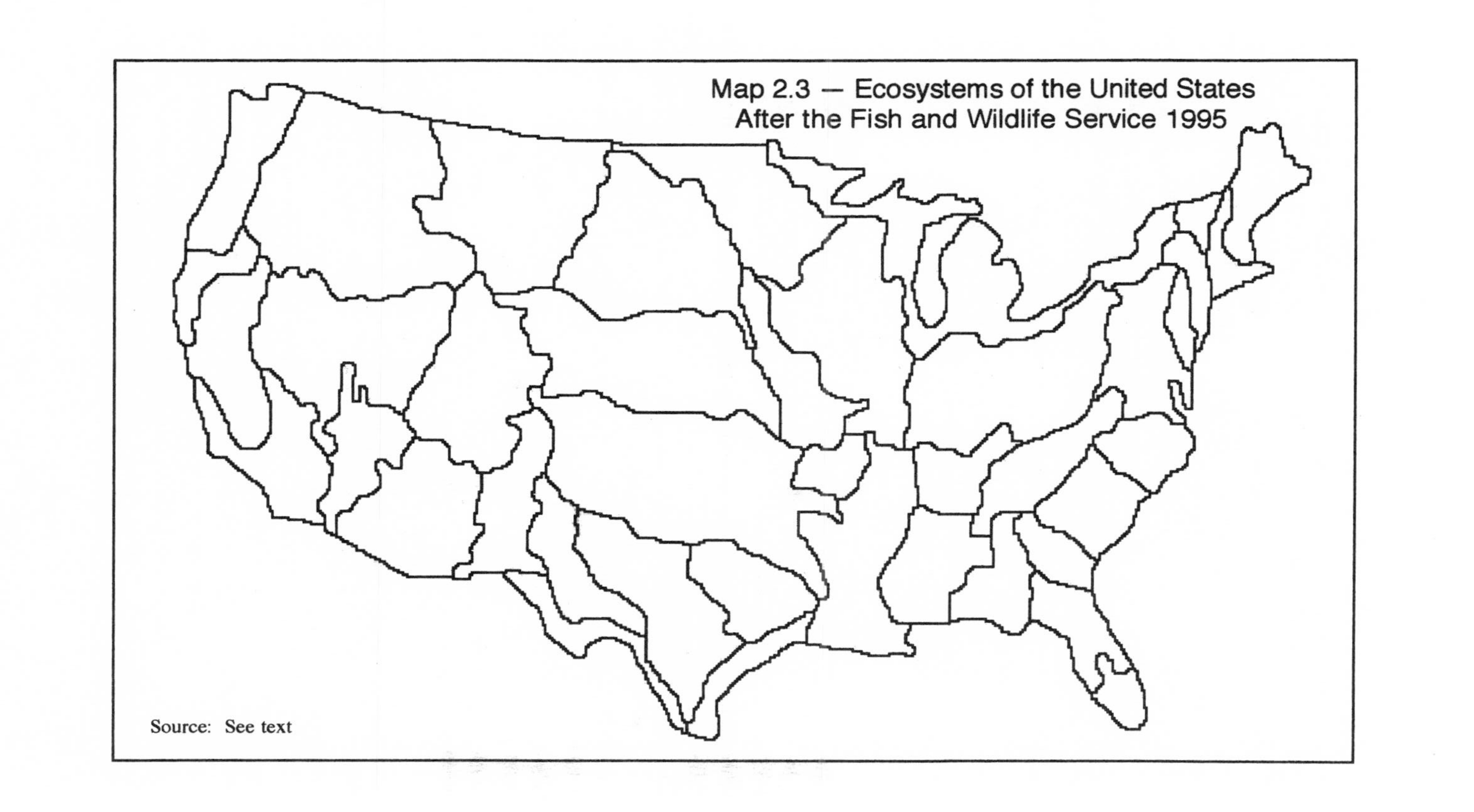
Map 2.3 — Ecosystems of the United States
After the Fish and Wildlife Service 1995

Source: See text

Perhaps the only practical advantage in using watersheds as the basis for delimiting ecosystems is that in the remaining two-thirds of the nation, their boundaries are much more easily located than are boundaries based on multiple criteria. And, of course, they do not shift with the seasons. While this has a certain management appeal, it represents little more than a return to river-basin planning that was fashionable in the 1930s.

Finally, the Sierra Club proposed an ecosystem map for the nation in 1994 (map 2.4).[54] This map, depicting sixteen ecoregions covering the contiguous states, is the basis for a major continuing national effort, The Critical Ecoregions Program. The club has organized throughout the nation in order to influence land and resource decision making so as to save our "home" as represented by these ecoregions. But where did this map come from? What data did the Sierra Club use in its construction? How and why did the club select this data? What means did the club employ to combine data to determine ecosystem boundaries? The club does not tell us.

The Sierra Club map is a classic example of what Bailey terms the "gestalt method" of identifying ecosystems. The club notes that the ecosystems they identify "will send those who need strict definitions into a tizzy. Nature has messy boundaries . . . and so do our ecoregions." To their credit, they acknowledge that their ecosystems often overlap and blend into one another. They define some regions "by their water," others by "geological features," and still others by "vegetation." This information is not particularly illuminating. The phrase "by their water," for example, is meaningless in a geographic or an ecological sense. As we have seen there are several ways to think about vegetation, so claiming to delimit ecosystems on the basis of vegetation without further explanation is not informative. The club's map appears to be little more than the result of some individuals sitting around a table and playing a new parlor game called "let's make up some ecosystems."

The club's method lacks any kind of scientific or public policy merit. It has no apparent analytic basis. Consequently, it should drive those "who need strict definitions into a tizzy." The universe of people that the club's approach to ecosystems tizzifies should include anyone calling themselves a scientist, all public officials, and all those who seek to base environmental and natural resource policymaking on reason and good science.

What can we learn from examining these four maps? Four organizations use four different sets of variables and four different means of melding those variables into a single national ecosystem map. They are free to do so because no ecological theory or methodology dictates otherwise. This is not to say that all the maps are equally valid; those done by Bailey and his colleagues and the Omernik map certainly have greater legitimacy as geographic portrayals of the ecosystem concept than either the FWS or the Sierra Club maps do, if for no

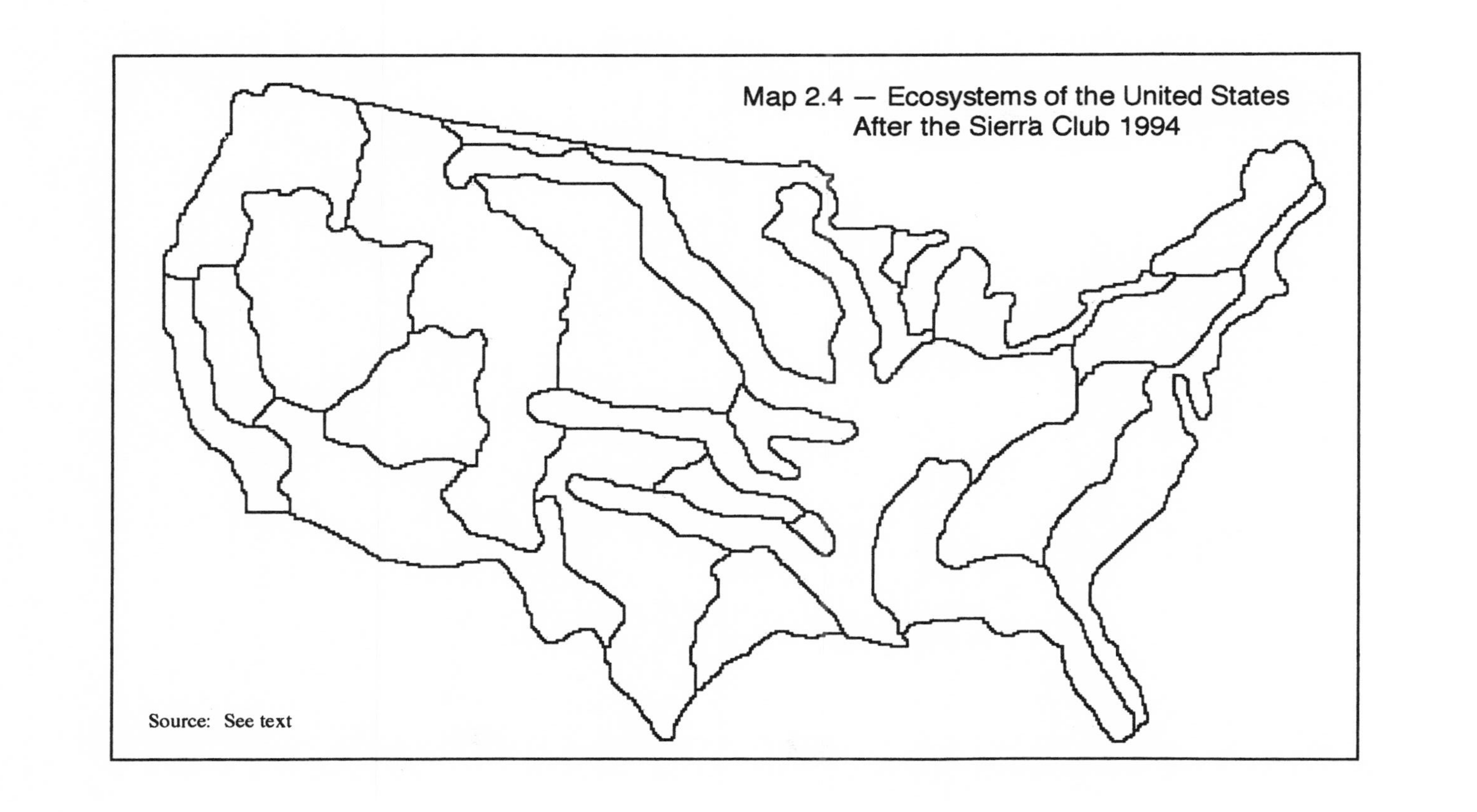

Map 2.4 — Ecosystems of the United States
After the Sierra Club 1994

Source: See text

other reason than that they include both biotic and abiotic features of the landscape in their compilation and do so in something of a systematic way. Taken together, the maps do not agree on the number, size, shape, or location of ecosystems for the nation. To be sure there are areas of similarity between some of the maps (e.g., the Bailey and Omernik maps for the Pacific Coast). On the other hand, Bailey divides Kansas into thirds vertically at the province level (seven jigsaw puzzle shapes at the section level), FWS cuts the state in half horizontally, and Omernik splits it into six irregular pieces.

The relevant public policy question becomes, Does one of these maps (or any other map)—to the exclusion of all other ecosystem maps—provide a rational geographic basis for identifying the ecosystems to be protected by the federal government under the new paradigm? The unequivocal answer is no. At best these maps are subjective cartographic representations of subjective classifications that rest on earlier subjective mapping and classifying activities. No one can verify them or independently reproduce them. At worst, they result from idiosyncratic speculation and mental meanderings without any attempt at employing precise and replicable methodologies or using what little scientific understanding exists regarding the spatial attributes of the ecosystem concept.

Thus far, I have looked at ecosystems in general. The time has come for a case study. The next chapter introduces the Greater Yellowstone Ecosystem and examines attempts to define and locate it as the effort to move to a new paradigm management regime for the region gets under way. Not surprisingly, each of the four ecosystem maps we have just reviewed treat the Yellowstone area differently. Omernik subsumes it within one or, perhaps, two ecosystems. The Sierra Club includes it within a single vast ecosystem reaching from New Mexico to beyond the Canadian border. It appears to be within one or two ecosystems at Bailey's province level and at least four separate ecosystems at the section level. Finally, the FWS assigns the area to two separate ecosystems. None of the maps is congruent, although Bailey at the province level is similar to Omernik. I turn now to the Greater Yellowstone Ecosystem.

Notes

1. U.S. General Accounting Office, *Ecosystem Management: Additional Actions Needed to Adequately Test a Promising Approach,* GAO/RCED-94–111 (Washington, D.C.: U.S. General Accounting Office, August, 1994), 42.

2. Mark Monmonier, *How to Lie with Maps,* 2nd ed. (Chicago: University of Chicago Press, 1992), 2.

3. Denis Wood, "The Power of Maps," *Scientific American,* May 1993, 90.

4. There are numerous texts available on cartography; for example, see Arthur H. Robinson, Joel L. Morrison, Philip C. Muehrcke, A. Jon Kimerling, and Stephen C.

Guptill, *Elements of Cartography,* 6th ed. (New York: John Wiley & Sons, 1995), and John Campbell, *Map Use and Analysis* (Dubuque, Iowa: Wm C. Brown, 1991).

5. Edward B. Espenshade Jr. and Joel L. Morrison, eds., *Goode's World Atlas,* 18th ed. (Chicago: Rand McNally, 1991), x–xii; John C. Garver Jr., "New Perspective on the World," *National Geographic* 174, no. 6 (December 1988): 910–13. There are a potentially unlimited number of projections, and analysts routinely propose new ones. Some become highly contentious, underscoring the importance of their biases and demonstrating the passions of their users. The debate of the so-called Peters' Projection is a case in point. Peters believes that it overcomes what he feels are the biases in favor of midlatitude developed nations present in the Mercator projection, which exaggerates size as you proceed poleward. According to Peters, the Mercator is rooted in "a colonial and racist mentality," and its continued use nurtures such views. Peters is quoted in Campbell, *Map Use and Analysis,* 287. A detailed account of the controversy can be found in Mark Monmonier, *Drawing the Line* (New York: Henry Holt, 1995), 9–43. See also Arthur H. Robinson, "Arno Peters and His New Geography," *The American Cartographer* 12, no. 2 (October 1985): 103–11; John Loxton, "The Peters Phenomenon," *The Cartographic Journal* 22, no. 2 (December 1985): 106–8; and German Cartographic Society, "The So-Called Peters Projection," *The Cartographic Journal* 22, no. 2 (December 1985): 108–10.

6. John C. Hudson, "Scale in Space and Time," in *Geography's Inner Worlds: Pervasive Themes in Contemporary American Geography,* ed. Ronald F. Abler, Melvin G. Marcus, and Judy M. Olson (Washington, D.C.: Association of American Geographers, 1992), 280–97.

7. Preston James, *All Possible Worlds: A History of Geographical Ideas* (Indianapolis: Bobbs-Merrill, 1972).

8. John Fraser Hart refers to a 1947 study commissioned by the Association of American Geographers that identified more than fifty separate types of regions. John Fraser Hart, "Presidential Address: The Highest Form of the Geographer's Art," *Annals of the Association of American Geographers* 72 no. 1 (March 1982): 1–29.

9. Espenshade and Morrison, eds., *Goode's World Atlas.*

10. C. Langdon White, Edwin J. Foscue, and Tom L. McKnight, *The Regional Geography of Anglo-American,* 5th ed. (Englewood Cliffs, N.J.: Prentice Hall, 1979), 98.

11. Tom McKnight served on my doctoral dissertation committee and first opened my eyes to the characteristics of regions during courses on Anglo-America.

12. Hart, "Presidential Address."

13. Richard Hartshorne, *Perspectives on the Nature of Geography* (Chicago: Rand McNally, 1959).

14. John Fraser Hart, "Presidential Address," 9; Derwent Whittlesey, "The Regional Concept and the Regional Method," in *American Geography Inventory and Prospect,* ed. Preston E. James and Clarence F. Jones (Syracuse, N.Y.: Association of American Geographers/Syracuse University Press, 1954), 19–69. The first subject matter chapter of this volume and its contemporary counterpart, *Geography's Inner Worlds,* deals with regions. The books' subsequent chapters bear little resemblance to each other.

15. Denis Wood, *The Power of Maps* (New York: Guilford Press, 1992).

16. Alan H. Strahler and Arthur N. Strahler, *Modern Physical Geography,* 4th ed. (New York: John Wiley & Sons, 1992), 141.

17. Paul E. Lydolf, *Weather and Climate* (Totowa, N.J.: Rowman and Allanheld, 1985), 82.

18. National Oceanic and Atmospheric Administration, *Climatic Atlas of the United States* (Washington, D.C.: U.S. Department of Commerce, 1983).

19. In the second century A.D., Ptolemy divided the world into zones based on the amount of daylight. Today climatological scholarship and research covers much more than classification. It includes efforts to develop theory and a deeper understanding of the processes and forces that create the various components of climate to enable us to make reasoned and accurate predictions of potential future climates. The general area of climate change has received an enormous amount of attention and research funding over the last two decades as concerns about the impact of human activities on climate have been raised in many quarters.

20. John E. Oliver and L. Wilson, "Climatic Classification," in *The Encyclopedia of Climatology,* ed. John E. Oliver and Rhodes W. Fairbridge (New York: Van Nostrand Reinhold, 1987), 221–37. See also Anastasios A. Tsonis, "Climate," in *Encyclopedia of Climate and Weather,* ed. Stephen H. Schneider (New York: Oxford University Press, 1996), 122–26; Richard C. J. Somerville and Diane Manuel, "Weather and Climate," in Schneider, ed., *Encyclopedia,* 127–31; John J. Hidore and John E. Oliver, *Climatology: An Atmospheric Science* (New York: Macmillan, 1993); Lydolf, *Weather and Climate;* and Glenn T. Trewartha and Lyle H. Horn, *An Introduction to Climatology,* 5th ed. (New York: McGraw-Hill, 1980).

21. Robert G. Bailey, *Description of the Ecoregions of the United States,* 2nd ed., Miscellaneous Publication No. 1391 (Washington, D.C.: USDA/Forest Service, 1995).

22. Oliver and Wilson, "Climatic Classification," 222.

23. Reed F. Noss, Edward T. LaRoe III, and J. Michael Scott, *Endangered Ecosystems of the United States: A Preliminary Inquiry,* Biological Report No. 28 (Washington, D.C.: U.S. Department of the Interior, National Biological Service, February 1995).

24. Paul Risser, "The Status of the Science Examining Ecotones," *BioScience* 45, no. 5 (May 1995): 318–25.

25. U.S. Forest Service, "Forest Type Groups of the United States," map, U.S. Forest Service, Washington, D.C., 1993.

26. See generally A. W. Küchler and I. S. Zonneveld, eds., *Vegetation Mapping* (Dordrecht: Kluwer Academic Publishers, 1988), and Robert H. Whittaker, ed., *Classification of Plant Communities* (The Hague: Dr. W. Junk, 1978).

27. A. W. Küchler, "The Classification of Vegetation," in Küchler and Zonneveld, eds., *Vegetation Mapping,* 67.

28. Federal Geographic Data Committee Vegetation Subcommittee, *FGDC Vegetation Classification and Information Standards* (Reston, Va.: U.S. Department of the Interior, U.S. Geological Survey, Federal Geographical Data Committee Secretariat, June 3, 1996).

29. H. Gyde Lund, executive secretary of the Federal Government Data Committee Vegetation Subcommittee, "Sorted Comments on FGDC Draft Vegetation Standards," letter to FGDC Vegetation Subcommittee members, October 23, 1996, downloaded from http://www.nbs.gov/fgdc.veg/comments.htm, November 4, 1996.

30. A. W. Küchler, "Potential Natural Vegetation," map, in *The National Atlas of the*

United States (Washington, D.C.: U.S. Department of the Interior, U.S. Geological Survey, 1985).

31. Küchler, "Potential Natural Vegetation," map (text on the back of the 1985 map).

32. William L. Baker and Gillian M. Walford, "Multiple Stable States and Models of Riparian Vegetation Succession on the Animas River, Colorado," *Annals of the Association of American Geographers* 85, no. 2 (June 1995): 320–38; Karl S. Zimmerer, "Human Geography and the 'New Ecology': The Prospect and Promise of Integration," *Annals of the Association of American Geographers* 84 no. 1 (March 1994): 108–25; Joel B. Hagen, *An Entangled Bank: The Origins of Ecosystem Ecology* (New Brunswick, N.J.: Rutgers University Press, 1992).

33. Thomas R. Loveland, James W. Merchant, Jesslyn F. Brown, Donald O. Ohlen, Bradley C. Reed, Paul Olson, and John Hutchinson, "Seasonal Land-Cover Regions of the United States," article and map supplement, *Annals of the Association of American Geographers* 85, no. 2 (June 1995): 339–55.

34. N. J. Mulder, "Digital Image Processing, Computer-Aided Classification and Mapping," in Küchler and Zonneveld, eds., *Vegetation Mapping.*

35. Isaak S. Zonneveld, "The Land Unit—a Fundamental Concept in Landscape Ecology, and Its Applications," *Landscape Ecology* 3, no. 2 (1989): 67–86; Michael Goodchild and Sucharita Gupta, "Preface," in *The Accuracy of Spatial Databases,* ed. Michael Goodchild and Sucharita Gupta (London: Taylor and Francis, 1989), xi–xv.

36. Delvin S. Fanning and Mary C. Fanning, *Soil: Morphology, Genesis, and Classification* (New York: John Wiley & Sons, 1989).

37. For a review of several soil classifications, see S. W. Boul, F. D. Hole, and R. J. McCracken, *Soil Genesis and Classification,* 3rd ed. (Ames: Iowa State University Press, 1989).

38. Soil Survey Staff, *Soil Taxonomy: A Basic System of Soil Classification for Making and Interpreting Soil Surveys,* U.S. Department of Agriculture Handbook No. 436 (Washington D.C.: USDA, 1975). The Soil Conservation Service was folded into the newly established Natural Resources Conservation Service within the U.S. Department of Agriculture in 1994.

39. Boul et al., *Soil Genesis and Classification,* 343.

40. Arthur H. Robinson, *Early Thematic Mapping in the History of Cartography* (Chicago: University of Chicago Press, 1982), 3.

41. Robert G. Bailey, "Ecoregions of the United States," map, U.S. Forest Service, U.S. Department of Agriculture, Ogden, Utah, 1976. The original map sheet measures thirty-two by forty-four inches and contains text describing the process for determining the ecoregions. Fuller descriptions of the ecosystems on this map are found in Robert G. Bailey, *Description of the Ecoregions of the United States,* Miscellaneous Publication No. 1391 (Washington, D.C.: U.S. Department of Agriculture, U.S. Forest Service, 1980), and Robert G. Bailey, "Delineation of Ecosystem Regions," *Environmental Management* 7, no. 4 (1983): 365–73. A second edition of *Description of the Ecoregions of the United States* was published in 1995 and was accompanied by the 1994 revision of the map "Ecoregions of the United States." Each was authored by Bailey. A third map, and the second one with a 1994 publication date, depicted a further geographic breakdown of ecoregions and was prepared by Robert G. Bailey, Peter E. Avers,

Thomas King, and Henry McNab, eds., "Ecoregions and Subregions of the United States," U.S. Forest Service, U.S. Department of Agriculture, Washington, D.C.,1994). See also ECOMAP, *National Hierarchical Framework of Ecological Units* (Washington D.C.: U.S. Department of Agriculture, U.S. Forest Service, October 29, 1993).

42. Bailey, "Ecoregions of the United States," map, 1976.

43. Robert E. Bailey, *Ecosystem Geography* (New York: Springer, 1996), 20.

44. Bailey identified fifty-two ecosystems at the province level in 1994, but some are discontinuous. In these cases I have tallied each separate occurrence as an ecosystem.

45. James M. Omernik, "Ecoregions of the Conterminous United States," *Annals of the Association of American Geographers* 77, no. 1 (March 1987): 118–25.

46. Omernik identified seventy-six ecosystems, but some are discontinuous. In these cases I have tallied each separate occurrence as an ecosystem.

47. Bailey, *Ecosystem Geography,* 28–29.

48. These included two maps used by Bailey: Küchler's map of potential natural vegetation and E. H. Hammond, "Classes of Land Surface Form," in *The National Atlas of the United States,* a map of land-surface form (physiography). Also used were J. R. Anderson, "Major Land Uses," in *The National Atlas of the United States of America,* and several soils maps.

49. U.S. Fish and Wildlife Service, *An Ecosystem Approach to Fish and Wildlife Conservation* (Washington, D.C.: Fish and Wildlife Service, February 1995).

50. Paul R. Seaber, F. Paul Kapinos, and George L. Knapp, *Hydrologic Unit Maps,* Water Supply Paper No. 2294 (Washington, D.C.: U.S. Department of the Interior, U.S. Geologic Survey, 1987).

51. In determining just how to aggregate watersheds, the FWS did consider vegetation, cover type, and physiography.

52. Tom L. McKnight, *Physical Geography,* 5th ed. (Upper Saddle River, N.J.: Prentice Hall, 1996); James Omernik and G. E. Griffith, "Ecological Regions versus Hydrologic Units: Frameworks for Managing Water Quality," *Journal of Soil and Water Conservation* 46 (1991): 224–340.

53. James M. Omernik and Robert G. Bailey, "Distinguishing between Watersheds and Ecoregions," *Journal of the American Water Resources Association* (formerly *Water Resources Bulletin*) 33, no. 5 (October 1997): 935–49.

54. Jane Elder, "At Home on the Planet," *Sierra,* March/April 1994, 53–56.

Chapter 3

Visions of the Greater Yellowstone Ecosystem

The overarching goal is to conserve the sense of naturalness and maintain ecosystem integrity.

—Greater Yellowstone Coordinating Committee (of federal agencies), "Vision for the Future: A Framework for Coordination in the Greater Yellowstone Area," 3–7

The draft [Vision] clearly does not go far enough . . . development activities must be precluded if they impair or destroy Ecosystem resources or values.

—Greater Yellowstone Coalition (of environmental groups), "Be Visionary: Send Your Comments on the GYCC 'Vision,'" *Ecoaction,* August 31, 1990, 1, 2

The "Vision document" . . . erroneously assumes that the region has been overrun by pollution, unsightly development, over population, and careless business and commercial activity.

—Wyoming Multiple Use Coalition (of business interests), "Review of Vision for the Future Document," 4

Many Americans think of Yellowstone National Park as a special place. As our first national park it is a testament to our societal commitment to sound environmental stewardship.[1] In the public's mind it is a land of grizzly bears, moose, and elk as well as the spectacular scenes of Old Faithful, the Grand Canyon of the Yellowstone River, and much more. To many environmentalists, scientists, and National Park Service personnel, however, it is the center of what they often refer to as "the largest essentially intact natural ecosystem in the temperate zone." Longstanding interest in the area plus the predominance of federal lands in the region—millions of acres of national forests surround the park—make the area a very attractive testing ground for new paradigm ideas. The effort to have the federal government adopt new land management policies in order to protect a large ecosystem was first attempted for the so-called Greater Yellowstone Ecosystem (GYE). This chapter examines the notion of the GYE and the attempts to make its protection the top priority in regional land use policy.

Government officials have studied the Yellowstone region since Brigadier General W. F. Reynolds explored the area in 1859–60. Second Lieutenant Gustavus Doane, a participant in an 1870 expedition, wrote in a well-publicized report of the time that "as a country for sightseers, it is without parallel; as a field for scientific research, it promises great results; in the branches of geology, mineralogy, botany, zoology, and ornithology, it is probably the greatest laboratory that nature furnishes on the surface of the globe."[2] When Congress set aside the over 2-million-acre Yellowstone National Park in 1872, interest in the region grew, but its isolation prevented more than a handful of people from visiting the area for many years. Over time, more people came to the park, and still others settled in the region, being drawn largely by forest and mineral resources and the area's agricultural potential. The region remains sparsely populated, but people have altered portions of the landscape with roads, fields, settlements, mines, and timber harvesting. Their actions and their policies changed conditions for many living things in the region.

Researchers interested in one of the region's most spectacular species, the grizzly bear, coined the term "Greater Yellowstone Ecosystem." John and Frank Craighead Jr. began to study the grizzly in Yellowstone and Grand Teton National Parks as well as sections of four adjacent national forests in 1959. Twenty years later, Frank Craighead Jr. included a map of what he termed the "greater Yellowstone ecosystem" in his book *Track of the Grizzly*.[3] The Craigheads announced a new region and produced the map to depict it (map 3.1). Yet Boise State University's John Freemuth observes that defining the boundary "is stunningly problematic."[4]

The Craighead's approach to defining the GYE is straightforward. They use only the range of the grizzly bear population to locate the boundary. It is simply "the extent of the grizzly bear's range in the region surrounding Yellowstone National Park," and the Craigheads fashion it by connecting individual points established by "bear sightings, kills, or radio fixes."[5] In other words, they connect the dots. The Craigheads define their GYE by using straight lines. Essentially, it is a polygon that "comprises some 5 million acres." Others seized on the concept of a GYE and came forward with their own ideas about how one should define and locate it. These subsequent views of the GYE are totally consistent with the arbitrary and imprecise nature of the ecosystem concept.

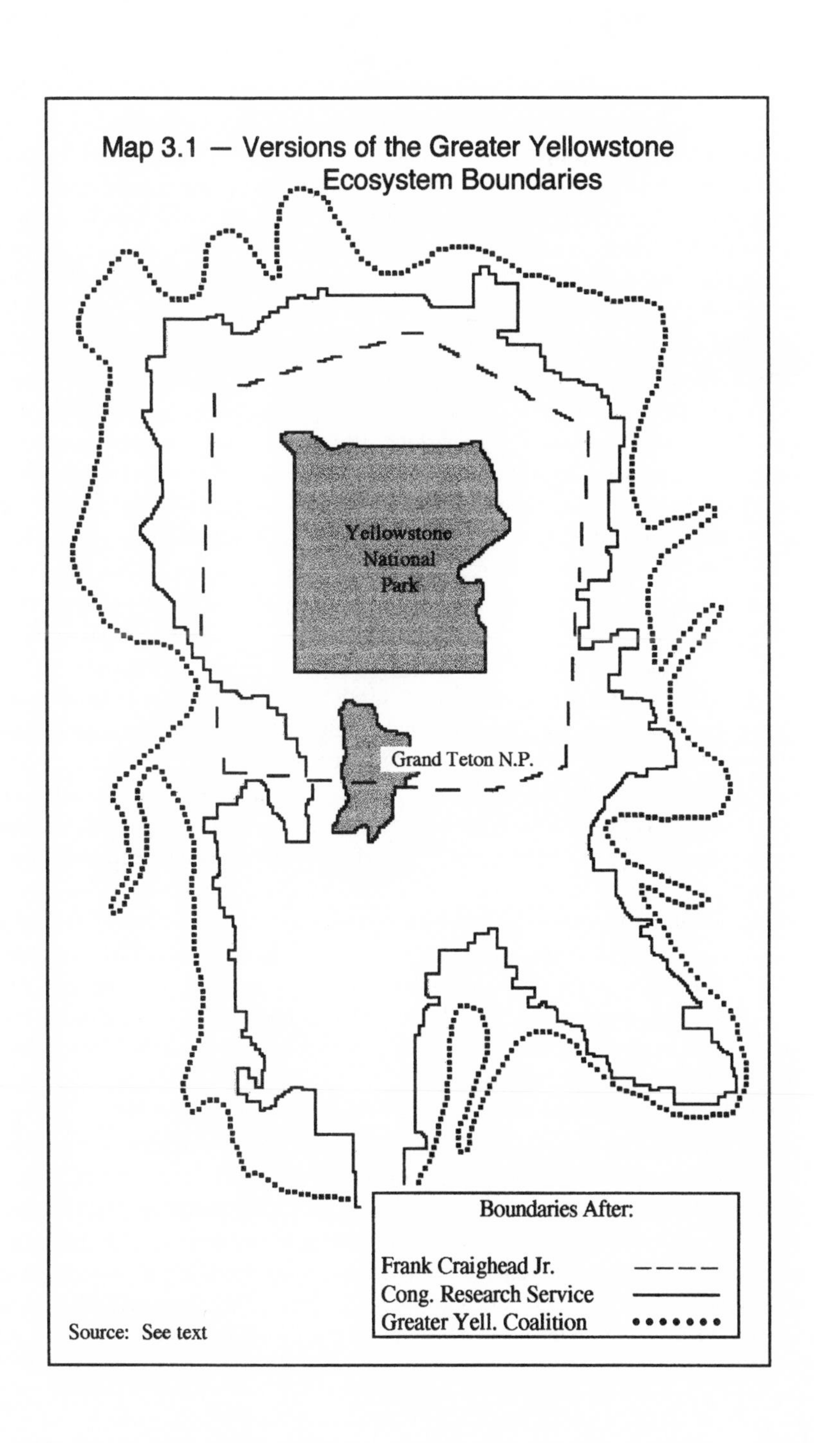
Map 3.1 — Versions of the Greater Yellowstone Ecosystem Boundaries
Yellowstone National Park
Grand Teton N.P.
Boundaries After:
Frank Craighead Jr.
Cong. Research Service
Greater Yell. Coalition
Source: See text

Rick Reese was one of the first to latch on to the idea of the GYE. He is the chief founder and first president of the Greater Yellowstone Coalition (GYC), the major umbrella group for environmental organizations seeking to influence land use policy in the region. While the Craigheads were generally motivated by a scientific concern with grizzly bears in proposing a GYE, something quite different pushes Reese and the GYC: nature worship. In the foreword to the second edition of Reese's *The Greater Yellowstone: The National Park & Adjacent Wildlands,* Terry Williams writes, "The Greater Yellowstone is the nation's medicine bundle. It must remain whole and intact. Let us send up our prayers, remembering the grizzly, remembering the wolf, remembering the white wings of the trumpeter swans. May we never forget each pilgrimage is holy."[6] In the first edition of *The Greater Yellowstone,* Reese estimates the holy ground to have "encompassed 8 to 10 million acres."[7] By the time the second edition was published in 1991, the GYE had grown to cover a core area of some 18 million acres plus additional acreage in transition zones to surrounding ecosystems.[8]

In *The Greater Yellowstone,* Reese writes that "how we define the geographic extent of the Greater Yellowstone depends on which of its characteristics we use as criteria for delineating boundaries."[9] He observes that wildlife biologists could use the combined ranges of all the wildlife centered on Yellowstone National Park and choose as a boundary the outermost barriers to movement. Geologists, on the other hand, might employ "major geological formations or landforms, or might include large areas of geothermal resources that lie" beyond park boundaries. He cites vague hydrographic criteria as another possible means of establishing the GYE's boundaries. Ultimately, he dismisses these approaches as lacking in the "precision, comprehensive scope, and objectivity we desire." In their place he opts for a GYE that finds its basis in a 1976 national ecosystem map that Robert Bailey produced that in turn is based on Küchler's map of potential natural vegetation.[10] Reese writes that "the concept of potential natural vegetation provides us with a definition of the Greater Yellowstone Ecosystem." Using the boundaries from Küchler's map, Bailey combines several of Küchler's classes of potential natural vegetation into larger units that he then labels "ecoregions."

On the Bailey-Küchler map, Reese's GYE is part of a much larger area labeled "M3110." It appears as a Rorschach shape gone mad, having tentacles reaching into northern New Mexico, diagonally across Utah, through Idaho, and more than halfway across Oregon. Reese does not explain how he extracts the GYE from this much larger region. One thing, however, is certain. We have already seen the speculative nature of Küchler's map, so if Reese is correct and this is the best we can do in terms of precision, comprehensiveness, and objectivity, then other approaches must be completely awash in guesswork from start to finish.

The GYC adopts the GYE map from the second edition of Reese's *The Greater Yellowstone* in its 1991 book *An Environmental Profile of the Greater Yellowstone Ecosystem,* but the GYC offers a totally different, and utterly insubstantial, explanation of how Reese crafted the map (map 3.1).[11] Editors Dennis Glick, Mary Carr, and Bert Harting acknowledge that "because of the dynamic nature of a complex ecosystem and the nearly limitless layers that exist within and without, the exact boundaries of a particular system may vary in time."[12] They assert that Reese uses many criteria to draw boundaries around the GYE; including climate, vegetation, geology, geothermal features, hydrography, topology, and the ranges and habitat needs of grizzlies, elk, deer, bison, mountain lions, bald eagles, trumpeter swans, and peregrine falcons. They curiously omit potential natural vegetation, which, according to Reese, is the basis for the map in the first place. Glick and his coeditors also refer to "critical functions," such as "pathways of material and energy flow, the space and other ingredients for successful interactions of the individual organisms that together define and sustain Greater Yellowstone."[13] Then, without a hint of how they determine the distributions of all these variables or how they subsequently blend them into the boundary of the GYE, Glick et al. present the same map used by Reese and proclaim that "complex relationships among plants, animals, climate, geology, hydrology, topography, and geothermal activities, rather than political boundaries, define this system." There is no geographic substance here, yet the authors of an article in the prestigious journal *Science* present this map as factual, giving ample witness to geographers' laments about the general cartographic ignorance of many map users.[14]

Regional officials from the National Park Service and the U.S. Forest Service established the Greater Yellowstone Coordinating Committee (GYCC) in the early 1960s. It stems from the realization that these two principal federal land managing agencies needed to coordinate their efforts more closely because over 10 million acres of national forests completely surround Yellowstone and Grand Teton National Parks. It is the GYCC that took the federal lead in addressing the matter of the GYE (which it prefers to all the Greater Yellowstone Area, or GYA), an effort that I can best describe as ethereal.[15]

In 1987 the GYCC published an extensive volume on the region.[16] The work, *The Greater Yellowstone Area: An Aggregation of National Park and National Forest Plans,* contains an enormous amount of data on biota, the physical environment, and socioeconomic factors. It is based on decades of study, countless analyses, and over a century of land management experience. The report contains fifty maps covering, for example, vegetation classification; noxious and exotic weeds; seasonal ranges of white-tailed deer, mountain goats, bighorn sheep, bison, moose, and antelope; habitats of several species; and general geology. It concludes that the GYE covers 19,906,000 acres that include national forests

(10,029,000 acres), national parks (2,567,000 acres), other federal agencies (907,000 acres), state lands (685,000 acres), private lands (4,838,000 acres), and Indian reservations (880,000 acres).[17] Given this level of detail and the amount of analysis that lay behind the report, it is reasonable to expect that the GYCC would produce an authoritative map of the GYE. It does not.

The GYCC's map of the GYE is a remarkable example of cartographic doublespeak because it does not show an ecosystem.[18] On pages 1–4, the authors of *The Greater Yellowstone Area: An Aggregation of National Park and National Forest Plans* present a map of what they call the "Greater Yellowstone Ecosystem," but the map contains no ecosystem boundary of any kind unless the edge of the map itself is supposed to serve that purpose, a most improbable supposition. The map simply depicts a large rectangular region extending some 160 miles east to west and 225 miles north to south and shows the lands managed by the Park Service, Forest Service, and Fish and Wildlife Service.[19] The federal lands are not all contiguous. It includes disconnected portions of three national forests. It displays major roads, settlements, rivers, streams, and lakes. The private, state, Indian, and other federal lands that the GYCC claims to be a part of the GYE do not appear on this map, although there is a map of land ownership elsewhere in the report.[20]

The authors of the report try to obscure the geographic shortcomings of the their work by proclaiming that "specific boundaries are not shown for the Greater Yellowstone Area. Rather . . . the resources and the area they cover define borders."[21] Figuring out how "the resources and the area they cover" do this is left up to the creativity of the reader. They do not suggest a method for selecting variables to use in crafting a boundary or offer any means to do so. The GYCC's effort is the rough equivalent of Omernik's publishing a blank map of the United States that he labels "Ecosystems of the United States" accompanied by maps of land use, potential natural vegetation, physiography, and soils and then telling the reader that these variables and "the areas they cover define ecosystem borders." The GYCC's approach to delimiting the GYE is implausible.

The Congressional Research Service (CRS) proposes another version of the GYE (map 3.1). Spurred by claims that people and industries were holding Yellowstone National Park under siege, the House of Representatives held hearings on the state of the park in 1985.[22] One result was *Yellowstone: Ecosystem, Resources, and Management,* published by the CRS in 1986.[23] The CRS concedes the impossibility of finding precise boundaries for the GYE and more generally says that "it is simply impossible to draw a line through an area and then assert that one independent ecosystem lies on one side of the line and another on the other side." Seemingly oblivious to the management significance of these conclusions, however, the CRS observes that "the ecosystem approach can . . . be useful as a management tool."[24] By eyeballing maps of topography, climate, and vegetation, the CRS produces a preliminary map of the GYE that they then compare to administrative boundaries of federal lands. When the boundary on the preliminary map was within a few miles of an administrative boundary, they adopt the latter as the GYE boundary. On the CRS map, straight lines connected at right angles predominantly compose the boundary of the GYE. It covers about 14 million acres and includes lands administered by federal and state agencies as well as tribal lands and some in private ownership.

Other researchers produce different 14-million-acre versions of the GYE. One exists in *The Greater Yellowstone Ecosystem,* edited by Robert Keiter, then at the University of Wyoming's Law School, and Mark Boyce of the university's Department of Zoology and Physiology.[25] Their map is similar to that of the GYCC in that it only identifies federal lands and contains no other boundaries, leaving it to the reader's imagination to decide whether the GYE is in fact congruent with the federal lands. Keiter notes that the GYE "cannot be easily defined by boundaries of any lasting significance."[26] Duncan T. Patton, writing in the same volume, agrees that the GYE "has no definite boundaries" but goes on to say that "it is bound by its ecological unity of cohesiveness."[27] Readers can be justifiably confused by a sentence that simultaneously asserts that something has no boundaries but is nonetheless bounded. Using a similar map with a basis in administrative boundaries, Seymour Fishbein, in his book for the National Geographic Society, *Yellowstone Country,* writes that "biology doesn't help much . . . in establishing firm [boundary] lines."[28] In contrast, Tim Clark and Dusty Zaunbrecher embrace the importance of biology when lamenting that "major administrative units do not coincide with actual biological boundaries of the ecosystem."[29] Despite this view, Clark and Zaunbrecher present a map that depends heavily on the same administrative boundaries found on other GYE maps. Confusion reigns.

In a thoughtful 1994 book, Tim Clark and Steven Minta present a 19-million-acre GYE that does not follow administrative boundaries.[30] They do not explain how they derive their boundary or even make it clear what variables they use in its creation, so their map is simply one more subjective and arbitrary view of the GYE. A major contribution of the book is the explicit recognition that "some of the definitions in currency are . . . clearly at odds with one another . . . there is disagreement on the location and extent of the GYE . . . [which simultaneously] cannot be 5 million and 19 million acres."[31] They find that differing views of the size and location of the GYE significantly impacts the perceptions of the land management issues facing the region as well as possible solutions. They are quite correct.

Analysts do not concur on just what constitutes the GYE two decades after the Craigheads first proposed it. No agreement on boundaries. No agreement on size. No agreement on shape. No agreement on the variables to use in defining the GYE. The only common characteristic of all the GYE maps is that they contain Yellowstone National Park, at least a portion of Grand Teton National Park, and some fraction of the national forests immediately adjacent to Yellowstone. This is not to say that the various maps are wholly dissimilar; similarities are plainly evident among some of the maps. Nonetheless, in the minds of various students of the Yellowstone area, the GYE can vary in size from 5 million to 19.6 million acres, a fourfold difference. Some believe that the boundary of the GYE is determined by a single criterion; others use multiple criteria. For some the GYE's boundary coincides with administrative boundaries of federal lands; for others it does not. For some the boundaries are chiefly straight lines; for others they are curvilinear. Some researchers see the GYE as a single continuous area; others view it as possessing discontinuous outliers. It may have a compact shape or, in Fishbein's words, be reminiscent "of a headless colossus, teetering on crumbling legs."[32] The GYCC sees federal lands accounting for 69 percent of the GYE, but the CRS considers them to occupy 92 percent of the GYE. The GYE contains some 4.8 million acres of private land according to the CYCC but only about 950,000 acres according to the CRS. In short, after an enormous investment of time and effort, "clear designation of the boundaries of the greater Yellowstone ecosystem remains somewhat vague."[33]

With no basic understanding of how to define the GYE or of its location, there is little wonder that there are significant differences among various stakeholders regarding how the government should manage this area. This controversy played out in the debate over the government's so-called Vision document for the GYE.

The "Vision Document"

Interest in management of the GYE took on new intensity in October 1985, when the House of Representatives held a hearing on federal land management in the GYE.[34] Several parties strongly criticized the management practices of the National Park Service and Forest Service. Stung, the two agencies moved to increase coordination by energizing the GYCC and entering into a new "Memorandum of Understanding," reaffirming earlier commitments to interagency cooperation.[35] The revitalized GYCC's first major project was to produce the noncontroversial document *The Greater Yellowstone Area: An Aggregation of National Park and National Forest Management Plans* in September 1987. The agencies viewed it as the precursor to developing what they termed the "Vision" for the future of the GYE.

The National Park Service and the Forest Service moved on to this next phase of the operation after they released *Aggregation*. They wanted to identify goals to guide future land use management in the GYE. Agency personnel discussed matters among themselves throughout 1988, but there was as yet no structured public involvement; indeed, a public participation plan was not approved until April 1989.[36] Between May and August 1989, informal discussions took place between GYCC staff and some interest groups. In September 1989 the affected park superintendents and forest supervisors wrote draft goals. The GYCC briefed the deputy director of the National Park Service, the chief of the Forest Service, interest groups, and congressional staff, then released the draft goals for public comment in December 1989. The public comment period continued until March 1990. The agencies then prepared the draft "Vision document." They released it in limited quantities on July 17, 1990, and more broadly on August 14.

There is an enormous difference between the *Aggregation* and the two subsequent works: the draft goals and the Vision document. Readers widely understood *Aggregation* to be a collection of data, and no one opposed the gathering of information and making it widely available to interested parties. The statement of goals and the Vision document, however, are quite different matters. Many readers, regardless of their views of what constitutes the correct future for the GYE, operated from the perspective that what they were witnessing was nothing less than a sweeping attempt to put a master land use plan for the GYE into place.

The agencies tried to have it both ways. On the one hand, the GYCC writes in the document that "the Vision is not a regional plan" but simply a "statement of principles."[37] The press statement accompanying the July 17 release of the document picked up this report-as-innocent-roadmap-to-a-better-future spin. The report quotes two senior agency officials overseeing development of the Vision document,—Lorraine Mintzmyer, the National Park Service's director of the Rocky Mountain region, and Gary Cargill, the Forest Service's regional forester for the Rocky Mountains—as saying that "the current draft document charts a new future for the Greater Yellowstone Area," but the release points out "that Cargill and Mintzmyer emphasized that the 'Vision document' is a statement of principles, and does not seek to make specific land allocation decisions, nor does it seek to change the separate missions of the national forests and national parks."

On the other hand, the lead officials and the Vision document itself made it abundantly clear that the goals of the report are to guide future land use management and that existing plans, guidelines, and policies should be brought into conformance with the newly prescribed vision of the GYE. Cargill and Mintzmyer say that "the principles provided by the completed 'Vision document' will be used as guidance in the creation of amendments to park and forest plans and provide a focus for action items carried in the 'Vision document.'"[38] The report notes that the "goals would describe desired future conditions and would be the centerpiece . . . that defined management philosophy for both agencies to use in developing agency plans."[39] It goes on to declare that "the final step in the . . . process is the completion of whatever amendments may be needed to Forest Service regional guides and forest service plans and to the parks' statements for management, general management plans, and resource management plans."[40] But what are the goals of the Vision document?

Ecosystem Management for the GYE: A Glimpse at a New Paradigm Future

The GYCC's vision for the future proclaims that "the overall mood of the GY[E] will be one of naturalness, a combination of ecological processes operating with little restraint and humans moderating their activities so that they become a reasonable part of, rather than encumbrance upon, those processes."[41] Accordingly, the GYCC announces that "the overarching goal is to conserve the sense of naturalness and maintain ecosystem integrity in the GY[E] through respect for ecological and geological processes and features that cross administrative boundaries."[42] This goal offers a revealing look at the new paradigm world of ecosystem management and provides a concrete example of this world's dependence on enviro-babble (mood of the landscape and sense of naturalness), indefinite ideas (maintenance of ecosystem integrity and respect for ecological process; see chapter 6), as well as its links to nature worship (subordinating human activity to the protection of nature as represented by the GYE itself; see chapter 5).

After stating the prime goal for land use management in the GYE, the agencies immediately try to explain what it means. They begin by addressing the idea of naturalness, but they never define the critical term *ecosystem integrity*. The GYCC acknowledges that naturalness carries many meanings that stem from "deep cultural, social, and religious value systems."[43] They reject the notion that one should define naturalness solely in terms of pristine—no humans or human activity—landscapes. Indeed, they specifically acknowledge that humans are part of natural settings. Of course, they could do nothing else since the law charges both the National Park Service and the Forest Service with managing the public's lands for human benefit. After claiming that the Vision document would recognize the "entire continuum of definitions," they nonetheless make it plain that nature is to be supreme and that humans must give way in the event of conflict. "A sense of naturalness," they write, "is defined as that state in which landscapes appear and ecological processes operate much as they would without the effects of modern man even though modern man and his activities occur."

The government officials do not leave much to the imagination here. The ideal ecological state, the perfect landscape, is one in which people have not affected nature with technologies beyond those of hunters, gatherers, or preindustrial farmers. The CYCC uses premodern Indian occupancy of a portion of Yellowstone National Park as an example of people operating in the area without unduly interfering with nature. They write that the park has "hosted very nearly the same fauna for the last 2,000 years," during which time "Prehistoric American Indians" occupied and used the area "without dramatic changes [occurring] in species diversity."[44] They were in harmony with nature because even though they were present for long periods, "most major ecosystem processes continued to occur."

For advocates of the Vision document, harmony became the buzzword for justifying subordination of people to nature. We twentieth-century humans are all right as long as we "harmonize with, rather than intrude upon, the landscape."[45] The federal land managers were to "be guided by the principle of harmonizing with ecological processes."[46] Thus, the CYCC sought to make subordination of human endeavors to nature protection the centerpiece of land management even though the Vision document contains a goal to "provide for sustainable economic opportunities in the GY[E] through the balanced and environmentally sensitive use of resources such as oil and gas, timber, forage, and recreation."[47]

When the authors of the Vision document clarified the land use pecking order, they made it clear that the economic opportunity goal was little more than a nod to human activity. "Responsibly managed recreation opportunities will continue in the GY[E] where they are compatible with other GY[E] goals and values," says the GYCC.[48] The phrase "where they are compatible with" subordinates recreation to the overarching goal of protecting naturalness and ecosystem integrity. The Vision document's discussion of the goal supposedly legitimizing economic activity repeats this same language for other kinds of economic uses of the land. "Responsible grazing of livestock on forest lands will continue where it is compatible with other GY[E] goals and values." "Responsible mineral development activities will continue on forest lands where it is compatible with other GY[E] goals and values." "Responsible timber harvesting will continue where it is compatible with other GY[E] goals and values." If that is not enough to undermine resource development, the GYCC's definition of sustainability completes the process. The sustainable use of a resource, it says, includes the notion that such use could not "impair" either the resource itself or undefined "ecosystem values."[49]

This is new paradigm thinking. It brings land use management into a never-never land of subjective and value-laden management. Adopting the protection of undefined "ecosystem values" and seriously asserting the existence of a "principle of harmonizing with ecological processes" that is to guide management actions has no basis in science (see chapter 6).

By indicating that various kinds of economic activities remained appropriate within the GYE yet making those activities subservient to protecting nature and ecosystems, the GYCC believed that they had found a middle ground. The GYCC wrote that "in short those GY[E] constituencies hoping for a so-called 'lockup' of the area's resources will be disappointed, but those elements hoping for a 'business-as-usual' approach to GY[E] industries will likewise be disappointed."[50] James Overbay, deputy chief of the Forest Service, echoes this view when he told the Department of Agriculture's assistant secretary for natural resources and environment—who oversees the Forest Service—that the following passage from the Vision document is an "applicable summary": "The GY[E] will continue to provide a diversity of livelihoods. Opportunities for recreation and commodity development, including timber harvesting, grazing, and minerals will be provided for on appropriate federal lands. The agencies will ensure that proposed developments are designed in harmony with resources of the GY[E]."[51] There were few takers for the agencies' interpretation of the Vision document, and reaction was swift and polarized.

Reaction to the Vision Document

The response from new paradigm supporters in the environmental community was generally along the lines of great-start-but-you-must-go-farther. The national heads of several prominent environmental organizations—National Parks and Conservation Association, Natural Resources Defense Council, National Wildlife Federation, Izaak Walton League, National Audubon Society, Environmental Defense Fund, Sierra Club, and Environmental Policy Institute—wrote that "[the] effort is a crucial opportunity to reorient the management of the Greater Yellowstone Ecosystem toward sustainable, ecologically sound management and away from the intensifying, incremental erosion of the region's natural and wildland qualities caused by recreation and commodity-based resource extraction."[52] These national environmental leaders applauded all the goals of the Vision document, whose "primary emphasis [is] on resource protection." All other goals, including any encouragement of "sustainable economic activities . . . must be subordinate to the protection-oriented goals." They wished to restrict or prevent recreational use by the public as well as commodity production in order to achieve the higher goal of ecosystem protection.

The GYC took this same course. Ed Lewis, the GYC's executive director, pointed with pride to the document's echoing of "many of the themes and sentiments the Coalition has been expressing for years."[53] But the GYC was vociferous in calling for the document's strengthening. "The current draft clearly does not go far enough to rein in environmentally damaging resource management activities in Greater Yellowstone," they proclaim in a newsletter urging members to write the GYCC commenting on the Vision document.[54] The GYC told recipients of the newsletter to endorse the overarching goal and to call for rewriting the document to "insist that development activities must be precluded if they impair or destroy Ecosystem resources or values." For the GYC, resources and values of the GYE include naturalness, air and water quality, soils, biological diversity, fish and wildlife, geothermal features, wild qualities, scenery, and the catchall category of "wonders." The GYC wanted the document to be absolutely unequivocal in its endorsement of the protect-the-ecosystem-first approach to land use management.

The National Parks and Conservation Association (NPCA) not only endorsed ecosystem management for the Greater Yellowstone but also saw in it an opportunity to spread ecosystem management across the nation. The NPCA's comments on the document left no question that protecting the ecosystem is of paramount importance. The NPCA called the Vision document one of "extraordinary importance" but indicated that its ecosystem-first philosophy should be made more explicit by noting that "*the Vision should clearly state, however, that activities that impair Ecosystem resources or values will be precluded*" [emphasis in the original].[55] Because of Yellowstone's special place in the national psyche, the NPCA saw the successful imposition of ecosystem management in this region as setting the tone for managing the lands around hundreds of other units of the national parks system. "Our vision for Yellowstone," they write, "reflects and shapes our vision for all national parks." In their view, parks are not "islands" and cannot survive if the government manages them as such. Instead, they are part of larger surrounding ecosystems whose "careful stewardship" is the key to the future for all national parks.

Criticism of a different kind came from many elected officials of the region as well as regional economic interests, ranging from bankers and builders to farmers and miners.[56] The congressional delegations of Wyoming, Montana, and Idaho were highly critical, and the governors of all three states requested that the document "be withdrawn or substantially revised."[57] Congressman Ron Marlenee of Montana told Lorraine Mintzmyer that "I won't mince words. . . . I'm opposed to the Vision document—period."[58] The Wyoming congressional delegation of Senators Malcolm Wallop and Al Simpson and Representative Craig Thomas told Secretary of the Interior Manuel Lujan that their concerns with the Vision document were "too numerous to outline" in a letter and thus that they wanted to meet with him in person.[59] These critics had several problems with the document.

At bottom was the belief that the document was a regionwide land use plan. The Wyoming delegation told Secretary Lujan that "the crux of our concerns rests with the worry that this document will ultimately be used as a 'master' management plan" and that it would "undermine the concept of multiple-use on public lands" in contravention of existing law. Congressman Marlenee, true to his self-description as one who "does not mince words," said that the document would bring "hunting, fishing, timber harvesting, grazing, mineral development, recreation and many other legitimate and vital uses of our public land to a screeching halt." An overstatement to be sure, but he certainly identifies the direction of the document. He indicated his desire to do "everything I can to make sure this Document is either scrapped or completely rewritten." Members of Congress and others felt that the document went well beyond congressional direction for the National Park Service and the Forest Service to improve cooperation and coordination in the region while posing a threat to the rights of the owners of the 4.8 million acres of private land within the GYCC's version of the GYE. Feelings were running high on all sides.

On September 11, 1991, the GYCC released the final version of the Vision document, retitled *A Framework for Coordination of National Parks and National Forests in the Greater Yellowstone Area.*[60] It reflects consideration of over 8,000 comments received during a nine-month public comment period. It is a markedly different document than the draft. It contains eleven pages, a far cry from the Vision's seventy-one. The GYE had shrunk from 19 million acres in multiple ownerships to 11.7 million acres of national forest and national parklands.

The *Framework* represented ecosystem management in an evolutionary, not revolutionary, form. Gone was the overarching goal of conserving a sense of naturalness and maintaining ecosystem integrity. The first principle of the *Framework* is to "maintain functional ecosystems," which could be done "through actions based on planning, research, education, and cooperation."[61] The *Framework* did not clarify what constituted a functional ecosystem as opposed to a nonfunctional ecosystem. Moreover, the new goal speaks to multiple ecosystems in the area rather than a single Greater Yellowstone Ecosystem. The GYCC notes that "both agencies depend on healthy ecosystems to fulfill their missions" but offers no explanation of why that is so or what ecosystem health means. Revision of management and planning documents for the affected federal lands was no longer a requirement of achieving this goal. Instead, it was tied to interagency cooperation, to working with state and local agencies and with private landowners, and to better research. Even the most suspicious of critics would be hard put to claim that the *Framework* is a regional land use management plan in disguise, but the document leaves ecosystem-based management, in all its inexplicability, on the table.

Framework also has a goal of encouraging "opportunities that are economically and environmentally sustainable."[62] It does not emasculate this goal via "compatibility" requirements or by defining sustainability in terms of protecting mysterious "ecosystem values." In fact, the *Framework* does not define sustainability at all, which does nothing to help the cause of providing rational direction to land managers. The *Framework*'s vision for the area fifty years into the future includes calls for clean air and water, productive soils, magnificent views, and rich biodiversity. It also includes people recreating, grazing livestock, cutting timber, and extracting minerals. The *Framework* strikes much more of a balance between environmental protection and human uses of public lands in the region. Mintzmyer and Cargill note that it emphasizes the "distinct missions of both agencies . . . because of widespread public concerns that the Forest Service's mission of 'multiple use' and the National Park Service's 'preservation and public use' mission were being blended."[63] New paradigm advocates generally lamented these changes, while many of the Vision document's critics in Congress and elsewhere approved of them.

Those who did not like the final outcome generally complained of many things: the GYCC's inability to define the problem correctly, the impossibility of instituting ecosystem management within the current institutional structure, poor outreach, and heavy-handed political interference with agency professionals.[64] It did not occur to dissenters that the underlying science is far too weak to justify major changes in land management policy or that they misunderstand what little science is available, as with the GYC's proclaiming that ecosystems "are inherently self-sustaining" entities that grow and "are held in a dynamic equilibrium."[65] New paradigmists failed to realize that many critics of the Vision document felt that its vagueness paved the way for a government and environmentalist power grab. In addition, they did not notice that the public was not willing to equate the welfare of cockroaches with that of fellow humans, a basic premise of new paradigm thought. Instead of critically examining the merits of their own position, champions of the Vision document generally cried foul and pointed fingers.

The Vision document was unfavorably changed, new paradigmists said, because of an inappropriate exercise of political power by the Bush administration. However, elected officials from throughout the region also faulted the Vision document, as did most of the public that submitted comments on the draft. New paradigmists conveniently overlooked all of this. Instead, they singled out White House Chief of Staff John Sununu as the loathsome mastermind, while they fingered Scott Sewell, deputy assistant secretary for fish, wildlife, and parks in the Department of the Interior, as the day-to-day henchman in charge of forcing changes on persecuted career professionals in the National Park Service and the Forest Service.[66] They made the issue a morality play. The GYCC played the role of Good trying to accomplish the Virtuous by subordinating human endeavors to the protection of Nature-as-Ecosystem. Republicans played the role of Evil intent on using Raw Power to protect Destroyers exemplified by Business. Evil won because the *Framework* represented business-as-usual when compared with what new paradigm proponents believed was the morally superior and scientifically enlightened Vision document.

Vice President Al Gore echoes this view of the first effort at new paradigm land use management when he bemoans the fact that the Vision document did not set management standards for the region. In a mistaken portrayal of events, he suggests that the Vision document properly represented the concerns of state and local governments, private landowners, public land users, businesses, and interest groups but nonetheless "certain interest groups successfully lobbied to keep the original plan from being implemented with little resistance from higher levels of the federal government."[67] If, as Gore infers, the Vision document fairly represented the interests of such a wide range of stakeholders, it would have been impossible to change it greatly; the governors of Idaho, Montana, and Wyoming would never have written the secretary of the interior asking him to withdraw the document, members of Congress would not have weighed in to call for major revisions, and the majority of the public would not have rendered a thumbs-down in their comments.

The experience at Yellowstone does not offer a compelling vision of a future in which the management and protection of ecosystems is the basis of federal environmental policy. Undaunted, however, new paradigm supporters are moving forward as quickly as possible, propelled by a sense of moral righteousness and a belief that their views are firmly grounded in sound science, all the while carrying a sign proclaiming "the end is near." The next chapter examines the state of the environment to see whether environmental calamity is on our doorstep.

Notes

1. John Ise, *Our National Park History* (Baltimore: The Johns Hopkins University Press, 1961), 13–50; John Muir, *Our National Parks* (Boston: Houghton Mifflin, 1909), 37–74.

2. Ise, *Our National Park History,* 15.

3. Frank Craighead Jr., *Track of the Grizzly* (San Francisco: Sierra Club Books, 1979).

4. John Freemuth, "Ecosystem Management and Its Place in the National Park Service," *Denver University Law Review* 74, no. 3 (1997): 721.

5. Craighead, *Track of the Grizzly,* 4.

6. Rick Reese, *Greater Yellowstone: The National Park and Adjacent Wildlands,*

2nd ed., Montana Geographic Series, no. 6 (Helena, Mont.: American and World Geographic Publishing, 1991).

7. Reese, *Greater Yellowstone,* 12.

8. In 1995, Louisa Wilcox, GYC program director, indicated in "The Yellowstone Experience," *BioScience* (Special Supplement on Science and Biodiversity Policy, June 1995): S79–83, that the GYE was approximately 20 million acres. No map was included, and it is unclear whether the 20-million-acre figure refers to the core or to the core plus transition zones.

9. Reese, *Greater Yellowstone,* 57.

10. Reese, *Greater Yellowstone,* 61. R. G. Bailey, U.S. Forest Service, *Ecoregions of the United States—RARE II Map B,* minor modifications by G. D. Davis (Washington, D.C.: U.S. Department of Agriculture, U.S. Forest Service, 1976).

11. Dennis Glick, Mary Carr, and Bert Harting, eds., *An Environmental Profile of the Greater Yellowstone Ecosystem* (Bozeman, Mont.: The Greater Yellowstone Coalition, 1991).

12. Glick et al., *An Environmental Profile,* 11.

13. Glick et al., *An Environmental Profile,* 13.

14. Joseph Alper, "Yellowstone Ecosystem: 'Win-Win' Solution," *Science* 255 (February 7, 1992), 685–86.

15. To avoid confusion, I will use GY[E] instead of GYA when quoting the GYCC.

16. Greater Yellowstone Coordinating Committee, *The Greater Yellowstone Area: An Aggregation on National Park and National Forest Plans* (N.p.: U.S. Department of the Interior, National Park Service, and U.S. Department of Agriculture, National Forest Service, September 1987).

17. Greater Yellowstone Coordinating Committee, *The Greater Yellowstone Area,* 2–8.

18. Greater Yellowstone Coordinating Committee, *The Greater Yellowstone Area,* map, p. 1–4.

19. Greater Yellowstone Coordinating Committee, *The Greater Yellowstone Area,* map, p. 1–4.

20. The report contains a map showing land ownership, pp. 2–10.

21. Greater Yellowstone Coordinating Committee, *The Greater Yellowstone Area,* 2–8.

22. U.S. House Committee on Interior and Insular Affairs, *Greater Yellowstone Ecosystem: Oversight Hearings before the Subcommittees on Public Lands, and National Parks and Recreation of the Committee on Interior and Insular Affairs,* 99th Cong., 1st sess., October 24, 1985, serial 99–18.

23. Congressional Research Service, *Yellowstone: Ecosystem, Resources, and Management,* 86–1037 ENR, by M. Lynn Corn and Ross W. Gorte (Washington, D.C.: Library of Congress, December 12, 1986).

24. Congressional Research Service, *Yellowstone,* 2.

25. Robert B. Keiter and Mark S. Boyce, eds., *The Greater Yellowstone Ecosystem* (New Haven, Conn.: Yale University Press, 1991).

26. Robert B. Keiter, "An Introduction to the Ecosystem Management Debate," in Keiter and Boyce, eds., *The Greater Yellowstone Ecosystem,* 5.

27. Duncan T. Patton, "Defining the Greater Yellowstone Ecosystem," in Keiter and Boyce, eds., *The Greater Yellowstone Ecosystem,* 19.

28. Seymour Fishbein, *Yellowstone Country* (Washington D.C.: National Geographic Society, 1989).

29. Tim W. Clark and Dusty Zaunbrecher, "The Greater Yellowstone Ecosystem: The Ecosystem Concept in Natural Resources Policy and Management," *Renewable Resources Journal* (summer 1987): 10.

30. Tim Clark and Steven Minta, *Greater Yellowstone's Future* (Moose, Wyo.: Homestead Publishing, 1994).

31. Clark and Minta, *Greater Yellowstone's Future,* 77.

32. Fishbein, *Yellowstone Country,* 28.

33. Jonathon Taylor and Nina Burkhardt, "Introduction: The Greater Yellowstone Ecosystem—Biosphere Reserves and Economics," *Society and Natural Resources* 6 (1993): 105.

34. U.S. House Committee on Interior and Insular Affairs, *Greater Yellowstone Ecosystem.*

35. This "Memorandum of Understanding" was signed on September 24, 1986, by the acting regional director for the Rocky Mountain region of the National Park Service and three regional foresters.

36. This chronology is developed from Greater Yellowstone Coordinating Committee, "Vision for the Future: A Framework for Coordination in the Greater Yellowstone Area"; its accompanying press release, "Greater Yellowstone Goals are Naturalness, Sustainable Economic Opportunities Where Human Activity Is to Harmonize with Landscape," July 17, 1990; and a briefing memorandum, "Draft Vision for the Future—a Framework for Coordination in the Greater Yellowstone Area," prepared for James R. Mosely, assistant secretary for natural resources and the environment, and John Beuter, deputy assistant secretary for natural resources and the environment, U.S. Department of Agriculture, by James C. Overbay, deputy chief of the U.S. Forest Service, January 29, 1991 (copies on file with the author).

37. Greater Yellowstone Coordinating Committee, Vision document, 1–6.

38. Greater Yellowstone Coordinating Committee, press release, July 17, 1990.

39. Greater Yellowstone Coordinating Committee, Vision document, 2–1.

40. Greater Yellowstone Coordinating Committee, Vision document, 1–6.

41. Greater Yellowstone Coordinating Committee, Vision document, 3–1.

42. Greater Yellowstone Coordinating Committee, Vision document, 3–7.

43. Greater Yellowstone Coordinating Committee, Vision document, 3–9.

44. Greater Yellowstone Coordinating Committee, Vision document, 3–10.

45. Greater Yellowstone Coordinating Committee, Vision document, 3–10.

46. Greater Yellowstone Coordinating Committee, Vision document, 3–10.

47. Greater Yellowstone Coordinating Committee, Vision document, 3–25.

48. Greater Yellowstone Coordinating Committee, Vision document, 3–31.

49. Greater Yellowstone Coordinating Committee, Vision document, G-5.

50. Greater Yellowstone Coordinating Committee, Vision document, 3–5.

51. "Draft Vision for the Future—a Framework for Coordination in the Greater Yellowstone Area," briefing memorandum prepared for James R. Mosely.

52. Paul C. Pritchard, National Parks and Conservation Association; John H. Adams, Natural Resources Defense Council; Jay D. Hair, National Wildlife Federation; Jack Lorenz, Izaak Walton League; Peter A. A. Berle, National Audubon Society; Frederic D. Krupp, Environmental Defense Fund; Michael L. Fischer, Sierra Club; and Michael S. Clark, for the Environmental Policy Institute, Friends of the Earth, and Oceanic Society, letter to Jack Troyer, team leader of the Greater Yellowstone Coordinating Committee, February 28, 1990 (copy on file with the author).

53. Ed Lewis, "The View from Greater Yellowstone," *Greater Yellowstone Report* (fall 1990): 3.

54. Greater Yellowstone Coalition, "Be Visionary: Send Your Comments on the GYCC 'Vision'"; see also Greater Yellowstone Coalition, "Be Visionary: Speak Out for the Yellowstone Ecosystem," undated mailing from the fall of 1989 (copy on file with the author).

55. Terri Martin, Rocky Mountain regional director of the National Parks and Conservation Association, letter to the team leader of the Greater Yellowstone Coordinating Committee, January 29, 1991 (copy on file with the author).

56. Wyoming Multiple Use Coalition, "Review of Vision for the Future Document"; Mark Kossler, chairman of the Range Subcommittee of the Montana State Rural Areas Development Committee, letter to the team leader of the Greater Yellowstone Coordinating Committee, January 26, 1991; Jess Cooper, Rocky Mountain Oil and Gas Association, letter to Jack Troyer, team leader of the Greater Yellowstone Coordinating Committee, January 30, 1991; and Jack Lyman, executive director, Idaho Mining Association, letter to Jack Troyer, Greater Yellowstone Coordinating Committee, January 21, 1991 (copies of letters on file with the author).

57. Pamela Lichtman and Tim W. Clark, "Rethinking the 'Vision'; Exercise in the Greater Yellowstone Ecosystem," *Society and Natural Resources* 7, no. 5 (1994): 463; Clark and Minta, *Greater Yellowstone's Future,* 98.

58. U.S. Representative Ron Marlenee, letter to Lorraine Mintzmyer, regional director of the Rocky Mountain region, National Park Service, March 19, 1991 (copy on file with the author).

59. U.S. Senators Al Simpson and Malcolm Wallop and U.S. Representative Craig Thomas, letter to Secretary of the Interior Manuel Lujan, August 22, 1990 (copy on file with the author).

60. Greater Yellowstone Coordinating Committee, *A Framework for Coordination of National Parks and National Forests in the Greater Yellowstone Area* (N.p.: U.S. Department of the Interior, National Park Service, and U.S. Department of Agriculture, National Forest Service, September 1991).

61. Greater Yellowstone Coordinating Committee, *A Framework,* 1.

62. Greater Yellowstone Coordinating Committee, *A Framework,* 8.

63. Greater Yellowstone Coordinating Committee, press release, September 9, 1991.

64. Clark and Minta, *Greater Yellowstone's Future,* 96–102; Bruce Goldstein, "The Struggle of Ecosystem Management at Yellowstone," *BioScience* 42, no. 3 (March 1992): 183–87.

65. Glick et al., *An Environmental Profile,* 113.

66. Tom Kenworthy, "Park, Forest Officials Allege GOP Pressure: Two Caught in

Crossfire over Public Lands," *Washington Post,* September, 23, 1991, A23; Bill McAllister, "Yellowstone Report Controversy Prompts House Inquiry," *Washington Post,* September 12, 1991, A21; Angus M. Thuermer Jr., "Mintzmyer Calls 'Vision' Paper a Political Fraud," *Jackson Hole News,* September 25, 1991, A1A.

67. Al Gore, *Creating a Government That Works Better and Costs Less: Reinventing Environmental Management* (Washington, D.C.: Office of the Vice President, September 1993), 13.

Chapter 4

Claims of Environmental Calamity

> Biological meltdown . . . [is caused by] 5.5 billion eating, manufacturing, warring, breeding, and real-estate-developing humans . . . the damage done in the United States is particularly well documented.
>
> —Dave Foreman, director of the Sierra Club and founder of Earth First!, "Missing Links," *Sierra,* September/October 1995, 52

For over three decades activists and others have bombarded Americans with claims that the environment is on the verge of destruction at the hands of humans. Claims of environmental destruction are passed on as accepted fact by politicians, journalists, and even some scientists, using phrases such as "By now, we are all familiar with . . ."[1] Vice President Al Gore writes in *The Earth in the Balance: Ecology and the Human Spirit* that "modern civilization . . . is colliding violently with our planet's ecological system. The ferocity of its assault on the earth is breathtaking."[2] Gore proclaims that "whether we realize it or not, we are now engaged in an epic battle to right the balance of our earth." Prophesiers want you to send money, proselytize the unwashed, and pressure elected officials to do the right thing by saving the world and nature from greedy profiteers. You must act immediately to save the planet. Consider some examples:

> You see, here at Greenpeace we refuse to bow down in silent acceptance of the destruction of our planet . . . the oil giants . . . the chemical and timber companies . . . the pulp and paper mills . . . the industrial fishing fleets . . . *these powers and super powers are destroying our Earth.* Who will challenge them? Who will hold them responsible?. . . . I'll tell you: Greenpeace!. . . . P.S. remember your gifts of . . . (Barbara Dudley, executive director, Greenpeace)[emphasis in the original][3]

> It is entirely possible that we may be the last generation of humans to know this wondrous earth as it was meant to be. . . . Our world is drowning in filth . . . [environmental] organizations form a "thin green line" between a healthy environment and disaster . . . [we bring] cases against callous corporate polluters and indifferent

government agencies. . . . The amount of your tax-deductible gift . . .(Vawter "Buck" Parker, executive director, Sierra Club Legal Defense Fund)[4]

In the midst of the environmental crisis. . . . The real challenge is how we as biologists can create that sense of urgency about biological diversity, climate change, and human population growth. . . . My thesis in part . . . [is] that environmental problems may grow so large, social chaos and quarreling over dwindling resources are likely to ensue, thwarting any possibility of remedial action. . . . We [biologists] need to act now. Now. (Thomas E. Lovejoy, counselor to the secretary of biodiversity and environmental affairs, Smithsonian Institution)[5]

Now we must admit that the communist and capitalist systems have jointly brought the world to the brink of ecological disaster. That imminent disaster, and the related problems of widespread poverty . . . racism, sexism, religious prejudice, xenophobia, and the use of war to maintain access to resources now stand as challenges to the long-term success of democracies with market-based economies . . . a population decline must be initiated as soon as possible . . . and maintaining the integrity of many extensive natural ecosystems should become a top priority of all government bodies . . . humanity must try to develop a symbiotic relationship with Earth as it heals. (Paul R. Ehrlich, professor of population studies and biological sciences, Stanford University, and Anne H. Ehrlich, senior research associate in biological sciences, Stanford University)[6]

Are we at a crisis point in the United States, as so many new paradigm advocates want us to believe? It all depends on what you use as a yardstick. If the standard is deviation from a totally natural landscape—that is, one that evolved without any human influence—then the answer is yes. It is also yes if the standard is variation from conditions as they existed in 1492, when the only people here were hunters and gathers and preindustrial agriculturalists who posited deities in natural objects. Many environmentalists romantically portray these residents as primitive ecologists living in harmony with nature despite engaging in practices that would appall most new paradigmists if done by Europeans. For example, Native Americans artificially maintained grasslands in much of the Midwest by deliberately burning vast tracts in Wisconsin, Illinois, Kansas, Nebraska, and Ohio over thousands of years, thereby preventing the establishment of forests. In the plains, other tribes drove herds of buffalo over cliffs, where most of the dead were left to rot and the maimed to die since the hunters could use only a small fraction of the available meat.[7]

If the standard that we use to judge the quality of the American environment is more objective; if it employs measurable phenomena; if it gets beyond romanticism, longing for a past that never existed, and nature worship, then you must conclude that overall conditions are good and getting better. Forest Service scientists Hal Salwasser, Douglas MacCleery, and Thomas Snellgrove write that "there is overwhelming evidence that, while some problems remain and others have emerged in the last few years, . . . the conditions of U.S. forests, wildlife, rangelands, agricultural lands, and related resources have improved dramatically during the last century."[8] Americans at the end of the twentieth century are neither raping nor pillaging the environment, nor are they presiding over the decay and decline of our physical world, as several authors recently demonstrate.[9] As *The Economist* puts it, "Forecasters of scarcity and doom are . . . invariably wrong."[10] Paul Portney, president of Resources for the Future (RFF), and Wallace Oates, a university fellow at RFF, agree: "No doubt many criticisms [of *The Economist*'s views] will be raised. It seems to us, however, that more than a germ of truth exists in what *The Economist* has to say. As a matter of fact, the prophets of environmental doom do have a very bad record. The forecasts have, as *The Economist* says, been 'invariably wrong.'"[11] Nonetheless, Americans can and should make further improvements in the quality of our environment. The best way to do so is by adopting more effective policies to achieve desired environmental goals and by becoming wiser and more efficient in our use of resources (see chapter 10).

Many new paradigm advocates appear to be caught in a time warp. Their portrayals of environmental conditions reflect the past, not the present. At various times and in various places, we were not good stewards of the environment. Forestry practices wasted timber. Farming techniques squandered soil. We poured excessive amounts of pollutants into the air until our eyes watered and our lungs burned. We put industrial and human refuse into our rivers and streams until they, too, burned and biota died. We placed hazardous materials into the ground. We killed off species through ignorance and wantonness. But we also looked about us, did not like what we saw, and made major changes in how we related to the physical world.

This is what we did. Federal, state, and local governments enacted thousands of laws protecting the environment. Laws such as the National Environmental Policy Act and its state equivalents require the government to consider the environmental impacts of their actions. Many statutes address the emission of pollutants into the air and water or the disposal of wastes into the ground. In response to these statutes, the U.S. Bureau of Economic Analysis tells us that annual spending for pollution abatement climbed steadily over the last twenty-five years and reached nearly $110 billion in 1993, an amount that exceeded the after-tax profit of our domestic agriculture, forestry, fishing, mining, construction, wholesale trade, retail trade, and service industries combined.[12] The Environmental Protection Agency (EPA) estimates that the cost of environmental protection measures would reach 2.5 percent of our gross national product by the year 2000, up from less than 1 percent in 1972.[13] This may be the highest rate of environmental spending of any nation in the world. Working at St. Louis' Washington University, Thomas Hopkins projects annual costs for environmental protection and risk reduction at $267 billion (1995 dollars) by the year 2000, up from $80 billion (1995 dollars) in 1977.[14] These costs would grow from 12 percent of total regulatory costs in 1977 to 37 percent by 2000. Other laws and policy changes address land use management. Lands associated with species threatened with extinction receive federal protection under the Endangered Species Act, and the federal government established new protective land management categories, such as federal wilderness areas, wild and scenic rivers, and national marine sanctuaries. The General Accounting Office finds that the amount of land that the federal government manages primarily or exclusively for conservation purposes rose from 66 million acres in 1964 to 272 million acres in 1994.[15] This exceeds the combined acreage of California, Florida, Indiana, Massachusetts, New York, Ohio, and Pennsylvania. At the same time Americans are developing and incorporating new technologies and responding to market signals that combine to lessen our impacts on the environment through greatly increased efficiencies.[16]

Despite all this societal effort, can things still be bad or getting worse? Judging by the weight of the evidence, the answer is a resounding, No!

Our Environment

The United States is the world's fourth-largest country behind Russia, Canada, and China. Examination of national land use patterns reveals that our country is not overrun with asphalt and concrete, we have placed a good deal of land in protected status, and we have not crowded the other species that we share the continent with onto a vanishingly small fraction of the country. Our total land area is 2,263,000 acres, of which 1,894,000 constitute the contiguous forty-eight states. As of 1992, analysts estimate cities and towns with populations of at least 2,500 cover 59 million acres (a little more than double the amount of 1960), or about 3 percent of the contiguous states.[17] This compares with 87 million acres of wilderness, parks, and other set aside lands in the forty-eight states and 229 million acres of reserved lands for the nation as a whole.

Another way to think of land is as habitat for species other than our own. Researchers at the U.S. Department of Agriculture have found that the total amount of developed acreage—cities, towns, roads, farmsteads, and so on—in the contiguous states at beginning of the 1990s was between 77 million and 90 million acres.[18] In other words, humans have built on about 5 percent of the lower forty-eight states. In these areas, other species have to adapt to often drastically altered environments in which some species thrive but many do not. Cropland represents further areas where species have to adapt to highly altered circumstances. Cropland takes up 20 to 22 percent of the land in the forty-eight states. The bottom line is that human actions have significantly changed approximately 25 to 27 percent of the landscape in the contiguous states by building cities and roads and by growing crops. When one includes Alaska, the built and cropped landscape represents about 22 percent of the nation's land. These are places where other species have to deal with highly altered environments. The hand of man rests more lightly on the remainder of the country. The overwhelming majority of the living things that share the land and water with us have adapted successfully to whatever changes we have brought about. A testimony to this is that so few species are known to have become extinct in the continental United States, as I point out later in this chapter.

Data gathered by the EPA conclusively demonstrates that the nation's air quality is steadily improving even though between 1970 and 1995 the U.S. population grew by 28 percent, we drove our cars more than twice as many miles in 1995 as in 1970, and our gross domestic product expanded by 99 percent during the same period.[19] The EPA tracks six so-called criteria pollutants to measure air quality: carbon monoxide, lead, nitrogen dioxide, ozone (ground level), particulate matter, and sulfur dioxide. Nationally, the concentrations and emissions of each of these pollutants declined between the mid-1980s and 1995, continuing the trend in improved air quality that began in the 1970s (table 4.1). The EPA established the National Ambient Air Quality Standards (NAAQS) as thresholds to protect human health. Concentrations of criteria pollutants below the NAAQSs do not adversely impact health. For the country as a whole, average concentrations of these pollutants are below these thresholds.

TABLE 4.1
National Trends in Air Quality, 1986–95

Pollutant	Change in Concentration	Change in Emissions	Concentration Below NAAQS in 1995?
Carbon monoxide	37% decrease	16% decrease	Yes
Lead	78% decrease	32% decrease	Yes
Nitrogen dioxide	14% decrease	3% decrease	Yes
Ozone	6% decrease	9% decrease	Yes
Particulates	22% decrease	17% decrease	Yes
Sulfur dioxide	37% decrease	18% decrease	Yes

Source: U.S. Environmental Protection Agency, *National Air Quality and Trends Report—1995* (Washington, D.C.: U.S. Government Printing Office, 1997).

National averages can mask regional and local problems, yet air quality is improving dramatically at these scales as well. The EPA reports that between 1986 and 1995, "the total number of 'unhealthful' days decreased 54 percent in Los Angeles, 35 percent in Riverside, California, and 58 percent in the remaining major cities across the United States."[20] Anyway you examine data against objective standards, the nation's air quality is improving, and it is good in most places on most days.

Let's consider American forests. Our relationship with the woods has been checkered over the years.[21] We have reviled and revered and consumed and conserved forests. Our present forests are extensive in area, and "by most criteria, U.S. forests are in excellent condition."[22]

When the colonists first arrived on the Atlantic seaboard, they were greeted by vast stretches of woodland. Yet it was hardly the primeval forest that nineteenth-century romanticists depicted. Native Americans routinely cleared woodland using fire and killed larger trees by cutting through the bark to destroy sap-carrying tissue. Several tribes burned forests to create fields and to change the vegetation from trees to meadow and scrubs to attract deer and other game that they used as food sources. Records from the Massachusetts Bay area, for example, paint a picture of "the desolation of the surrounding woods caused by repeated Indian burnings."[23] One account from 1630 regarding the land around Salem, Massachusetts, notes that "much ground is cleared by the Indians . . . and I am told that about three miles from us a man may stand on a little hilly place and see diverse thousands of acres of ground . . . and not a Tree on the same."[24] Arriving Europeans quickly adopted fire as a forest management tool, using it and other measures to greatly extend human impacts on our woodlands.

For the next 300 years the forests lost ground. Forest acreage in the contiguous forty-eight states declined from about 950 million acres in 1600 to around 600 million acres in 1920, the least amount of woodland in the nation's history.[25] Farmers cleared land for agriculture. People turned trees into a wide range of wood products, from furniture to railroad ties and props in underground mines. Moreover, for many of us in the current world of fossil fuels, nuclear power, and solar panels, it is easy to forget that wood was the nation's chief fuel until well into the last half of the 1800s.

Today's forests have rebounded nicely from past insults and careless practices so that they presently cover about 650 million acres, or nearly 70 percent of what existed in 1600. We can attribute the gains to several factors. The nation replaced wood with other energy sources. Increases in agricultural productivity allow today's farmers to produce enough food for over 270 million Americans as well as many tens of millions abroad from the same amount of land used in 1920 to feed a domestic population of 106 million. We have adopted a wide range of forestry conservation measures and embraced scientific practices, such as replanting with genetically improved seedlings. Net forest growth began exceeding harvests in the 1950s. In the 1990s forest growth surpassed harvests by about one-third, and net annual growth was nearly four times what is was in 1920.[26] Ironically, accumulation of biomass in some forests has become so extensive that it has created an unusual fire hazard, particularly in the West.

Forest recovery is more than just acreage and productivity. A wide range of wildlife benefits greatly by our improved care of woodlands and other measures. Forest-dependent species whose populations were low at the turn of the twentieth century but are now doing well include black bear, white-tailed deer, beaver (including the family that dined on a willow that I was nurturing at our suburban Virginia home), Rocky Mountain elk, wild turkey, wood duck, and whistling swans. Not all woodland species are thriving. The northern spotted owl and red-cockcaded woodpecker are cases in point, but they are exceptions. Many also express concerns regarding aquatic species inhabiting streams in forest environments altered by human use.[27]

As a further testament to the condition of our forests, people are enjoying the woods in record numbers. Visitors spent some 345 million days in the national forest system alone in 1995, up from 234 million in 1980. Such increased popularity is not based on a morbid fascination to experience impoverished or devastated landscapes. Quite to the contrary, visitors to the national forests come to enjoy the scenery, hunt, fish, camp, hike, view wildlife, canoe, climb rocks, and pursue other recreations found in wooded landscapes. While all this is going on, managers are protecting watersheds, livestock grazes, companies produce oil and natural gas, and various other beneficial uses of woodlands go forward.

Still, not all agree that our forests are in good condition. We hear claims of a forest health crisis requiring the pursuit of sustainable forestry (the issue of sustainability is addressed in chapter 6). By and large, critics claim that we so adversely impact biological diversity that we need to make major changes in how we manage forestlands.[28] The Sierra Club's April 1996 position statement, calling for the termination of commercial logging on public lands throughout the United States, represents this view.[29]

Biological Diversity

Clamoring for the protection of biodiversity is a cause célèbre among a cadre of activists, politicians, and scientists. The faithful present saving biodiversity basically as a moral and ethical imperative while making ancillary references to the utilitarian benefits of doing so in order to enhance the appeal of their position with politicians and the public (see chapter 5). Laurie Ann MacDonald, who chaired the Sierra Club's Biodiversity Resource Group, told Congress that "to impoverish the earth's biological diversity is like cutting the heart out of us. It takes away our lives, and it takes away our soul."[30] MacDonald's no-science emotionalism is typical of many of those asserting that the end is near. However, many scientists also raise the issue of a loss of biological diversity. When we rationally define biodiversity, the possibility of its decline has enough scientific legitimacy to be a reasonable subject for policy debates. But the data for the United States point to a lesser, rather than a greater, problem and certainly not to anything rising to the level of a biological, moral, spiritual, or economic crisis.

Biological diversity is enigmatic even within the scientific community.[31] Like beauty, "one usually has no difficulty in recognizing diversity when one sees it, . . . [but] in most cases one is hard pressed to characterize it with adequate precision."[32] It means so many different things that the eminent ecologist Robert MacArthur once proposed that "perhaps the word 'diversity' like many of the words in the early vocabulary of ecologists . . . should be eliminated from our vocabularies as doing more harm than good."[33] Niles Eldredge of the American Museum of Natural History writes that "we have several different kinds of phenomena masquerading under the general rubric of *biodiversity*" [emphasis in the original].[34] He emphasizes that for systematists—those biologists who identify and classify organisms into species and who are concerned with their lineages—diversity means the number of species within a taxon (a level within a classification), not the number of species in an ecosystem.[35] In this view, biodiversity has roots in genealogy, not geography. This stands in contrast to those who generally regard biodiversity as species per area and multiple ecosystems on the landscape. In turn, British scholars John Harper and David Hawksworth argue that we should not consider ecosystems a part of biological diversity because the ecosystem concept includes the nonliving components of the environment that by definition cannot be biologically diverse.[36] In its place they offer either community or ecological diversity because these concepts are restricted to combinations of biota. In addition, they find that *species* is too restrictive and favor *organismal* in its place in order to "embrace taxonomic categories above species rank." Of course, this gets away from tangible entities—organisms—and into the artificial planes of genus and family within taxonomic classification that biologists use to catalog living things.

Taxonomic classifications have the same characteristics of other classifications. They are artificial. The Chinese, some say, once employed a system dividing animals into categories such as "belonging to the emperor," "embalmed," "tame," "frenzied," "innumerable," and "that from a long way off look like flies."[37] Classifications change through time; the idea of two kingdoms (plants and animals) gave way to five (plants, animals, bacteria, fungi, and protists, which include algae, protozoa, slime molds, and so on). One proposed classification contains thirteen kingdoms as advances in new technologies, such as electron microscopy, reveal differences at subcellular levels.

Scientists vigorously debate classifications and approaches to their formulation. The science is unsettled. "There is no general agreement on the definition of systematics, and of related words, such as taxonomy, biological system and classification," according to Alessandro Minelli.[38] In *Evolution and Classification,* Mark Ridley identifies three major approaches to classification: phenetic, which depends on observed sets of selected characteristics; phylogenetic, which has a foundation in ancestral relationships; and evolutionary, which combines the other two. Within each of these schools are further divisions because of the vast number of attributes that organisms possess and the multiple ways in which scientists can describe, measure, and aggregate those individual attributes to form the basis for categorizing biota. Lynn Margulis and Karlene Schwartz recognize the limits of biological knowledge by noting that their classification is the best that it can be considering the "varying, fragmented, and often inconsistent literature" that typifies scientific writing on the earth's past and present living things.[39] There is neither a definitive classification of the world's biota nor a consensus regarding the best approach to prepare it.

The species concept lies uneasily at the heart of biological classifications. Systematists aggregate similar species into genera, genera into families, families into orders, orders into classes, classes into phyla (or divisions), and phyla into kingdoms.[40] This follows the lead of Swedish botanist Carl Linnaeus, whose work dates from the mid-1700s. Linnaeus reasoned that species are specific creations of God, be they mice or moths, and that they are knowable and unchanging. Since Linnaeus's time, the species idea has become increasingly complicated, and some researchers challenge the very notion that species exist. Species, concluded a prominent researcher in 1908, have "no actual existence in nature . . . they are mental constructs and nothing more."[41]

Within the scientific community, "The Species Problem," to borrow the title of a 1957 book from the American Association for the Advancement of Science, is omnipresent and the subject of an enormous amount of sometimes acrimonious literature extending over more than 200 years.[42] In his preface to the book, editor Ernst Myer, writes,

> Few biological problems have remained as consistently challenging through the past two centuries as the species problem. Time after time attempts were made to . . . declare the species problem resolved either by asserting dogmatically that species did not exist or by defining, equally dogmatically, the precise characteristics of species . . . these pseudo solutions were obviously unsatisfactory. [Ignoring the problem] has been tried, but the consequences were confusion and chaos.[43]

the infinite (ecosystems). They heap the tangible and intangible together. No wonder the authors of the preface for the American Institute of Biological Science's special publication on biodiversity write that "the various dimensions of biodiversity are complex and difficult to quantify and convey to others."[61]

Biodiversity lacks any agreed-on currency or calculus for its measurement or assessment by specialists, policymakers, or land managers.[62] "Until we have decided how to measure 'biodiversity' we cannot begin to mobilize serious science into answering" any number of scientific and policy-relevant questions, such as the relationships, if any, between biodiversity and stability, productivity, or sustainability.[63] While scientists quantify biological diversity in many ways and on multiple levels, the resulting numbers or indices are of limited application because of inherent weaknesses.[64] When scientists tally species per area as a measure of biological diversity, the underlying assumption is that every species contributes equally to an area's biodiversity, yet a systematist would claim that the diversity of an area with five species would be greater if each species represented a separate taxonomic family. This is the equivalent to saying that there is greater transportation diversity represented in five items that included one airplane, one locomotive, one car, one bicycle, and one snowmobile compared with diversity found among one Chevy, one Ford, one Dodge, one Honda, and one Toyota. Moreover, scientists know that the presence of some species in some locations facilitates the presence of other kinds of living things. Trees, as a rule, provide more opportunities than do annual plants. Enormous difficulties exist in trying to find a biodiversity currency even when we limit biological diversity to consideration of tangible living things and their constituent genetic materials. Broadening the term to encompass arbitrary landscape units, such as ecosystems, may be emotionally fulfilling and politically expedient but will not facilitate development of sound scientific measurement of biological diversity.

Current definitions move biological diversity from the real world into that of the surreal and confound rational policymaking. Not only do they foreclose the development of useful measures of biodiversity, but they render its protection meaningless as a basis for public policies. New paradigmists exhort us to use the power of the federal government to protect ecosystems as a means of safeguarding biological diversity, yet these same individuals define ecosystems as part of biological diversity, so we face strident calls to protect ecosystems in order to protect ecosystems.

Biological Diversity in the United States

The United States is a biologically rich country. Scholars do not know how many species reside in the United States, although estimates are between

250,000 and 750,000 species.[65] The International Union for the Conservation of Nature (IUCN) estimates that since the year 1600, eighty-seven species of mammals, birds, reptiles, amphibians, fishes, and invertebrates have become extinct in the continental United States, with an additional 114 species lost in the Hawaiian Islands.[66] The IUCN also estimates that twenty-two (out of approximately 16,000) species of vascular plants (ferns, conifers, and flowering plants) are now extinct in the United States.[67] In contrast, the Defenders of Wildlife claim that approximately 500 species of plants and animals became extinct in the last 500 years.[68] Even the Defenders' figure represents only .1 to .2 percent of our estimated species endowment.

The U.S. Fish and Wildlife Service (FWS) lists 1,136 species of plants and animals as threatened or endangered as of the end of May 1998.[69] The list exaggerates the actual number of threatened or endangered species by including subspecies (geographically separated populations of the same species having only minor differences) and hybrids (crosses between distinct species). Grizzly bears appear as an endangered species on the basis of threats to the bear population centered on Yellowstone National Park. The United States, however, is in no danger of becoming void of grizzlies, as they abound in Alaska. The FWS lists the red wolf as an endangered species whose survival was in doubt until the FWS began a million-dollar-per-year captive breeding and release program. Yet the red wolf appears to be a hybrid, the product of mating between coyotes and gray wolves.[70] If every species on the list became extinct tomorrow, our supply of species would drop by less than 1 percent, hardly an indication of biological disintegration. This is not justification for careless stewardship of our natural resources or wanton disregard for other species, but it is important to keep the impacts of our actions to date in perspective because it means that we can safely reject efforts to infect the national psyche with an enormous ecological guilt trip.

Humans can increase biological diversity as well as detract from it. We have added more species to our biological stocks than we have eliminated. The Congress' Office of Technology Assessment reports that a minimum of 4,500 species from outside the United States have established resident populations here owing to human activities.[71] Humans value some of these species highly, having deliberately introduced many of them, including virtually all our food crops and domestic animals. Others arrived unbidden and unwanted. The Asian gypsy moth, boll weevil, and Mediterranean fruit fly are highly unwelcome newcomers, but they add to our overall biodiversity. Simple arithmetic suggests that since the coming of Europeans, the nation's biological diversity has increased, not decreased, because the number of known introductions exceeds the number of known extinctions by a wide margin. Indeed, known introductions exceed the total of known extinctions plus the total number of threatened and endangered (but not extinct) species by a factor of three. The data put prophets of biological meltdown in an awkward position, but you would never know it from their rhetoric.

New paradigmists continue to beat the drum of biodisaster. Testifying on behalf of several groups, the Environmental Defense Fund's Michael Bean told Congress that "our nation is experiencing a dramatic loss of its biological diversity" and that "endangered species are slipping over the brink of extinction."[72] Researchers estimate that between 1940 and 1990, thirty-nine species of vertebrates became extinct in the United States.[73] An extinction rate of less than one species per year hardly qualifies as a "dramatic loss." As for endangered species "slipping over the brink of extinction," the FWS notes that through 1993 less than 1 percent of the species listed under the Endangered Species Act had become extinct.[74] One way in which new paradigmists get around these inopportune data is to use accounting gimmicks, such as not counting exotics as part of our biological stock and inventing ecosystems that they can then claim people are threatening or destroying.

When assessing the biological diversity of the United States, why should we count buffalo but not cattle? Why should we tally bluestem but not cheat grass? Cattle clearly meet Wilson's definition of a species as "a population of organisms that interbreed freely with one another under natural conditions."[75] So does cheat grass and thousands of other exotic species brought here via human agency. In the new paradigm view of the world, we cannot count them because their association with humans taints them. Their presence in the landscape is not natural, so they are biotic pariahs. "When dealing with endangered species our concern is to prevent extinction, with invasive exotics our concern is to cause extinction," writes the University of Connecticut's Kent Holsinger.[76]

Even when exotics perform what many widely regard as useful functions, new paradigmists may still judge them unworthy. At a June 1995 conference on biological diversity held by the American Institute of Biological Science, the Smithsonian Institution's Thomas Lovejoy extols the American oyster's ability to cleanse the waters of the Chesapeake Bay as a highly important beneficial subsidy of nature and complains about our failure to take this into account when valuing oysters. He regrets that because of a declining population it now takes the oysters a year to process the bay's waters, whereas in times past there were enough of them to do the job in a week. At the same conference, however, Holsinger says that one of the more negative consequences of the rapidly growing population of exotic zebra mussels in Lake St. Clair is their ability to "completely filter all its water in 24 days." In Lake Erie, "measures of water transparency have almost doubled," but activists consider this an adverse impact. In the new paradigm view, if a native species filters water, that is good, but if an exotic species filters water, that is bad.

Rejection of exotic species as contributors to the nation's biological diversity is grounded in emotion, not science. We may not like the impacts of an exotic species, but that is no reason to discount them as part of our biological stocks.[77] The arrival of a new species adds to an area's biological diversity. Species deploy over the earth by numerous means. Ranges of plants, animals, and other biota vary over time in response to changing conditions. Mechanisms for dispersal vary. Vertebrates fly, walk, or swim to new homes. Winds and currents disperse seeds, and birds carry seeds afield in their guts. Seeds cling to fur and feathers, to be brushed off in new locations. The means by which a species arrives at a new location is irrelevant to the fact that its presence adds to biological diversity. There is no ecological distinction between a seed arriving at a new location by hitchhiking on a wolf or on the wolf's hide when worn by a hunter. When the first finches arrived on the Galapagos Islands, the overall species tally increased by one. It did not matter whether they were carried from the mainland by storm winds, as ecologists suggest, or whether they escaped from a ship at anchor. Once on land, their mode of transport did not matter to the finches, to the preexisting biota, or to later-arriving life forms. Finches were a new species on the island. They were now a part of the islands' biodiversity. So it is with the thousands of exotic species that now share the land and waters of the United States with us. The unwillingness of new paradigmists to count exotics as part of our biological inventory is driven by a reverence for nature rather than an interest in the size and variety of the U.S. gene pool.

Many who claim that the United States is in the midst of a biological diversity crisis rely on false assertions that Americans are destroying ecosystems as though they were real entities, such as grizzly bears or bald eagles. This is an accounting trick whose public acceptance depends on widespread ignorance of the ecosystem concept. We know that anyone can invent as many ecosystems as they choose and then go on to define the supposed degradation and death of these imaginary entities in wholly arbitrary ways. The idiosyncratic nature of ecosystems is perfect for doomsayers because there are no independent standards or means by which to evaluate their claims of destruction. If you accept their definitions and standards, then you must accept their conclusions.

Claimants of ecosystem destruction in America use as a standard the deviation of present landscapes from those that existed prior to the arrival of Europeans or their descendants (or deviation from landscapes that they imagine would exist today had not Europeans come to North America). Authors of the 1995 Department of the Interior report *Endangered Ecosystems of the United States: A Preliminary Assessment of Loss and Degradation* define ecosystem decline as the "destruction, conversion to other land uses, or significant degradation of ecological structure, function, or composition since European settlement."[78] According to them, a "critically endangered" ecosystem is one that has suffered at least a 98 percent decline, an 85 to 98 percent decline places an ecosystem in the "endangered" category, while declines thought to be in the 70 to 85 percent range were assigned the label "threatened." The Defenders of Wildlife adopt a similar deviation-from-nature standard for ecosystem degradation and destruction in their book *Endangered Ecosystems: A Status Report on America's Vanishing Habitat and Wildlife*.[79] They tell us that "since the time of European colonization, millions of acres of North America's natural ecosystems have been destroyed, degraded and replaced by agriculture and concrete" and complain that more than 90 percent of our streams cannot qualify as wild or scenic under federal law.[80] Neither the public nor policymakers have a compelling reason to accept either this standard or the related claims of biodisaster. Quite to the contrary, we should reject both if we require that standards have a rational basis and if we have an understanding of the evolution of landscapes.

There is no scientific basis for anyone to adopt a change-from-1492 standard to judge today's landscapes. What an area looked like or what species happened to occupy it at the time of European settlement has no inherent ecological importance compared to conditions five, ten, or a hundred years earlier or later. The moment that Europeans arrived was simply another instant in the constant evolution of the landscape. Indeed, it was not a moment at all but a gradual occupation as hunters, trappers, explorers, traders, miners, farmers, ranchers, and others slowly moved inland from the first coastal towns. This complicates fixing a baseline. Do you select the year before the first European of any kind arrived in the area, or do you choose some later time, as when villages began to appear or when the first farm or ranch was established? Science gives us no rules to aid us in making such decisions.

For the sake of argument, assume that researchers agree on an exact date for the baseline against which to assess the status of a current landscape. How do you then determine the landscape conditions of the agreed-on time? Geographers and others can develop general pictures of long-ago landscapes by reference to several sources whose availability is geographically uneven and of varying quality. They can use historical documents, such as explorers' journals and letters. Occasionally, early travelers made biological surveys that provide partial inventories of biota, but coverage is limited to where the explorers actually went and what they, as individuals, chose to record. Old inventories done by scientific expeditions are likewise restricted in spatial coverage and reflect the interests of the observers and their level of biological acumen. Researchers can obtain valuable information via examination of the fossil record, but it provides only point data on a limited number of biota and does not give information down to the species level in some cases, as noted by the University of Minnesota's Dwight Brown in his study of midcontinent grasslands in the 1800s.[81]

When scholars amalgamate these data, they produce qualitative regional sketches rather than detailed reconstructions of past conditions that offer a rigorous scientific baseline against which to measure landscape change. We cannot expect otherwise. Since we have yet to inventory any portion of United States for all its living things and we cannot agree on a boundary for the Greater Yellowstone Ecosystem after more than a decade of effort, why should anyone seriously believe that precise depictions of conditions that existed 500 years ago are possible?[82]

Since I am making heroic assumptions for the sake of argument, let me make some more. Assume that 1) scholars agree on a single set of ecosystems covering the nation just prior to European settlement, 2) these ecosystems are geographically unambiguous and that detailed and accurate descriptions are available, 3) the present state of each of these ecosystems is fully known, and 4) the difference in conditions between today and the past can be computed with an agreed-on calculus. Even if we get this far, we still do not know how much of the observed change is due to human action and how much is attributable to natural causes.

Brown writes that "contradictions and ambiguities are evident in the facts and theory that bear upon the nineteenth-century biogeography of the Plains grasslands."[83] Important difficulties are by no means limited to grasslands or to the nineteenth century. If biogeographers and ecologists cannot work out the cause of observed changes occurring over the last 200 years, how can we expect cogent findings for a 500-year period? Embedded in the last 500 years are major local, regional, and nationwide natural events and changes that affect the distribution of living things. Disturbance events (e.g., fires, hurricanes, and floods) alter biotic communities. Diseases shift the makeup of plant and animal communities, as do multiyear swings in precipitation and temperature. Longer-term climate changes affect the distribution of individual species. Over the last 500 years there have been significant long-term regional changes in climate that are poorly understood and that vary among regions.[84] If this were not enough, individual species are not uniformly affected by either human or natural factors. Brown concludes that the westward expansion of tall grasses in the plains cannot be explained by climate, while Melissa Savage finds that "climate was an important catalyst in the southwest Ponderosa pine regeneration pulse of this century."[85]

The new paradigm advocates' assumption that differences between current landscapes and those of pre-European times provides a scientific, objective, or meaningfully quantifiable standard by which to judge the condition of today's environment or to guide public policy is inescapably erroneous. Instead, it is rooted in a reverence for nature. I pursue this subject in greater depth in the next chapter but address it in the next section in the context of judging landscape change.

Landscape Ideals and Claims of Calamity

Landscapes evolve over time through human actions and natural processes. From prehistory to the present day, human actions affect the land and other living things. We do so as a species without regard to location, race, creed, or economic, political, and legal system. Writing over 2,000 years ago, Plato and Eratosthenes describe landscape changes in Attica and on the island of Cyprus caused by the felling of forests, while the Chinese philosopher Mencius recalls similar actions in his homeland.[86] The late geographer Carl Sauer of the University of California observes that "every human population, at all times, has needed to evaluate the economic potential of its inhabited area, to organize its life about its natural environment in terms of the skills available to it and the values which it accepted. . . . Wherever men live, they have operated to alter the aspect of the Earth, be it to their boon or bane."[87] Sauer further notes that the degree to which an area is deflected from its "pristine, or prehuman," condition is a function of length of human occupancy, population size, and technological level. Humans cannot live on the earth without altering it.

How people judge those alterations depends on the standards used. From Greek times to the eighteenth and nineteenth centuries, human manipulation of the landscape was generally viewed positively in Western societies. Planting, clearing, draining, irrigating, harvesting, and similar activities were regarded by most everyone as rightfully tending nature's garden as stewards of creation.[88] But signs of the scope of human activities continued to mount over the centuries and across the globe, and in 1864 George Perkins Marsh published his classic work *Man and Nature: Physical Geography as Modified by Human Action,* "the first great work of synthesis in the modern period to examine in detail man's alteration of the face of the globe."[89] Scholarly references to the book abound to this day. Marsh criticizes careless and wasteful practices. He urges moderation, reclamation of exhausted areas, and "the restoration of disturbed harmonies."[90] The purpose of his urgings, however, is to ensure that human-induced landscape change is rational and constructive so as to better serve humans, the earth's "noblest inhabitant."[91] For Marsh, change is a "boon or bane," depending on its consequences for humans. His standard is anthropocentric.

New paradigm advocates use a different standard when viewing human interaction with the environment. For them, nature unblemished by human actions is the ideal. The pristine landscape is nature perfected, and wilderness is its purest form. Wilderness becomes sacred space, the "model of perfection" whose defense is a "sacred charge" with new paradigm organizations such as the Wilderness Society.[92] "Wilderness," writes the University of Wisconsin's William Cronin, "serves as the unexamined foundation on which so many of the quasi-religious values of modern environmentalism rest . . . [it is] the standard against which to measure the failings of our human world. Wilderness is the natural unfallen antithesis of an unnatural civilization that has lost its soul."[93] Nature is profound, man is profane.

The sacredness of wilderness landscapes for advocates of ecosystem management flows from their lack of visible signs of modern man. It is unrelated to environmental attributes, such as the presence of rare, threatened, unusually productive or so-called fragile ecosystems. Bald eagles do not make a place sacred, but the absence of roads and buildings may. Activists defend wilderness areas against modern intrusions because they desecrate the temple of nature just as people protect churches, mosques, and synagogues against arsonists and vandals. Wilderness represents the ideal, yet nature is found in other landscapes. Forests, fields, suburbs, and cities all contain nature in some form or another; it is simply nature in less than the ideal state. New paradigm supporters defend nature in these landscapes in order to prevent further retreat from the ideal or to move toward it.

Nature is viewed by champions of ecosystem management as a good and beneficent god deserving of our adoration. This perspective allows new paradigmists to view all human-caused landscape change as an affliction and drives their efforts to prevent or limit human use of the earth. As long as humans exist, a landscape-evaluation standard based on nature worship provides new paradigmists with an unlimited number of sins to point to and allows them to utter never-ending cries of Repent! Nature worship and the new paradigm is the subject of the next chapter.

Notes

1. Ruth Kaplan, *Our Earth, Ourselves* (New York: Bantam, 1990), 3.

2. Al Gore, *Earth in the Balance: Ecology and the Human Spirit* (New York: Penguin, 1993), 269.

3. Barbara Dudley, executive director, Greenpeace, fund-raising letter, April 1993 (copy on file with the author).

4. Vawter "Buck" Parker, executive director, Sierra Club Legal Defense Fund, fund-raising letter, fall 1992 (copy on file with the author).

5. Thomas E. Lovejoy, "Will Expectedly the Top Blow Off?" *BioScience* (Special Supplement on Science and Biodiversity Policy, June 1995): S5–6.

6. Paul R. and Anne H. Ehrlich, *Healing the Planet* (Reading, Mass.: Addison-Wesley, 1991): 285–86.

7. Karl W. Butzer, "The Indian Legacy in the American Landscape," in *The Making of the American Landscape,* ed. Michael P. Conzen (New York: Routledge, 1994), 27–50. The *Annals of the Association of American Geographers* 82, no. 3 (September 1992), contained several articles on the condition of the Americas at the time of Columbus; see Karl

W. Butzer, "The Americas before and after 1494: An Introduction to Current Geographical Research," 345–68; William M. Denevan, "The Pristine Myth: The Landscape of the Americas in 1492," 369–85; and William E. Doolittle, "Agriculture in North America on the Eve of Contact: A Reassessment," 386–401. See also Ralph H. Brown, *The Historical Geography of the United States* (New York: Harcourt Brace, 1948). The state of Montana has preserved an example of a site where Native Americans stampeded buffalo over a precipice at the Madison Buffalo Jump State Park west of Bozeman.

8. Hal Salwasser, Douglas W. MacCleery, and Thomas A. Snellgrove, "The Pollyannas vs. the Chicken Littles—Enough Already," *Conservation Biology* 11, no. 1 (February 1997): 283.

9. Julian L. Simon, *The Ultimate Resource 2* (Princeton, N.J.: Princeton University Press, 1996); Ronald Bailey, ed., *The True State of the Planet* (New York: The Free Press, 1995); Greg Easterbrook, *A Moment on Earth: The Coming Age of Environmental Optimism* (New York: Penguin, 1995); Stephen Budiansky, *Nature's Keepers* (New York: The Free Press, 1995); Joseph L. Bast, Peter J. Hill, and Richard C. Rue, *Eco-Sanity: A Common-Sense Guide to Environmentalism* (Lanham, Md.: Madison, 1994); Ben Bloch and Harold Lyons, *Apocalypse Not: Science, Economics, and Environmentalism* (Washington, D.C.: Cato Institute, 1993); and Dixie Lee Ray and Lou Guzzo, *Environmental Overkill: Whatever Happened to Common Sense?* (Washington, D.C.: Gateway, 1993).

10. "Environmental Scares, Plenty of Doom," *The Economist* 345, no. 8048 (December 20, 1997): 19.

11. Paul R. Portney and Wallace E. Oates, "On Prophecies of Environmental Doom," *Resources,* no. 131 (spring 1998): 17.

12. U.S. Department of Commerce, *Statistical Abstract of the United States,* 116th ed. (Washington, D.C.: U.S. Department of Commerce, October 1996), tables 384, 828.

13. A. Carlin and the Environmental Law Institute, *Environmental Investment: The Cost of a Clean Environment—a Summary,* EPA-230–12-90–083 (Washington, D.C.: U.S. Environmental Protection Agency, December 1990); Alan Carlin, Paul F. Scodari, and Don H. Garner, "Environmental Investments: The Costs of Cleaning Up," *Environment* 34, no. 2 (1992): 12–20, 38–44.

14. Thomas D. Hopkins, *Regulatory Costs in Profile,* Policy Study No. 132 (St. Louis: Center for the Study of American Business, Washington University, 1996). Such estimates generally understate the overall costs to society of our actions to protect the environment. For example, not included are jobs lost or not created, increased prices of products, forgone economic growth, or the decline in the value of properties defined by the government as wetlands or as habitat for endangered species. See Michael Hazilla and Raymond J. Kopp, "Social Costs of Environmental Quality Regulations: A General Equilibrium Analysis," *Journal of Political Economy* 98, no. 4 (1990): 853–73.

15. U.S. General Accounting Office, *Land Ownership: Information on the Acreage, Management, and Use of Federal and Other Lands,* GAO/RCED-96–40 (Washington, D.C.: U.S. General Accounting Office, March 1996). See generally *NWI Resource* 6, no. 1 (spring 1995).

16. Indur M. Goklany, "Factors Affecting Environmental Impacts: The Effect of Technology on Long-Term Trends in Cropland, Air Pollution and Water-Related Diseases," *Ambio* 25, no. 8 (December 1996): 497–503.

17. Arthur B. Daugherty, *Major Land Uses of the United States, 1992,* Agricultural Economic Report No. 723 (Washington, D.C.: U.S. Department of Agriculture, Economic Research Service, 1995).

18. Calculated from Arthur B. Daugherty, *Major Land Uses,* and Soil Conservation Service and Iowa State University Statistical Laboratory, *Summary Report 1987 National Resources Inventory,* Statistical Bulletin No. 790 (Washington, D.C.: U.S. Government Printing Office, 1990).

19. U.S. Environmental Protection Agency, *National Air Quality and Emissions Trends Report, 1995* (Washington, D.C.: U.S. Government Printing Office, 1997).

20. U.S. Environmental Protection Agency, *National Air Quality and Emissions Trends Report, 1995,* Executive Summary, 4.

21. See generally Douglas MacCleery, *American Forests: A History of Resiliency and Recovery* (Durham, N.C.: Forest History Society, 1994); Michael Williams, *Americans and Their Forests: An Historical Geography* (Cambridge: Cambridge University Press, 1989); and Marion Clawson, "Forests in the Long Sweep of American History," *Science* 204 (June 15, 1979): 1168–74.

22. Roger A. Sedjo, "Forest Resources: Resilient and Serviceable," in *America's Renewable Resources: Historical Trends and Current Challenges*, ed. Kenneth D. Frederick and Roger A. Sedjo (Washington, D.C.: Resources for the Future, 1991), 113.

23. Brown, *The Historical Geography of the United States,* 15.

24. Brown, *The Historical Geography of the United States,* 15.

25. Estimates of forest acreage can be found in Daugherty, *Major Land Uses;* Sedjo, "Forest Resources"; and MacCleery, *American Forests.*

26. MacCleery, *American Forests.*

27. Robert T. Lackey, "Pacific Salmon, Ecological Health, and Public Policy," *Ecosystem Health* 2, no. 1 (March 1996): 61–68; Peter A. Bisson, Thomas P. Quinn, Gordon H. Reeves, and Stanley V. Gregory, "Best Management Practices, Cumulative Effects, and Long-Term Trends in Fish Abundance in Pacific Northwest River Systems," in *Watershed Management,* ed. Robert J. Naiman (New York: Springer-Verlag, 1992), 189–232; a particularly critical perspective on human impacts on aquatic organisms is provided by Christopher A. Frissell and David Bayles, "Ecosystem Management and the Conservation of Aquatic Biodiversity and Ecological Integrity," *Water Resources Bulletin* 32, no. 2 (April 1996): 229–40.

28. There are numerous published works on sustainable forestry, for example, Chris Maser, *Sustainable Forestry: Philosophy, Science, and Economics* (Delray, Fla.: St. Lucie Press, 1994), and Gregory H. Aplet, Nels Johnson, Jeffrey T. Olson, and V. Alaric Sample, eds., *Defining Sustainable Forestry* (Washington, D.C.: Island Press, 1994).

29. Robbie Cox (president) and Carl Pope (executive director), "Statement of the Passage of Sierra Club Forestry Ballot," downloaded from the Sierra Club's Web site, http://www.sierraclub.org, April 23, 1996.

30. Laurie Ann MacDonald, testimony on S. 58, The National Biological Diversity Conservation and Environmental Research Act, before Congress, Senate Subcommittee on Environmental Protection of the Committee on Environment and Public Works, 102nd Cong., 1st sess., July 26, 1991, Committee Print 102–188, 199, 37.

31. Boris Zeide, "Assessing Biodiversity," *Environmental Monitoring and Assessment* 48 (1997): 249–60.

32. Ian Tattersall, "Systematic versus Ecological Diversity: The Example of Malagasy Primates," in *Systematics, Ecology, and the Biodiversity Crisis,* ed. Niles Eldredge (New York: Columbia University Press, 1992), 25.

33. Robert H. MacArthur, *Geographical Ecology* (Princeton, N.J.: Princeton University Press, 1972), 197.

34. Niles Eldredge, "Where the Twain Meet: Causal Intersections between Genealogical and Ecological Realms," in Eldredge, ed., *Systematics, Ecology, and the Biodiversity Crisis,* 1–14.

35. Niles Eldredge, "Introduction," in Eldredge, ed., *Systematics, Ecology, and the Biodiversity Crisis,* 1–19.

36. John L. Harper and David L. Hawksworth, "Preface," in *Biodiversity Measurement and Estimation,* ed. D. L. Hawksworth (Oxford: The Royal Society, 1995), pp. ii-xiv.

37. Peter J. Bowler, *The Environmental Sciences* (New York: W. W. Norton, 1992), 10.

38. A detailed discussion of taxonomy is far beyond the purpose of this book. See Lance Grande and Olivier Rieppel, eds., *Interpreting the Hierarchy of Nature* (San Diego: Academic Press, 1994); Alessandro Minelli, *Biological Systematics: The State of the Art* (London: Chapman & Hall, 1993), text quote from p. 9; Andrew Cockburn, *An Introduction to Evolutionary Ecology* (Oxford: Blackwell Scientific Publications, 1991); Lynn Margulis and Karlene V. Schwartz, *Five Kingdoms: An Illustrated Guide to the Phyla of Life on Earth* (New York: W. H. Freeman, 1988); and Mark Ridley, *Evolution and Classification* (London: Longman, 1986). For a less technical discussion, see Charles C. Mann and Mark L. Plummer, *Noah's Choice* (New York: Alfred Knopf, 1996).

39. Margulis and Schwartz, *Five Kingdoms,* 3.

40. The prefixes *sub* and *supra* can be added to any level as determined by those doing the classifying. Species, for example, can be divided into subspecies when populations become isolated from one another.

41. Charles E. Bessey, 1908 president of the American Association for the Advancement of Science, as quoted in Mann and Plummer, *Noah's Choice,* 37.

42. Ernst Mayr, ed., *The Species Problem* (1957; reprint, North Stratford, N.H.: Arno Press, 1974; Publication No. 50 of the American Association for the Advancement of Science, Washington, D.C.).

43. Mayr, *The Species Problem,* iii.

44. Edward O. Wilson, "Biodiversity: Challenge, Science, and Opportunity," *American Zoologist* 34 (1994): 5–11.

45. Daniel P. Faith, "Phylogenetic Pattern and the Quantification of Organismal Biodiversity," in Hawksworth, ed., *Biodiversity Measurement and Estimation,* 48.

46. Wilson, "Biodiversity: Challenge, Science, and Opportunity," 5.

47. Anthony G. O'Donnell, Michael Goodfellow, and David L. Hawksworth, "Theoretical and Practical Aspects of the Quantification of Biodiversity among Microorganisms," in Hawksworth, ed., *Biodiversity Measurement and Estimation,* 65–73.

48. Grande and Rieppel, eds., *Interpreting the Hierarchy of Nature,* 280.

49. Minelli, *Biological Systematics,* 64.

50. See Edward O. Wilson, "Toward Renewed Reverence for Life," *Technology Review* 95, no. 2 (November/December 1992): 72, and generally Hawksworth, ed., *Biodiversity Measurement and Estimation.*

51. Michael L. Rosenzweig, *Species Diversity in Space and Time* (Cambridge: Cambridge University Press, 1995).

52. National Research Council, *A Biological Survey for the Nation* (Washington, D.C.: National Academy Press, 1993).

53. Rosenzweig, *Species Diversity,* 263; K. S. Schrader-Frechette and E. D. McCoy, *Method in Ecology* (Cambridge: Cambridge University Press, 1993), 77; Joel Cracraft, "Species Diversity, Biogeography, and the Evolution of Biotas," *American Zoologist* 34 (1994): 33–47; Peter S. Ashton, "Species Richness in Plant Communities," in *Conservation Biology,* ed. Peggy L. Fielder and Subodh K. Jain (New York: Chapman & Hall, 1992). The theory was proposed by Robert H. MacArthur and Edward O. Wilson, *The Theory of Island Biogeography* (Princeton, N.J.: Princeton University Press, 1967). See also Daniel Simberloff, "Biogeographic Approaches and the New Conservation Biology," in *The Ecological Basis for Conservation,* ed. S. T. A. Pickett, R. S. Ostfeld, M. Shachak, and G. E. Likens (New York: Chapman & Hall, 1997), 274–84.

54. Lawrence B. Slobodkin, "Islands of Peril and Pleasure," *Nature* 381, no. 6579 (May 16, 1996), 205.

55. Harper and Hawksworth, "Preface," in Hawksworth, ed., *Biodiversity Measurement and Estimation,* and G. T. Prance, "Biodiversity," in *Encyclopedia of Environmental Biology,* vol. 1, editor-in-chief, William Nierenberg (San Diego: Academic Press, 1995), 183–93.

56. Conference papers were published as E. O. Wilson, ed., *Biodiversity* (Washington, D.C.: National Academy Press, 1988).

57. Harper and Hawksworth, "Preface," in Hawksworth, ed., *Biodiversity Measurement and Estimation.*

58. Council on Environmental Quality, "Ecology and Living Resources—Biological Diversity," in *Environmental Quality: 11th Annual Environmental Quality Report* (Washington, D.C.: U.S. Government Printing Office, 1980), 31–80; Council on Environmental Quality and the U.S. Department of State, *The Global 2000 Report to the President of the U.S.,* Gerald O. Barney, study director (New York: Pergamon Press, 1980), 149–55.

59. United Nations, *Convention on Biological Diversity* (New York: United Nations, 1992), Article 2.

60. Keystone Center, *Report of a Keystone Policy Dialogue—Biological Diversity on Federal Land* (Keystone, Colo.: Keystone Center, 1991), 6.

61. Hal Mooney and Clifford J. Gabriel, "Preface," *BioScience* (Special Supplement on Science and Biodiversity Policy, June 1995).

62. Andrew Metrick and Martin L. Weitzman, "Patterns of Behavior in Endangered Species Preservation," *Land Economics* 72, no. 1 (February 1996): 1–16; Andrew Solow, Stephen Polasky, and James Broadus, "On the Measurement of Biological Diversity," *Journal of Environmental Economics and Management* 24, no. 1 (1993): 60–68.

63. John L. Harper and David L. Hawksworth, "Preface," in Hawksworth, ed., *Biodiversity Measurement and Estimation,* 5.

64. Robert M. May, "Conceptual Aspects of the Quantification of the Extent of Biological Diversity," in Hawksworth, ed., *Biodiversity Measurement and Estimation,* 13–20.

65. Estimates are rather flexible. Peter Raven of the Missouri Botanical Garden used a figure of 250,000 in 1991 congressional testimony but put the number of species at 750,000 in a personal communication with David Pimentel that Pimentel adopted in a 1997 article in *BioScience* after he had used a figure of 500,000 in a 1992 *BioScience* article. Peter Raven, testimony on H.R. 585 and H.R. 2082, "National Biological Diversity Conservation" before Congress, House Subcommittee on Environment of the Committee on Science, Space, and Technology, 102nd Cong., 1st sess., May 23, 1991, Committee Print 63, 3; David Pimentel, Ulrich Stachow, David A. Takacs, Hans W. Brubaker, Amy R. Dumas, John J. Meaney, John A. S. O'Neil, Douglas E. Onsi, and David B. Corzilius, "Conserving Biodiversity in Agricultural/Forestry Systems," *BioScience* 42, no. 5 (May 1992): 354–62; and David Pimentel, Christa Wilson, Christine McCullum, Rachel Huang, Paulette Dwen, Jessica Flack, Quynh Tran, Tamara Saltman, and Barbara Cliff, "Economic and Environmental Benefits of Biodiversity," *BioScience* 47, no. 11 (December 1997): 747–57.

66. IUCN (International Union for the Conservation of Nature), *1994 IUCN Red List of Threatened Animals* (Gland, Switzerland: IUCN, 1993). For discussions of issues related to extinction rates, see John H. Lawton and Robert M. May, eds., *Extinction Rates* (Oxford: Oxford University Press, 1995); Stuart L. Pimm, Gareth J. Russell, John L. Gittleman, and Thomas, M. Brooks, "The Future of Biodiversity," *Science* 269 (July 21, 1995): 347–50; Julian L. Simon and Aaron Wildavsky, "Species Loss Revisited," *Society* 30 (November/December 1992): 41–46; and Charles C. Mann, "Extinction: Are Ecologists Crying Wolf?" *Science* 253 (August 16, 1991): 736–37. The journal *Nature* published an informative series of letters on extinction rate estimates. See Fraser D. M. Smith, Robert M. May, Robin Pellew, Timothy H. Johnson, and Kerry S. Walter, "Estimating Extinction Rates," *Nature* 364 (August 5, 1993), 494; Alfredo D. Cuarón, "Extinction Rate Estimates," *Nature* 366 (November 11, 1993), 118; Nigel E. Stork and Christopher H. C. Lyal, "Extinction or Co-Extinction rates?" *Nature* 366 (November 25, 1993), 307; Vernon H. Heywood, Georgina M. Mace, Robert M. May, and S. N. Stuart, "Uncertainties in Extinction Rates," *Nature* 368 (March 10, 1994), 105; and a letter from Roman Dial and one from Stephen Budiansky, each published under the heading "Extinction or Miscalculation?" *Nature* 370 (July 14, 1994), 104.

67. IUCN (International Union for the Conservation of Nature), *1997 IUCN Red List of Threatened Plants* (Gland, Switzerland: IUCN, 1998).

68. D. H. Chadwick, "Mission for the 1990s," Defenders of Wildlife Special Report, Washington, D.C.

69. U.S. Fish and Wildlife Service, Division of Endangered Species, "Listed Species and Recovery Plans as of May 31, 1998," downloaded from the FWS's Web site at http://www.fws.gov/r9endspp/boxscore.html, August 11, 1998.

70. Robert K. Wayne and John L. Gittleman, "The Problematic Red Wolf," *Scientific American* 273, no. 1 (July 1995), 36–39.

71. U.S. Congress, Office of Technology Assessment, *Harmful Non-Indigenous Species in the United States,* OTA-F-565 (Washington, D.C.: U.S. Government Printing Office, 1993).

72. Michael J. Bean, testimony on behalf of the Environmental Defense Fund, American Association of Zoological Parks and Aquariums, American Society for Plant Physiologists, Center for Marine Conservation, Humane Society of the United States, National Audubon Society, Natural Resources Defense Council, and Society for Animal Protective Legislation, on H.R. 585 and H.R. 2082, "National Biological Diversity Conservation" before Congress, House Subcommittee on Environment of the Committee on Science, Space, and Technology, 102nd Cong., 1st sess., May 23, 1991, Committee Print 63, 148.

73. James D. Williams and Ronald M. Nowak, "Vanishing Species in Our Own Backyard: Extinct Fish and Wildlife of the United States and Canada," in *The Last Extinction,* ed. Les Kaufman and Kenneth Mallory (Cambridge: MIT Press, 1993), 115–48.

74. U.S. Fish and Wildlife Service, Division of Endangered Species, "Endangered Species Home Page," downloaded from the FWS's Web site at http://www.fws.gov/r9endspp/faqrecov.html), January 15, 1998.

75. Wilson, "Biodiversity: Challenge, Science, and Opportunity," 5.

76. Kent E. Holsinger, "Population Biology for Policy Makers," *BioScience* (Special Supplement on Science and Biodiversity Policy, June 1995): S16.

77. Ecologists often express a desire to do away with exotic species on the grounds that they harm endemics; see Elizabeth Culotta, "Biological Immigrants under Fire," *Science* 254, no. 5037 (December 6, 1991): 1444–46.

78. Reed F. Noss, Edward T. LaRoe III, and J. Michael Scott, *Endangered Ecosystems of the United States: A Preliminary Assessment of Loss and Degradation,* Biological Report No. 28 (Washington, D.C.: U.S. Department of the Interior, National Biological Service, February 1995), 50.

79. Reed F. Noss and Robert L. Peters, *Endangered Ecosystems: A Status Report on America's Vanishing Habitat and Wildlife* (Washington, D.C.: Defenders of Wildlife, 1995).

80. Noss and Peters, *Endangered Ecosystems,* viii.

81. Dwight A. Brown, "Early Nineteenth Century Grasslands of the Midcontinent Plains," *Annals of the Association of American Geographers* 83, no. 4 (December 1993): 589–612.

82. Carol Yoon, "Counting Creatures Great and Small," *Science* 260 (April 30, 1993): 620–22.

83. Brown, "Early Nineteenth Century Grasslands," 589.

84. Raymond S. Bradley and Philip D. Jones, eds., *Climate since A.D. 1500* (New York: Routledge, 1992).

85. Melissa Savage, "Structural Dynamics of a Southwestern Pine Forest under Chronic Human Influence," *Annals of the Association of American Geographers* 81, no. 2 (June 1991): 287.

86. Clarence J. Glacken, "Changing Ideas of the Habitable World," in *Man's Role in Changing the Face of the Earth,* ed. William L. Thomas Jr. (Chicago: University of Chicago Press, 1956), 70–92.

87. Carl O. Sauer, "The Agency of Man on Earth," in Thomas, ed., *Man's Role in Changing the Face of the Earth,* 49. Even though this volume was produced forty years ago, it remains an indispensable collection of readings for those interested in the impacts of humans on the landscape through the corridors of time and across cultures.

88. A classic work on man and nature in Western thought is Clarence Glacken, *Traces on the Rhodian Shore* (Berkeley: University of California Press, 1967).

89. George Perkins Marsh, *Man and Nature; or Physical Geography as Modified by Human Action* (New York: Charles Scribner, 1864), republished in 1965, ed. David Lowenthal (Cambridge, Mass.: Belknap Press); William L. Thomas, "Introductory," in Thomas, ed., *Man's Role in Changing the Face of the Earth,* xxix.

90. Marsh, *Man and Nature,* 3.

91. Marsh, *Man and Nature,* 42.

92. Dyan Zaslowsky and the Wilderness Society, *These American Lands* (New York: Henry Holt, 1986), 213; J. Donald Hughes and Jim Swan, "How Much of the Earth Is Sacred Space?" *Environmental Review* 10, no. 4 (winter 1986): 247–59; and Linda Graber, *Wilderness as Sacred Space* (Washington, D.C.: Association of American Geographers, 1976), 13.

93. William Cronin, "The Trouble with Wilderness; or, Getting Back to the Wrong Nature," in *Uncommon Ground: Toward Reinventing Nature,* ed. William Cronin (New York: W. W. Norton, 1995), 80.

Chapter 5

Nature Worship and the New Paradigm

Nature was his church . . . and . . . protection of Nature became a Holy War.
—Roderick Nash, on Sierra Club founder John Muir, *The Rights of Nature,* p. 41

"Reverence for the ecosystem," concludes David Kinsley of MacMaster University, "is often expressed in terms of reverence of Mother Earth, or the planet as a living being, or the ecosystem as an organic whole with rights of its own that are equal to, or transcend, the rights of human beings or even the human species as a whole."[1] For many of its advocates, the new paradigm is inevitably intertwined with the worship and veneration of nature. For them it is a stalking horse to impose elements of their nature-worshiping faith on what is mainly a Christian nation.[2] In the United States, people are free to worship whatever they wish. If someone chooses to venerate nature or Mother Earth they may, but doing so is inconsistent with two millennia of Christian teaching. Yet activists are increasingly successful in their efforts to make reverence for nature an element of religious life in the United States and elsewhere. By doing so, new paradigmists enlist powerful allies in support of ecosystem management, which they see as the policy vehicle to limit human use of the earth in defense of Mother Nature. In this chapter, I explore the religious underpinnings of the new paradigm. I give particular attention to biocentrism (or ecocentrism) and to the success that new paradigmists have had in luring mainline churches into preaching the mantra of nature worship. I demonstrate the extent to which Congress has already incorporated veneration of nature into federal law. Finally, the chapter considers the nature of nature. Is it a beneficent god dispensing blessings from the wilderness, or is it something else?

Researchers have long recognized that environmentalism is akin to a nature-based religion. Linda Graber, in her book *Wilderness as Sacred Space,* observes that "since the 1880s until the present [1976] writers have noticed the crypto-religious behavior of the conservation movement." University College (of

London) geographer David Lowenthal is more direct, saying, "A new religion is in the making. Worshipers of nature exhort us from the pulpits of countless conservation societies and Audubon clubs. . . . Nature is wonderful, we are told, pay homage to it in the Wilderness."[3] Martin Lewis finds that religion is much more than an inert component of environmentalism. In his conclusion to *Green Delusions: An Environmentalist Critique of Radical Environmentalism,* he writes that the religious aspects of "eco-radicalism" cannot be overemphasized. For true believers, "the modern world is thoroughly derelict, and it will either perish for its sins or we will collectively find eco-salvation."[4] Some environmentalists acknowledge their religion.[5] Victor Scheffer argues that environmentalism "found fertile ground on college campuses . . . [because] it offered a religion to young people who, through a liberal education, were moving beyond dependence on gods. They were looking for guidance from natural, not supernatural, wisdom."[6] For Scheffer and others of this ilk, traditional Judeo-Christian faiths represent the supernatural. As Alston Chase notes in his book *In a Dark Wood,* "The Judeo-Christian tradition was out, and the ecosystem was in."[7]

New paradigmists use several interlinked lines of reasoning to conclude that nature is worthy of adoration. For some, God and nature are one and the same. Environmentalist doyen David Brower (former Sierra Club executive director and founder of Friends of the Earth) says that "to me, God and Nature are synonymous."[8] Others seize on the long-held view in Western thought and religion that nature is a creation of God and conclude that it is worthy of reverence on that account.[9] As I show later in this chapter, by convincing church leaders that Americans are reckless stewards of God's creation, new paradigmists enlist many churches into their cause. Another new paradigm route to nature worship lies in neoanimism. Here, advocates place gods in natural phenomena. In his foreword to the book *The Big Outside* (authored by Earth First! cofounders Dave Foreman and Howie Wolke), Michael Frome speaks of the wisdom of "the Dolphin-God, or of the many gods in sky, rocks, trees, and water."[10] Political scientist and environmentalist Lynton Caldwell of the University of Indiana writes admiringly of "neo-paganism" as a sophisticated form of nature worship based on a respect for all life. Natural religion, he concludes, is the rational route to a "harmonious relation of man to Earth."[11] Biocentrists, who are the driving force in today's environmental movement, argue not only that all life should be respected but that all species—including humans—are of equivalent value and that nonliving things possess full rights as coinhabitants of a planet that they consider an "organic whole." Biocentric philosophers Bill Devall and George Sessions write "that all organisms and entities in the ecosphere, as parts of the interrelated whole, are equal in intrinsic worth."[12] For public policy purposes, the path that new paradigmists take to nature worship is less important than their insistence on the rest of us adopting their faith.

New paradigmists see themselves as saviors of Mother Earth. They are self-anointed and zealous in pursuit of their creed. Scheffer claims that "the opponents seemed not to understand that environmentalism is a morality of life and death for the human race."[13] "*No More Compromise!*" writes Dave Foreman, we must enter into a "humble joining with Earth, becoming the rain forest, the desert, the mountain" and defend Earth "against hostile humanism . . . against extinction for what is sacred and right: the Great Dance of Life."[14] As is often the case with the self-righteous, advocates of the new paradigm have little patience with nonbelievers. Arguing that the earth is in the balance, activists accuse scientists, researchers, clergy, journalists, politicians, and others who fail to see the innate wisdom and moral superiority of the new paradigm of everything from religious bigotry to a "*Betrayal of Science and Reason,*" the title of a book by Stanford University biologists and activists Paul and Anne Ehrlich. They label as "brownlash" the "efforts being made to minimize the seriousness of environmental problems."[15] Secretary of the Interior Bruce Babbitt also routinely denounces those who disagree. He calls efforts to make badly needed reforms to federal environmental laws a plan "that is hostile to God's creation and determined to undermine the very legal tools . . . that allow us to restore that creation."[16] Efforts of reformers, he says, can "only drive us toward destruction" and away from "our moral obligation as stewards of God's creation." New paradigmists see those who do not share their views as being more than just advocates of wrongheaded policies—they see them as morally impoverished.

Prophesying for Nature

Scholars do not have to look far to find nature worship in environmentalism. John Muir, "the most rapt of all prophets of our out-of-door gospel," constantly evokes religious imagery in his writings.[17] During the early twentieth-century debate over converting the Hetch Hetchy Valley in Yosemite National Park into a drinking-water reservoir for the city of San Francisco, Muir rails against those in favor of the project: "These temple destroyers, devotees of ravaging commercialism, seem to have a perfect contempt for Nature, and, instead of lifting their eyes to the God of the mountains, lift them to the Almighty Dollar. Dam the Hetch Hetchy! As well dam for water-tanks the people's cathedrals and churches, for no holier temple has ever been consecrated by the heart of man."[18] Environmental historian Roderick Nash says that, in Muir's view, "wild nature was replete with 'Divine beauty' and 'harmony'," with "spiritual power" emanating from the landscape.[19]

Muir's continuous invocation of god and his use of religious rhetoric blinds many from fully understanding that his god is not that of Abraham or Isaac. It is not part of the Christian Trinity. Instead, Muir finds god and nature "indivisible concepts," according to historian Daniel Payne.[20] Muir's god is a "loving, intelligent spirit" controlling the universe from dwelling places in rocks and trees and animals. In 1871 Ralph Waldo Emerson visited Yosemite Valley. In a note to Emerson, imploring him to remain longer in Yosemite, Muir writes,

> The spirits of these rocks and waters hail you as a kinsmen . . . I invite you to join me in a month's worship with Nature in the high temples of the great Sierra Crown beyond our holy Yosemite . . . in the name of all the spirit creatures of these rocks and of this whole spiritual atmosphere . . . I am yours in Nature.[21]

To use the subtitle of Thurman Wilkins's biography of Muir, he is an "Apostle of Nature."[22] Another biographer, Stephen Fox, argues that "Muir made a permanent break with Christianity" and departed "the faith of Jesus."[23] Indeed, he "had no use for Christianity, either as institution or belief system."[24] No longer a Christian, he adopted a "pantheistic spiritualism . . . [that] makes transcendentalism seem pale in comparison."[25] Muir believed that spirits dwelt in the objects of nature and so "felt closer to pagan European and native American religious traditions" than to Western theology.[26] In rejecting Judeo-Christian teaching on the relations between God, humans, and the rest of creation and in its place arguing that all things were of value as part of an organic, cosmic whole, Muir advocates biocentrism before the term came into existence.[27] "Again and again," write Devall and Sessions, "the reform and deep ecology movements have returned to the insights expressed by John Muir." They take particular note of Muir's view that "Nature's object in making animals and plants might possibly be first of all the happiness of each one of them, not the creation of all for the happiness of one."[28]

Aldo Leopold is another of nature's messengers and "the spiritual father of ecosystem management," in the view of forestry professor Boris Zeide.[29] In his classic work *Wilderness and the American Mind,* Nash titles the Leopold chapter "Aldo Leopold: Prophet." With a degree in forestry from Yale University and extensive experience in resource management, Leopold's views reflect more scientific understanding than do Muir's and contain less religiosity. Even so, his reputation is built on nonscientific essays, and many who write about him employ terms used by acolytes in awe of a high priest. Commenting on the importance of *A Sand County Almanac,* Leopold's most famous work, writer Wallace Stegner opines that it is "one of the prophetic books, the utterance of an American Isaiah." It is "The Holy Writ of American Conservation," according to ecologist René Dubos, and "one of [the environmental movement's] new testament gospels." The volume is "the most important book ever written," in the judgment of Dave Foreman. Historian Donald Fleming sees Leopold as "the Moses of the New Conservation impulse of the 1960s and 1970s, who handed down the Tablets of the Law."[30]

New paradigmists give Leopold the status of a prophet because he embraces what later became biocentrism. The earth, in Leopold's view, is "a living being."[31] Devall and Sessions write that "in the 1920s and 1930s, Aldo Leopold underwent a dramatic conversion from the 'stewardship' resources management mentality to what he called *ecological conscience*" [emphasis in the original].[32] They praise his work as "radical," "subversive," and a "landmark in the development of the biocentric position." "His enduring achievement," according to biographer Susan Flader, was his melding of "ecological and evolutionary thought" with "the conservation imperative" into "a compelling ethic for our time," which philosopher J. Baird Callicott calls "a brand new nature ethic."[33] Leopold's reputation as a seminal thinker in American environmentalism is based heavily on his essay "The Land Ethic," which appears in *A Sand County Almanac*.[34] In it, Leopold argues that people must expand their circle of ethical concern beyond that of fellow humans "to include soils, waters, plants, and animals, or collectively: the land." He says that people must think of ourselves as simply "plain" members and citizens of the land community and that we must fully respect not only the other nonhuman citizens but also the community itself as a separate entity. The land ethic "cannot prevent the alteration and use of 'resources'," Leopold writes, "but it does affirm the right" of soil, water, plants, animals, and communities "to continued existence, and, at least in spots, their continued existence in a natural state."

But how will Americans know when we are doing what is ethically correct? New paradigmists answer by citing Leopold's precept that "a thing is right when it tends to preserve the integrity, stability, and beauty of the biotic community. It is wrong when it tends otherwise."[35] Callicott claims that what made Leopold unique among the "celestial chorus of voices crying in and for the wilderness" is his provision of "a sound *scientific* foundation for a land or environmental ethic" [emphasis in the original], but there is no science in Leopold's standard of environmental correctness.[36] As I show in chapter 6, researchers are unable to give substantive meaning to the term *biotic integrity,* and ecologists today overwhelmingly reject the idea of biotic stability. Scientists cannot offer objective standards for Leopold's wholly subjective notion of biotic beauty. Science is equally absent from other new paradigmist Leopold favorites, such as "Thinking Like a Mountain," the title of an essay in which Leopold holds that "only the mountain has lived long enough to listen objectively to the howl of a wolf" and that "to keep every cog and wheel is the first precaution of intelligent tinkering."

Biocentrism is basically a form of nature worship blended with enough (often times dubious) science to broaden its appeal and mask its foundation. New paradigmists expend a good deal of effort to portray biocentrism as anything but nature worship. This is not new. Muir and Leopold, for example, hid the more radical nature-worshiping components of their thinking from the public because they realized that the United States is mainly a Christian nation and its people are unlikely to embrace a philosophy wherein humans and other life forms, to say nothing of inanimate objects, are on equal spiritual footing.[37]

The situation may be changing as new paradigmists are increasingly successful in moving nature worship into contemporary religious teachings by mainline churches.[38] Traditional faiths are "vital to environmentalism," says philosopher Max Oelschlaeger, because they need the "moral legitimacy" of these religions to achieve their ultimate purpose of transforming "society's relations to nature."[39] Thus, new paradigmists incessantly campaign to convince church leaders that we are poor stewards of Creation and hoodwink church leaders into embracing nature worship cloaked as caring for the Garden.

In June 1991 two dozen U.S. church leaders issued a "Statement by Religious Leaders at the Summit on Environment."[40] The leaders woodenly chant the new paradigmist litany. Americans are killing ecosystems, global warming not only threatens ecosystem integrity but will create "millions of environmental refugees," and 20 percent of the world's species may be lost in the next thirty years. Americans must, in the view of these church officials, maintain Creation as God made it. They see "the cause of environmental integrity . . . [as having] the potential to unify and renew religious life." These church officials wax eloquent and proclaim, "in the strongest possible terms, the indivisibility of social justice and ecological integrity" and go on to declare that "economic equity, racial justice, gender equity and environmental well-being are interconnected and are all essential to peace." Their rhetoric neatly reflects radical ecology. In the introduction to her book *Radical Ecology,* Carolyn Merchant writes, "Radical ecology emerges from a sense of crisis . . . on a new perception that the domination of nature entails the domination of human beings along lines of race, class, and gender."[41]

The church leaders offer grand pronouncements but demonstrate little ecological understanding or knowledge about the practical relationship between humans and the environment. They seem, for example, to be oblivious to ecologists' failure to give the idea of environmental integrity substantive meaning, the reality that ecosystems are not alive and that we cannot keep Creation as God made it because it is constantly changing with or without human activity. Consider their bold proclamation of the "indivisibility of social justice and ecological integrity." As I point out in the next chapter, most researchers seeking to give meaning to ecological integrity argue that the ecosystems with the greatest degree of integrity are natural ecosystems, places where people do not live or where human impacts on the land are minimal. The church leaders do not tell us what they mean by social justice, but at minimum it would include life's basic necessities of food, clothing, and shelter. Yet people wrestle these items from the physical environment by changing it and thus reducing its ecological integrity as many understand the term. The religious leaders wish to have their cake and eat it too.

Their claim of an interconnectedness of economic equity, racial justice, gender equity, environmental well-being, and peace is unsupportable. The church leaders do not explain these linkages because they cannot explain them. Peace, for example, prevailed in much of Eastern Europe and the Union of Soviet Socialist Republics in most of the last half of the twentieth century; however, there is little evidence of what anyone would call a high level of environmental well-being in that region. Similarly, I would ask the officials to explain their proclaimed link between clean air or protection of wilderness areas and racial tolerance and harmony. How does water quality affect gender understanding? How does the existence of Yellowstone National Park improve economic equity? The soaring rhetoric of these church officials does not reflect reality.

Christian churches have not fully clarified their new view of human/nature relationships in their drift toward nature worship. Douglas Chial of the Presbyterian Church of the United States writes, "The ecumenical movement has only recently acknowledged the need to encourage eco-centric spirituality that embraces our dependence on the earth and God's real presence in all creation." When Chial asks, "What does it mean to do theology from the perspective of birds, water, air, trees and mountains?" he moves closer to Muir's pantheism than to the teachings of Jesus.[42] Ecotheologian Sallie McFague sees the earth as God's body; God dwells in creation, and so the line between Creator and creation blurs if it is not eliminated entirely.[43] The World Council of Churches (WCC) goes still further. A group of WCC-sponsored theologians write that "we not only reject a view in which the cosmos does not share in the sacred . . . we also repudiate hard lines drawn between animate and inanimate and human and non-human." They continue that we are accountable "to the Spirit, one another, and the rest of the biotic community."[44] In other words, Mother Earth will judge us in addition to God. Catholic ecotheologian Thomas Berry ignores God entirely in his essay "The Viable Human." Berry writes that "in creating the planet Earth, its living forms, and its human intelligence, the universe has found . . . the most elaborate manifestation of its deepest mystery. Here, in its human form, the universe is able to reflect on and celebrate itself in a unique mode of self-awareness."[45] In this view, Nature (the universe), not God, creates itself as well as humans.

What standards will Mother Earth use to judge us? Christians have a good idea of God's standards of judgment, the Ten Commandments as passed to Moses and reaffirmed by Jesus.[46] But nature provides people with no comparable rules of conduct, so new paradigm advocates propose to speak for her. Thus, they tell us that "a thing is right when it tends to preserve the integrity, stability, and beauty of the biotic community. It is wrong when it tends otherwise." According to philosopher Paul Taylor, we cannot show any "bias in favor of some [species] over others" and must exhibit "an attitude of respect toward the individual organisms, species-populations, and biotic communities of the Earth's natural ecosystems."[47] Chial says, "Christians need to . . . live in harmony with creation and thereby heal the brokenness that surrounds us." This sort of ethereal mysticism is far removed from the precision of the Ten Commandments. One need not be a theologian to grasp God's meaning when he says, "You shall not steal." On the other hand, no amount of training—theological or ecological—can give substance to notions such as "the integrity, stability, and beauty of the biotic community." Nonetheless, many theologians find such language enticing. Pope John Paul II, venturing outside the realm of theology, writes of the "requirements of the order and harmony which govern nature itself" and discusses "the ecosystem" as though it were real.[48]

Pope John Paul II does not, however, wander into nature worship. In his message "The Ecological Crisis: A Common Responsibility," he emphasizes God's creation of "the heavens, the sea, the earth, and all it contains" and God's entrusting "the whole of creation to the man and the woman."[49] The pope points out that "Adam and Eve's call to share in the unfolding of God's plan . . . distinguish the human being from all other creatures." He writes, "[when] man is no longer able to see himself as 'mysteriously different' from other earthly creatures; [when] he regards himself merely as one more living being, as an organism which, at most, has reached a very high stage of perfection . . . [he loses his sense of God and] no longer grasps the 'transcendent' character of his 'existence as man.' "[50] Humans, according the pope, are to exercise dominion over the earth guided by "wisdom and love." People must, he says, have a "due respect for nature." But God and nature are wholly distinct; God created the earth, but he is found in heaven, not in the heuristic device of ecosystems.

One reason for new paradigmist success in their greening of Christian churches is their ability to lay an ecological guilt trip on the faithful and their leaders. Historian Lynn White made the Judeo-Christian-theology-is-to-blame-for-impending-environmental-disaster theme fashionable with his 1967 article in *Science,* titled "The Historical Roots of Our Ecological Crisis."[51] He faults Western religion for teaching that humans are different from other living things and have a special place in God's creation. "The roots of our trouble," he writes, "are so largely religious, that the remedy must also be religious." Robert Fowler finds that the "universal constant" among a "fascinating, diverse, and fiercely controversial proliferation of eco-theologies" is "the conviction that creation is in crisis."[52] Jesuit Father Albert Fritsch begins his book *Eco-Church: An Action Manual* with "Today, God's Earth is in deep trouble . . . the litany of woes is almost endless. Truly, there is an urgency to do something to save the Earth from destruction."[53] Dominican Father Matthew Fox goes much further. The highly controversial Fox proclaims that "the killing of Mother Earth in our time is the number one ethical, moral, spiritual, and human issue of our time."[54] Biocentrically inclined or suddenly guilt-ridden theologians and others jump on the ecological crisis bandwagon and hurry to reread the Bible through green glasses, utter mea culpas, or generally bash Christianity, as with Wendall Berry's rant that "Christianity connives directly in the murder of Creation."[55]

Many church leaders join hands to kneel before nature in guilt-driven rituals that reflect ecological ignorance and theological confusion and that are awash in gross oversimplifications and distortions. Typifying this trend is "A Service of Worship: The Earth Is the Lord's—a Liturgy of Celebration, Confession, Thanksgiving, and Commitment," by the National Council of the Churches of Christ (NCCC).[56] The service begins with a "call to worship" that acknowledges that "the Earth is the Lord's"; so far so good, but in the next portion—"the act of confession"—it goes quickly off track. Participants are told from the pulpit that "we are responsible for massive pollution of earth, water, and sky." Guilt! An environmentally informed member of the congregation may well ask, "But what of all the data that show our air is clean and getting cleaner, that our waters are cleaner than they have been in decades; what about all the controls we have on toxic and hazardous wastes as well as on just plain garbage, and $100-plus-billion we spend every year to improve environmental quality and right past wrongs?" At this point in the service, the entire congregation is to confess, "We are killing the earth." Guilt! An analytical member of the congregation could pause, speculating, "How can we kill the earth since it is not alive in the first place?" From the pulpit the parade of presumed environmental horribles resumes—including such nonsense as, "We use trees faster than they can regrow," and the congregation is to respond, "We are killing the waters" and "We are killing the skies." Guilt! Knowledgeable members of the congregation might ask in despair, "Where are the clergy finding all this propaganda?" They would quickly answer their own question: from environmentalists.

The service proceeds. In the "prayer of grace" the congregation asks forgiveness not only from God but also "from the earth and from future generations." According to the National Council of Churches, the earth is empowered to forgive sins, but that is inconsistent with church dogma. Christian teaching provides that God can forgive sins; clergy, acting in his name, can forgive sins; and it calls on Christians to forgive those who sin against them, but it does not bestow on the earth authority to forgive sin. The notion that the earth can forgive sins is nature worship. At this point in the service, those familiar with the Bible might wonder, "How can I quietly depart from this druidic ceremony?"

Clergy expose members of the congregation to new paradigmists' romantic views during the service. In the liturgy's "act of thanksgiving," for example, the minister thanks God for the "passion of the children and youth . . . who push us to recognize the urgency of the environmental crisis." But educators such as Larry Gigliotti of Cornell University note that students too often come from an environmental education system that produces people who are "emotionally charged but woefully lacking in basic ecological knowledge," and Professor Jim Bowyer finds that his students at the University of Minnesota "consistently indicate the environmental situation to be worse than it really is."[57] One reason, as Stanford University climatologist Stephen Schneider points out, is that teachers too often expose them to "flaky" educational materials.[58] The Independent Commission on Environmental Education concurs. After reviewing K-12 teaching materials, it concludes that "factual errors are common in many environmental education materials and textbooks," "many high school environmental science textbooks have serious flaws . . . [including a propensity to] mix science with advocacy," and "environmental educational materials often fail to prepare students to deal with controversial environmental issues . . . [or] to help students understand tradeoffs in addressing environmental problems."[59]

The U.S. Catholic Conference also dons sackcloth and seeks to enhance the church's green credentials by offering such ecologically and theologically bewildering assertions as "other species, ecosystems, and even distinctive landscapes give glory to God."[60] Ecosystems, we know, are but arbitrary constructs of the human mind, so how can they "give glory to God?" What is a distinctive landscape? Manhattan? Yosemite Valley? A wheat field? A junkyard? What qualities make landscapes distinctive, and how do they allow such a landscape to glorify God? Are undistinctive landscapes unable to give glory? How is it that they fall into this heathen category? Can they be rehabilitated and somehow be made worthy?

The Bishops encourage parishes to take up environmental matters. They "invite the Catholic community to join with us and others of good will in a continuing effort to understand and act on the moral and ethical dimension" of the environmental crisis.[61] Parishes should hold meetings and get involved in issues. Like the NCCC, the Bishops erroneously believe that teenagers have "current knowledge of environmental conditions" and therefore should be especially encouraged to participate. The Bishops demonstrate more naivete when they suggest that parishioners invite environmentalists—who are often major sources of environmental myths and misinformation in the first place (see chapter 4)—into the parishes "as good sources of information" and that "materials from these organizations be displayed on tables" to educate parishioners.[62] This is akin to an altruistic farmer not only asking the fox to guard the hen house but also giving him a badge and uniform to enhance his authority and credibility with the hens.

Many theologians and scholars fault the contemporary movement toward nature worship. Colin Russell, in the Templeton Lectures on religion and science at

Cambridge University, sees "the current trend toward a vague deification of the Earth" as "flying in the face" of 2,000 years of Christian teaching and points out that the Old Testament is filled with examples of the prophets trying to protect God's chosen people, Israel, from nature-worshiping neighbors while in the New Testament "the teaching and life of Jesus can leave us in no possible doubt as to the transcendence of the God whom he called 'Father'." He says of modern-day pantheism that this "recourse to pre-Christian notions of a divine universe . . . is both an aberration of theology and an abandonment of science."[63] In the final analysis, says Russell, "at no point in Scripture are we ever encouraged to view the Earth as coterminous with God (or part of him)."[64] In his monumental work on man and nature in Western thought, *Traces on the Rhodian Shore,* geographer Clarence Glacken observes that in Christian theology the idea of worshiping the Creator, not creation (or creatures), is constantly recurring. "The works of God can be discerned in creation, but God is transcendent," writes Glacken, "the creation is by him but not of him, and is only a partial teacher. One can see His ways in it but worship is for the Creator alone."[65] Philosopher Stephen Clark, who believes that we are in the midst of an environmental crisis, nonetheless cautions that "we must not retreat to a romantic paganism, pantheism or animism."[66]

The difficulty that ecotheologians face in their efforts to fashion a biocentric Christian faith—wherein people, rats, and slime molds are morally equivalent and have equal standing before God—is that their position runs afoul of the fundamental source for Christian dogma: the Bible. The Scriptures can certainly be cryptic, and it is possible for theologians to selectively extract quotes to support divergent opinions, but on some points it is very clear.[67] Father Robert Sirico observes that "the Genesis account of creation provides enough theological evidence to counter the greening of theology."[68] In Genesis, it is written that "God created man in his image; in the divine image he created them; male and female he created them."[69] God further distinguishes humans from the remainder of creation in Genesis 1:28–31:

> Be fertile and multiply; fill the earth and subdue it. Have dominion over the fish of the sea, the birds of the air, and all the living things that move on the earth. . . . I give you every seed-bearing plant all over the earth and every tree that has seed-bearing fruit on it to be your food; and to all the animals of the land, all the birds of the air and all the living creatures that crawl on the ground, I give all the green plants for food . . . God looked at everything he had made and found it was very good.

Not surprisingly, biocentrically inclined theologians revile this text. Roger Gottlieb sees in this passage "an essential source for the havoc wreaked by Western societies upon the earth."[70] That Gottlieb is wrong in his claims that Christian societies have ruined the earth is beside the point; the fact remains that in

Genesis, God calls on beings he created in his own image to have dominion over the remainder of his Creation. People may worship Mother Nature, ecosystems, or various components of the environment if they wish; they are free to believe that all living and non-living things share a moral equivalency with humans; but they cannot do so within the confines of 2,000 years of Christian thought and understanding of God's teachings as reflected in the Bible.

Sirico criticizes the move to nature worship inherent in many of the new liturgies proposed by Protestant and Catholic theologians and organizations. The United Nations Environmental Program trumpets a liturgy for their "Environmental Sabbath."[71] In one prayer ministers exhort the faithful to "restore the Earth . . . [because] we cannot let our mother die. We must love and replenish her."[72] Sirico observes that "describing the earth as our living mother either constitutes a pagan form of earth worship or comes dangerously close to it."[73] He faults the Catholic Bishops for lending support to a movement that—contrary to Christian teaching—"seems to view human beings as the earth's most undesirable part of creation." Green Peace's Paul Watson sees humans "as the AIDS of the Earth . . . a viral epidemic to the Earth" and who, in committing crimes against nature, will be judged and punished by "Holy Mother Gaia."[74] Theologian Thomas Derr is more general in his rejection of green theology. He writes that instead of scrambling to say "Me, too," Christianity should reject biocentrism because the "historic and traditional" Christian valuation of "Creation is a perfectly sufficient guide to sound ecology."[75] Genesis 2:15, for example, recalls that "the Lord God then took the man and settled him in the garden of Eden, to cultivate and care for it." Christian understanding has us accountable to God for our dominion over his Creation, and there is scant biblical evidence that he is looking for plunderers rather than for stewards.

In chapter 4, I highlight the depth of commitment Americans have to sound stewardship. One does not have to be a new paradigmist to be in favor of clean air or clean water. One does not have to worship Mother Earth to support protecting other species, wanting well-managed national parks, or backing the existence of the national wilderness system. What separates many new paradigmists from others on environmental policy matters is their insistence on the elevation of nature protection over all other considerations in decision making. They reject balance in favor of a nature-first approach to land use management that is driven by their veneration of Mother Earth.

Nature Worship in Federal Law

Suppose that the head of the Conference of Catholic Bishops in the United States (or the leader of any other major religious denomination) sent the following letter to the president and Congress:

Dear —

The bishops and I just returned from an extensive tour of the nation. We visited every part of the country and are appalled by what we saw. Our God is under assault throughout the United States. We find sacrilege at every turn. We find widespread desecration and unbelief. We demand that the government take immediate and bold actions to inculcate our faith into federal law. Your actions should include establishing large regions wherein the worship and protection of our God must take precedence over all other human endeavors and use of federal power to force unbelievers to increasingly adhere to precepts of our faith.

Sincerely, Cardinal —

Politicians, evening-news anchors, editors, and constitutional scholars would rush to point out that Article I of the Constitution directs that "Congress make no law respecting an establishment of religion" and that the Supreme Court interprets this provision very broadly. The cardinal's suggestion that the government create special places for the practice of Catholicism and that it coerce non-Catholics into behaviors more consistent with Catholic teaching would ignite screams of protest and be seen by many as a threat to the very foundations of the country. However, leaders of the eco-church successfully make precisely these kinds of demands of our presidents and congresses.

Congress passed the Wilderness Act in 1964. The act is not an overt attempt to establish eco-church lands, but in practice it has that effect. Congress initially set aside 9 million acres of federal lands as statutory wilderness areas and provides specific management guidelines.[76] It does not contain any limit on the amount of land that may be placed in the system, but during congressional debates even staunch wilderness preservationists "pointed out that the most land the system would ever include was about fifty million acres."[77] Once Congress passed the act, however, environmentalists abandoned the notion of limits on wilderness acreage and began a ceaseless campaign to expand the system so that by 1997 it included 104 million acres in thirty-eight states. They continue to press for major new additions and support the Wildlands Project, which would put some 50 percent (about 950 million acres) of the contiguous states in wilderness or near-wilderness status (see chapter 8).[78] The new paradigmists' religiously motivated there-can-never-be-too-much-wilderness zealotry distinguishes them from American society at large in that, while deriving benefits from wilderness areas that are unrelated to nature worship, the general public opts for a more balanced apportionment of the nation's real estate. But are lands in the national wilderness preservation system really eco-church lands? Are they reasonably comparable to holy places where the faithful of traditional churches gather to worship God?

The Wilderness Act is the first place we should look to answer these questions. The act declares that it is the policy of Congress "to secure for the American people of present and future generations the benefits of an enduring resource of wilderness." Congress does not spell out these presumed benefits in the act; rather, they simply assume that it is in the national interest to set aside lands "for preservation and protection in their natural condition." They tell us that wilderness is "an area where the earth and its community of life are untrammeled by man, where man himself is a visitor who does not remain." Congress says that wilderness is a "primeval" landscape dominated "by the forces of nature." Almost as an afterthought, the act notes that wilderness areas "*may also* contain ecological, geological, or other features of scientific, educational, scenic or historical value" [emphasis added]. For Congress, what makes a landscape special and worthy of government protection and preservation as a wilderness is a condition of undisturbed nature. Scientific, educational, recreational, ecological, or other possible attributes of the landscape are not determinative when Congress decides on an area that should be set aside as a wilderness according to the act.

Undisturbed nature is precisely what makes landscapes sacred space for new paradigmists. It is in wilderness that environmentalists find their goddess—nature—in her most perfect form. Wilderness is both the physical manifestation of the goddess and a place for worship. "In wildness," wrote Thoreau, "is the preservation of the world."[79] If "Cardinal Green" of the eco-church sent the letter introducing this section, then Congress began complying with his wishes when it created the National Wilderness Preservation System because, as a practical matter, the Wilderness Act sets aside large regions for the protection and worship of the cardinal's goddess. Moreover, Congress often looks favorably on the cardinal's request for the expansion of church lands.

For "Cardinal Green," setting aside areas dedicated to the eco-church is insufficient. Human actions, when on sacred ground, must conform to the precepts of the church. Again Congress complies. The Wilderness Act stipulates that wilderness areas "shall be administered for the use and enjoyment of the American people in such manner as will leave them unimpaired for future use and enjoyment as wilderness." People are welcome in small numbers and only if they abide by the teachings of the eco-church, which holds that twentieth-century humans desecrate the temple and so the trappings of contemporary society must be forbidden from sacred ground. The Bureau of Land Management's wilderness regulations are typical of other federal land managing agencies. They provide that "wilderness areas shall be managed to promote, perpetuate and, where necessary, restore the wilderness character of the land."[80] The regulations prohibit, among other things, "temporary or permanent roads; aircraft landing strips, heliports, or helispots; use of motorized equipment, motor vehicles, motorboats, or other forms of mechanical transport; landing of aircraft; dropping of materials, supplies, or persons from aircraft; [and] structures or installations [including youth camps or communication towers]."[81] Their regulations emphasize limiting defilement of nature by human action. Congress not only establishes and protects the cathedrals of the earth goddess but also sees to their maintenance and requires those who visit them to do so on the terms dictated by her worshipers. The fact that most Americans may concur with Congress's directions for wilderness management does not change the reality that "Cardinal Green" is obtaining much of what he requests in his letter, but he gets even more.

The cardinal wants the government to protect his goddess outside her cathedrals as well as inside. Congress obliges in legislation such as the Endangered Species Act (ESA) and Section 404 of the Clean Water Act (CWA). These acts direct that protection of nature take precedence over human needs, and, unlike the Wilderness Act, their geographic reach extends beyond public land to cover privately owned land. The government appoints regulators to ensure that nonbelievers behave in ways consistent with the teachings of the eco-church in matters relating to endangered species or wetlands: Thou shalt not cause distress to an endangered species, Thou shalt not disturb an ecosystem favored by an endangered species, and Thou shalt not destroy a wetland ecosystem.

The public is raising strong protests over the nature-first absolutism of these statutes as their provisions become more widely known and the lives of more and more people are adversely impacted. New paradigmists, motivated by religious fervor and fearful of losing their legislative advantage, oppose efforts to amend these laws in order to put consideration of human well-being on an equal footing with nature protection. Because so many people now demand that Congress introduce balance into the ESA and the CWA's Section 404, the voice of church leaders on their behalf is vital if new paradigmists are to prevail in the current public policy debate on these statutes.

Some church leaders join new paradigmists in opposing a leveling of the playing field because they believe that Americans must do everything humanly possible to prevent the extinction of every species lest we fall short of God's call for stewardship. A coalition consisting of the Evangelical Environmental Network, the Coalition on the Environment and Jewish Life, and the National Council of Churches of Christ, USA, write the Senate Committee on Environment and Public Works that the ESA must seek to preserve all species for all time.[82] They liken ESA to Noah's Ark, "preserving and nurturing the remnants of God's creation." Their vision of a tattered Garden is a remarkable victory for new paradigmist propaganda, for, as I show in chapter 4, America's environment is in good condition and generally improving.

The church leaders' public policy position betrays substantial ecological ignorance. Their belief that human endeavor can prevent extinction has no grounding in reality, as scholars tell us that the overwhelming majority of the various kinds of living things ever to inhabit the Garden are extinct and became so through natural processes. Extinction is obviously part of God's plan. That does not mean that Americans should wantonly destroy other life forms. But what Christians understand as a biblically imposed stewardship responsibility does not appear to require Americans to adopt programs that cause significant human harm or foreclose consideration of the human implications of policies intended to protect other species. If I am wrong, then we should immediately begin to reintroduce large predators into populated areas (e.g., returning grizzly bears to much of California), restore conditions favorable to disease causing biota (e.g., re-creating swamps in Washington, D.C.), and stop our efforts to wipe out infectious diseases (e.g., polio and typhoid fever). Finally, their litmus test for amending the ESA asks, "Will the condition of life and habitat be more or less 'good'?" Again, the clergy demonstrate no grasp of ecology. Researchers understand that environmental conditions are in constant flux, that species populations wax and wane across the landscape, and that what is "good" for one species will invariably be "bad" for another. The clerics' standard of legislative judgment is incoherent, for neither they nor scientifically inclined analysts can give it substantive meaning (see chapters 6 and 7).

New paradigmists offer a range of proposals to enshrine ecosystem protection as the prime objective of federal environmental and land use policy and regulation (see chapter 8). If they succeed, they will fulfill "Cardinal Green's" wildest dreams.

The Nature of Nature

If we are to venerate nature, it is worthwhile to examine the makeup of the god that many new paradigmists wish to foist on us. Scientists tell us that Mother Earth is about 4.5 billion years old and, for the first 500 million of those years, that she lacked living things of any kind anywhere. Microscopic organisms first developed in the oceans some 4 billion years ago. Life remained in the water until some 450 million years ago, when it began to venture forth onto the land. For most of her history, then, Mother Earth did not provide conditions under which life could survive on terra firma.[83]

Nature is violent. In the sweep of time, land masses form and continents drift across the surface of the planet. Mountains, lakes, and oceans come and go. The atmosphere and climate change. In these comings and goings there are droughts, floods, volcanic activity, rising and falling of sea levels, and advances and retreats of mountain and continental glaciers. All these processes continue. Researchers, for example, are not sure whether we are at the end of the period of Pleistocene glaciation, which at its peak covered one-third of the land surface of the earth with ice, or simply in another interglacial time wherein glaciers simply retreat, only to advance again. After all, 10 percent of the land surface of the planet remains covered by ice, and permafrost lies beneath another 22 percent.[84] Glaciation and other natural processes are hard on living things. According to Harvard University's E. O. Wilson, "Almost all of the species that ever lived are extinct," and it is Mother Earth herself, without help from humans, that did nearly all the killing.[85] The 1976 earthquake in Tangshan, China, took 750,000 human lives. Hurricane Andrew wiped out entire neighborhoods in southern Florida and did $25 billion in damage in 1992.[86] The new paradigmist portrait of nature as a loving, nurturing goddess caring for her offspring bears little resemblance to reality. On the other hand, these same processes lay the foundation for an earth teeming with more life than we have been able to catalog, making it unique in the universe, as far as we know. The earth both supports and destroys life.

Humans sort through the raw stuff of nature to find things to improve their well-being while defending themselves against those parts that are harmful or deadly. From nature's wild plants comes the stock we fashion into highly beneficial foodstuffs, such as today's wheats, rices, maizes, and tubers, but nature also harbors plant diseases and insects eager to harvest what they do not sow. Nature offers rains to sustain crops and soil to grow it in but also sends droughts to wither plants and floods that destroy plants and remove soil. Nature provides a gazelle for a hunter's table as well as a child for the tiger's. Nature is neither god nor demon, neither good nor bad—it is just there. Nature is neutral.

The land we call the United States possesses a variety of different environments generally hospitable to humans. The earliest people arriving in the conterminous states (at least 12,000 years ago) found numerous regions amenable to their hunting-and-gathering way of life.[87] They also discovered an area in considerable flux. Glaciers were retreating, new patterns of lakes and surface drainage were emerging, and new vegetation complexes were forming (e.g., what was to become the eastern deciduous forests), and the fauna mix was changing as mammoths, mastodons, and camels slipped into extinction. By the time of European contact, Native Americans had diversified their lifestyles and taken advantage of what nature provided (or allowed). They practiced extensive agriculture over much of the eastern third of the nation; a mix of supplementary agriculture, hunting, and gathering in the middle third, and chiefly hunting and gathering in the western third, except for concentrations of agriculture largely in what is now Arizona and New Mexico. While nature offered opportunity, she also wreaked havoc. Changes in rainfall patterns are generally viewed as a major cause of the abandonment of large settlements in the Southwest, and entirely natural diseases (plague, smallpox, measles, scarlet fever, and whooping cough) introduced by Europeans killed large segments of indigenous populations. Scholars estimate the Native American population of the conterminous states at about 2.5 million to 3 million in 1500. The Census Bureau says the U.S. population surpassed 270 million in 1999. Yet current inhabitants of the United States live far longer and at a far higher standard of living than did earlier occupants of the continent. What happened?

Contemporary Americans are far better off than our predecessors because, using constantly expanding levels of technology and knowledge, we are able to dampen the impacts of nature's swings and expand the number of things found in the environment that help improve the quality of human life. We build dams to lessen flooding, provide water in times of diminished rainfall, and generate power. We develop new seeds to increase production per acre and to resist nature's pests. We use new building techniques and materials to make structures more resistant to earthquakes and other of nature's attacks. Through electricity, we convert coal to images on a computer screen that may be a diagnosis of disease or a picture of the moon used by a sixth-grade student doing a research project. We convert old vegetation that nature did not completely recycle—crude oil—into trips to the grocery store, Yellowstone National Park, and the workplace. And on and on, ad nauseum.

Nonetheless, we have not and cannot conquer the forces of nature, as any witness to the damage brought by earthquakes, hurricanes, floods, forest fires, tornadoes, and other natural phenomena can attest. In their foreword to *Restless Earth: Disasters of Nature,* Harm de Blij and Richard Williams Jr. ask, "Who among us has never come face to face with nature's fury?" and observe that "overshadowing our daily routines is the latent possibility that an unexpected act of nature will change our lives forever. It has happened before and will happen again."[88] The evidence fully supports their views. Between 1981 and 1991 nature sent the United States an average of seventy-four tornadoes per year that each caused over $500,000 in property damage while giving us floods that averaged $2.3 billion a year in damage for the same period.[89] In addition to dramatic acts featured on the evening news, nature continues to find new ways to bring us pain and death, as with ever-changing disease agents. Rising concerns in the public health and medical communities brought researchers from all over the world to Atlanta in March of 1998 to discuss their findings on new, reemerging, and drug-resistant infectious diseases at a conference sponsored by the Center for Disease Control and others.

Nature—Mother Earth—the goddess of the new paradigm, would just as soon send us a disease as show us a cure and would just as soon see us starve as see us prosper. It does not matter to her. To the contrary, Christians understand their God to be loving and caring.

New paradigmists interweave religion and science, but despite their success in commingling nature worship with mainline churches, they continue to emphasize science in public policy debates. They fully understand, as did Muir and Leopold, that the public would be far less likely to embrace their policy prescriptions if it understood them to be based on the veneration of Mother Earth. "The certificate of science," writes historian Anna Bramwell, "is crucial to the effectiveness of the ecological argument," so new paradigmist writers "give their work a scientific gloss."[90] She notes that "greens do base their new religion on science," which may be "biased, unreliable, [and] alarmist." I examine the state of ecosystem science in the next chapter.

Notes

1. David Kinsley, *Ecology and Religion: Ecological Spirituality in a Cross-Cultural Perspective* (Englewood Cliffs, N.J.: Prentice Hall, 1995), 162.
2. I am a Christian writing from that perspective.
3. David Lowenthal, "Is Wilderness 'Paradise Enow'? Images of Nature in America," *Columbia University Press Forum* 7, no. 2 (spring 1964), as quoted in Graber, *Wilderness as Sacred Space* (Washington, D.C.: Association of American Geographers, 1976), 8.
4. Martin Lewis, *Green Delusions: An Environmental Critique of Radical Environmentalism* (Durham, N.C.: Duke University Press, 1992), 248.
5. Robert H. Nelson, "Environmental Calvinism: The Judeo-Christian Roots of Ecotheology," in *Taking the Environment Seriously,* ed. Roger E. Meiners and Bruce Yandle (Lanham, Md.: Rowman & Littlefield, 1993), 233–55.
6. Victor Scheffer, *The Shaping of Environmentalism in America* (Seattle: University of Washington Press, 1990), 8.
7. Alston Chase, *In a Dark Wood: The Fight over Forests and the Coming Tyranny of Ecology* (Boston: Houghton Mifflin, 1995), 129.
8. David Brower, *Let the Mountains Talk, Let the Rivers Run* (New York: HarperCollinsWest, 1995), 176.
9. For a history of the idea of nature in Western thought, see Clarence J. Glacken, *Traces on the Rhodian Shore* (Berkeley: University of California Press, 1967).
10. Dave Foreman and Howie Wolke, *The Big Outside* (New York: Harmony Books, 1992), ix.
11. Lynton Keith Caldwell, *Between Two Worlds: Science, the Environmental Movement, and Policy Choice* (Cambridge: Cambridge University Press, 1990), 74.
12. Bill Devall and George Sessions, *Deep Ecology: Living as if Nature Mattered* (Salt Lake City: Peregrine Smith Books, 1985), 67.
13. Scheffer, *The Shaping of Environmentalism in America,* 9.
14. Dave Foreman, *Confessions of an Eco-Warrior* (New York: Harmony Books, 1991), 9.
15. Paul R. Ehrlich and Anne H. Ehrlich, *Betrayal of Science and Reason: How Anti-Environmental Rhetoric Threatens Our Future* (Washington, D.C.: Island Press, 1996), 1. See also John D. Echeverria and Raymond Booth Eby, eds., *Let the People Judge: Wise Use and Property Rights Movement* (Washington, D.C.: Island Press, 1995), and David Helvarg, *The War against the Greens* (San Francisco: Sierra Club Books, 1994).

16. Bruce Babbitt, "Between the Flood and the Rainbow: Stewards of Creation," *Christian Century* 113, no. 16 (May 8, 1996), 500.

17. Roderick Nash, *Wilderness and the American Mind* (New Haven, Conn.: Yale University Press, 1967), 122, quoted from "About the Yosemite," *American Review of Reviews* 45 (1912): 766–67.

18. John Muir, *The Yosemite* (1912; reprinted in *John Muir: Nature Writings* [New York: Library of America, 1997], 817).

19. Nash, *Wilderness and the American Mind,* 124.

20. Daniel G. Payne, *Voices in the Wilderness: American Nature Writing and Environmental Politics* (Hanover, N.H.: University Press of New England, 1996), 96.

21. Stephen Fox, *John Muir and His Legacy: The American Conservation Movement* (Boston: Little, Brown, 1981), 5.

22. Thurman Wilkins, *John Muir: Apostle of Nature* (Norman: University of Oklahoma Press, 1995); an extensive collection of Muir's work is found in John Muir, *Nature Writings,* ed. William Cronin (New York: Penguin, 1997).

23. Fox, *John Muir and His Legacy,* 50.

24. Fox, *John Muir and His Legacy,* 79.

25. Payne, *Voices in the Wilderness,* 84.

26. Fox, *John Muir and His Legacy,* 80.

27. Muir's biocentrism is noted by Payne, *Voices in the Wilderness,* 85; Wilkins, *John Muir,* 266; and Roderick Nash, *The Rights of Nature* (Madison: University of Wisconsin Press, 1989), 41.

28. Devall and Sessions, *Deep Ecology,* 104.

29. Boris Zeide, "Another Look at Leopold's Land Ethic," *Journal of Forestry* 96, no. 1 (January 1998): 14.

30. As quoted in Nash, *The Rights of Nature,* 63.

31. As quoted by Roderick Nash, "Aldo Leopold and the Limits of American Liberalism," in *Aldo Leopold: The Man and His Legacy,* ed. Thomas Tanner (Ankeny, Iowa: Soil Conservation Society of America, 1987), 72.

32. Devall and Sessions, *Deep Ecology,* 85.

33. Susan L. Flader, *Thinking Like a Mountain* (Madison: University of Wisconsin Press, 1994), 5; J. Baird Callicott, "The Scientific Substance of the Land Ethic," in Tanner, ed., *Aldo Leopold,* 90.

34. Aldo Leopold, *A Sand County Almanac: With Essays on Conservation from Round River* (New York: Ballantine, 1966), 237–64. *A Sand County Almanac* was first published in 1949.

35. Leopold, *A Sand County Almanac,* 262.

36. Callicott, "The Scientific Substance of the Land Ethic," 87.

37. Susan L. Flader, "Evolution of a Land Ethic," in Tanner, ed., *Aldo Leopold,* 9–10; Nash, *The Rights of Nature,* 41; Payne, *Voices in the Wilderness,* 92.

38. "Godliness and Greenness: Thou Shalt Not Covet the Earth," *The Economist* 341, no. 7997 (December 21, 1996): 108–10; Max Oelschlaeger, *Caring for Creation* (New Haven, Conn.: Yale University Press, 1994); Nash, *Rights of Nature;* Jon Naar, "The Green Cathedral," *The Amicus Journal* 14, no. 4 (winter 1993): 22–28; Michael Brown, "Earth Worship or Black Magic?" *The Amicus Journal* 14, no. 4 (winter 1993): 32–34.

39. Oelschlaeger, *Caring for Creation,* 61.

40. Reprinted in Roger S. Gottlieb, ed., *This Sacred Earth: Religion, Nature, Environment* (New York: Routledge, 1996), 636–39.

41. Carolyn Merchant, *Radical Ecology* (New York: Routledge, 1992), 1.

42. Douglas L. Chial, "The Ecological Crisis: A Survey of the WWC's Recent Responses," *Ecumenical Review* 48 (January 1996): 59.

43. Sallie McFague, "The Scope of the Body: The Cosmic Christ," in Gottlieb, ed., *This Sacred Earth,* 286–96.

44. From "Report of a Pre-Assembly Consultation on Subtheme I—'Giver of Life, Sustain Your Creation'," published as "Implications of Subtheme I," *Ecumenical Review* 42 (July/October 1990), 316.

45. Thomas Berry, "The Viable Human," in *Deep Ecology for the 21st Century: Readings in the Philosophy and Practice of the New Environmentalism,* ed. George Sessions (Boston: Shambala, 1995), 8.

46. Exodus 20:1–17, Matthew 19:18–19, and Matthew 22:37–40. All biblical citations are from the *New American Bible* (Nashville: Catholic Bible Press, 1987).

47. Paul W. Taylor, *Respect for Nature: A Theory of Environmental Ethics* (Princeton, N.J.: Princeton University Press, 1986), 45–46.

48. Pope John Paul II, "The Ecological Crisis: A Common Responsibility," A Message of His Holiness Pope John Paul II for the Celebration of the World Day of Peace, January 1, 1990 (Vatican City, December 8, 1989), 2.

49. Pope John Paul II, "The Ecological Crisis," 1.

50. Pope John Paul II, *The Gospel of Life* (New York: Times Books, 1995), 39.

51. Lynn White, "The Historical Roots of Our Ecological Crisis," *Science* 155, no. 3767 (March 10, 1967): 1203–7. See Victor Ferkiss, *Nature, Technology and Society: Cultural Roots of the Current Environmental Crisis* (New York: New York University Press, 1993).

52. Robert Booth Fowler, *The Greening of Protestant Thought* (Chapel Hill: University of North Carolina Press, 1995), 91.

53. Albert Fritsch with Angela Ladavaia-Cox, *Eco-Church: An Action Manual* (San Jose, Calif.: Resource Publications, 1992).

54. Matthew Fox, *The Coming of the Cosmic Christ: The Healing of Mother Earth and the Birth of a Global Renaissance* (San Francisco: HarperSanFrancisco, 1988), 144.

55. For writings in this vein, see Matthew Fox and Rupert Sheldrake, *Natural Grace* (New York: Doubleday, 1996); Elizabeth Breuilly and Martin Palmer, eds., *Christianity and Ecology* (London: Cassell Publishers, 1992); Charles Birch, William Eakin, and Jay B. McDaniel, eds., *Liberating Life: Contemporary Approaches to Ecological Theology* (Maryknoll, N.Y.: Orbis Books, 1990); and Fox, *The Coming of the Cosmic Christ.* See also Wendall Berry, "Christianity and the Survival of Creation," in *Sacred Trusts: Essays on Stewardship and Responsibility,* ed. Michael Katakis (San Francisco: Mercury House, 1993), 52.

56. National Council of the Churches of Christ, "A Service of Worship: The Earth Is the Lord's—a Liturgy of Celebration, Confession, Thanksgiving, and Commitment," in Gottlieb, ed., *This Sacred Earth,* 480–83.

57. Larry Gigliotti, "Environmental Education: What Went Wrong? What Can Be

Done?" *Environmental Educator* 22, no. 1 (fall 1990): 9; Jim Bowyer, "Fact vs. Perception," *Forest Products Journal* (November/December 1995), 31–36.

58. Stephen Schneider, as quoted in Karen F. Schmidt, "Green Education under Fire," *Science* 274, no. 5294 (December 13, 1996): 1828–30.

59. Independent Commission on Environmental Education, *Are We Building Environmental Literacy?* (Washington, D.C.: ICEE, 1997), 3.

60. U.S. Catholic Conference, "Renewing the Earth," A Pastoral Statement of the U.S. Catholic Conference, Washington, D.C., November 14, 1991, 7.

61. U.S. Catholic Conference, "Renewing the Earth," 1.

62. U.S. Catholic Conference, *Renewing the Face of the Earth* (Washington, D.C.: U.S. Catholic Conference, 1994), 30–31.

63. Colin A. Russell, *The Earth, Humanity and God* (London: UCL Press, 1994), 131. See also Robert Whelan, Joseph Kirwan, and Paul Haffner, *The Cross and the Rainforest: A Critique of Radical Green Spirituality* (Grand Rapids, Mich.: William B. Eerdmans, 1996).

64. Russell, *The Earth, Humanity and God,* 133.

65. Glacken, *Traces on the Rhodian Shore,* 161.

66. Stephen R. L. Clark, "Global Religion," in *Philosophy and the Natural Environment,* ed. Robin Attfield and Andrew Belsey (Cambridge: Cambridge University Press, 1994), 125.

67. For a variety of biblical quotes supporting different viewpoints, see Fowler, *The Greening of Protestant Thought,* chap. 2.

68. Robert Sirico, "Despoiler or Problem-Solver," *National Catholic Register* (October 23, 1994), 5.

69. Genesis 1:27.

70. Roger S. Gottlieb, "Introduction: Religion in an Age of Environmental Crisis," in Gottlieb, ed., *This Sacred Earth,* 9.

71. The UN Environmental Program (UNEP) grew out of the 1972 UN Stockholm Conference on the Human Environment. Its mission is to represent environmental concerns within the UN system. The "Environmental Sabbath" is an effort to use churches as a means of advancing the UNEP agenda that the earth faces a cataclysmal environmental crisis.

72. Reprinted in Fritsch, *Eco-Church,* 29.

73. Sirico, "Despoiler or Problem-Solver," 5.

74. Paul Watson, "On the Precedence of Natural Law," *Journal of Environmental Law and Litigation* 3 (1988): 79–90.

75. Thomas S. Derr, "The Challenge of Biocentrism," in *Creation at Risk,* ed. Michael Cromarie (Grand Rapids, Mich.: William B. Eerdmans, 1995), 104.

76. 16 U.S.C. (1994 ed.) § 1131.

77. Nash, *Wilderness and the American Mind,* 223.

78. Reed Noss, "The Wildlands Project: Land Conservation Strategy," *Wild Earth* (special issue, 1992): 13; Foreman and Wolke, *The Big Outside.*

79. Henry David Thoreau, as quoted in Nash, *Wilderness and the American Mind,* 84.

80. 43 C.F.R. 8560.0–6.

81. 43 C.F.R. 8560.1–2.

82. "Testimony of the Evangelical Environmental Network, Coalition on the Environment and Jewish Life, and the National Council of Churches of Christ, USA on S. 1180," The Endangered Species Recovery Act of 1997, September 23, 1997 (copy on file with the author).

83. E. O. Wilson, *The Diversity of Life* (New York: Workman Publishing, 1991), and William K. Hartmann and Ron Miller, *The History of Earth* (New York: Workman Publishing, 1991).

84. Tom McKnight, *Physical Geography: A Landscape Appreciation* (Upper Saddle River, N.J.: Prentice Hall, 1996).

85. Wilson, *The Diversity of Life,* 216.

86. H. J. De Blij, *Nature on the Rampage* (Washington, D.C.: Smithsonian Institution, 1994).

87. This account follows Karl W. Butzer, "The Indian Legacy in the American Landscape," in *The Making of the American Landscape,* ed. Michael P. Conzen (New York: Routledge, 1994), 27–50.

88. H. J. de Blij, Michael H. Glantz, Stephen L. Harris, Patrick Hughes, Richard Lipkin, Jeff Rosenfeld, and Richard S. Williams, Jr., *Restless Earth: Disasters of Nature* (Washington, D.C.: National Geographic Society, 1997): 8, 11.

89. U.S. Bureau of the Census, *The American Almanac—Statistical Abstract of the United States: 1993–1994* (Austin, Tex.: Reference Press, 1993), table 381.

90. Anna Bramwell, *The Fading of the Greens* (New Haven, Conn.: Yale University Press, 1994), 165–66.

Chapter 6

Science, Ecosystems, and the Emperor's New Clothes

> Ecosystem ecology for too long has operated in a dream world with few hypotheses and even fewer data.
>
> —Stuart Pimm, "An American Tale," *Nature* 379 (July 21, 1994), 188

Advocates of ecosystem management claim the mantle of science for their approach to environmental and land use policy. Ecosystem management, they say, is grounded in science presumably because they trace many of its ideas to scientists and to the discipline of ecology. The public and policymakers seem unaware of the deep divisions within contemporary ecology and erroneously assume that many of the concepts they associate with ecosystem management carry the imprimatur of a unified discipline and are widely accepted and understood within it. How many people fully appreciate that many ecologists find it "difficult" to associate the ecosystem concept "with any coherent theory or set of principles?"[1] Ecologists Stewart Pickett, Jurek Kolasa, and Clive Jones write that "public knowledge of ecology is out of date in some important ways" and "perhaps the most troubling aspect of this situation is that the large metaphors the public uses to carry ecological knowledge into its discourse reinforce the gap between public and scientific understanding of ecology."[2] For example, when a citizen at a public hearing, a politician in a policy discussion, or a Saturday afternoon cartoon character refer to the balance of nature, they reveal a poor grasp of scientific knowledge.

In this chapter, I look at the ecosystem concept in science and examine the ideas of ecosystem health, integrity, and sustainability as used by ecologists as well as the increasingly fashionable notion of ecosystem services. I look at new paradigmists' assumptions that there is a harmony or balance of nature and that sustainable economies (and societies) are dependent on sustainable ecosystems.

My purpose in this chapter is not to denigrate ecology or ecologists. Ecologists do important work, and ecology is a vital field of knowledge. Yet one cannot escape the conclusion that the key ideas new paradigmists import from ecology remain precarious at best within the discipline and frequently take on the aura of notions derived from faith. Regarding biological diversity (whose preservation he supports), Jack Ward Thomas nonetheless writes that "high priests swoop in from afar to impart knowledge and wisdom" to the uninitiated in rituals "complete with a Latin service—*in situ, ex situ, alpha, beta,* and *gamma*—amen." The wise men offer transparency after transparency that reveal "hieroglyphics and other mysterious writings worthy of the ancient tombs of the Pharaohs" and proclaim "new covenants between humans and the earth" while chanting in a "complex and convoluted" language worthy of the "Tower of Babel."[3] All disciplines possess a jargon, word shortcuts that members of the fraternity use to convey broader substantive understandings. The jargon that new paradigmists extract from ecology to advance their ecosystem protection agenda in the public policy arena, however, does not rest on substance; rather, it is more akin to buzzwords, for ecologists are unable to give them widely agreed-on meanings.

Fields of knowledge change over time. Scholars put forward ideas and concepts for debate and analysis by their peers who modify, discard, or accept them as evidence mounts and researchers gain additional insight and understanding. Theories come and go, with a few eventually achieving the status of laws after lengthy and rigorous scrutiny by experts in the discipline. Within a given discipline scholars organize themselves into various schools of thought, advance multiple lines of inquiry, and argue about methods and meaning. Ecology shares with other disciplines the comings and goings of ideas, the development of schools, and debates over subjects and methods, but after a century of existence and its share of heated exchanges, it is a fragmented discipline, in which many basic ideas remain vague and problematic and that has yet to produce theories with significant predictive power.

Ecologists looking at their discipline see multiple and serious problems. Robert Peters offers a lengthy critique in an effort to stimulate its growth.[4] "To many contemporary ecologists," he writes, "the weakness of ecology is patent and the problem needs little elaboration."[5] Among the difficulties he sees are a "lack of scientific rigor," "weak predictive capability," "lack of testable theory," and "a tendency of ecologists to demagogy and polemics." Peters views ecology as a less coherent science in part because it contains "many constructs of dubious merit." Robert McIntosh of the University of Notre Dame concludes that "after three decades of anarchic development in ecology, ecologists have lost the feeling belonging to a unified discipline."[6] He notes that the discipline is characterized by a succession of "new" ecologies, each one becoming the "old" ecology in its time, resulting in discomfort, disappointment, and confusion among ecologists. "Far too often ecology seems to have been dominated by one or another of various theoretical bandwagons," opines Claudia Pahl-Wostl. She sees fragmentation as preventing "ecology from defining its own identity as a science" so that "fruitful discussion is often prevented by the lack of conceptual guidelines," thereby promoting "the vagueness of concepts."[7]

More than confused ideas hinder productive discussions among practitioners. Frequently, ecologists are impatient with other points of view. Members of the discipline's schools have, says Joel Hagen, "been talking past one another for almost a generation," and not all the talk is gentlemanly.[8] Robert May pointedly notes that "over the past 20 years or so, ecology has been ill-served by some rancorous debates about how one or other versions of 'The Scientific Method' should govern everything we do."[9]

Over and over again, scholars point out the vagueness of key concepts in ecology (I note the questions regarding biological diversity and species in chapter 4). Critics see ecosystem ecology as particularly susceptible to the charge of conceptual fuzziness. Peters, for example, identifies no fewer than forty separate ideas associated with the longtime staple of stability.[10] These ideas grade into one another in a hodgepodge of jargon whose terms lack substance. Researchers attribute so many different meanings to *stability,* for example, that they find themselves in a "terminological and conceptual morass," according to Peters.[11] Stuart Pimm believes that his colleagues use *stability* to refer to at least five conditions.[12] Ecologists dispute the meaning of each of these five conditions—ability to return to equilibrium, resilience, variability, persistence, and resistance to change—in turn. It is little wonder that after reviewing its history in ecology, K. S. Shrader-Freschette and E. D. McCoy conclude that the "stability concept is problematic."[13] Lionel Johnson goes further. He feels that "concepts like 'stable' and 'stability'" are "virtually useless unless strongly qualified."[14] Another notion commonly used by some ecologists is that of keystone species (species that play a dominant role in an ecosystem). But after thirty years the term remains ambiguous, and ecologists have yet to give it an operational definition.[15] Consequently, write ten scientists in a 1996 article in *BioScience,* "ecologists still lack the empirical basis needed to detect, interpret, and predict general patterns in the occurrence of keystone species or to apply the concept to management."[16]

Ecosystem ecologists cannot make general predictions as a matter of course.[17] The National Research Council recommends that we "should work toward developing predictive models to facilitate" management of the environment.[18] Yet it is hard for anyone to develop general models possessing significant predictive power when so many of the concepts in the area of study are suspect and the linkages between those same concepts are unclear. Scholars leave us with metaphors. Researchers James Kay and Eric Schneider, for example, write that "ecosystem behavior . . . is like a large musical piece such as a symphony" and that the role of the researcher is to understand how the ecosystem symphony is composed.[19] They concede that we should not expect "a science of ecology which allows us to predict what the next note will be." Within ecosystem ecology there is movement away from an understanding of science as a relatively value-free pursuit of knowledge based on theory, confirmed observation, and verifiable experimentation and to a world wherein work that is distinctly value laden and marked by vague ideas is deemed science.[20] "We do not consider our work to be scientific in the narrow sense that it consists of value-free descriptions and explanatory hypotheses," write ecologists Bryan Norton and Robert Ulanowicz, "we, on the contrary, believe that conservation biology is a normative science . . . guided most basically by a commitment to important social values."[21] I wonder whether this approach to research is what decision makers have in mind when they call for policy-based sound science? Leaders must be cautious because the line between science and advocacy can become blurred. University of Arizona public policy professors Helen Ingram and H. Brinton Milward, along with World Wildlife Fund researcher Wendy Laird, write that in their desire to influence public policy, some science activists reach "considerably beyond their data to suggest appropriate courses for public action."[22]

Ecosystems Again

Sir Arthur Tansley proposed the idea of the ecosystem in 1935 as part of an intense debate occupying ecologists during the early decades of the twentieth century. Researchers have told the story of the emergence of the ecosystem concept many times so I will but summarize the tale.[23] Frank Golley writes that

> Tansley formulated the ecosystem concept as a solution to a conceptual argument that divided plant community ecology into two opposing camps. One group emphasized the significance of the individual stand of vegetation and organized these stands into hierarchies of community organization. The other hypothesized that vegetation was a complex organism that developed, matured, and became senescent.[24]

In the early twentieth century, American ecologist Frederic Clements sought to explain the patterns of species assemblages that researchers found on the landscape. To do so he employed organismal metaphors, which were commonly used in intellectual discourse at the time. Clements, however, goes beyond metaphor. Biological historian Joel Hagen observes that "for Clements the plant community really was a 'complex organism'" whose life history could be traced by studying ecological succession.[25] In his view they have a life cycle of birth, growth, and development that proceeded along an orderly path toward a climax, a state wherein the vegetation community would remain in a dynamic equilibrium, adjusting to environmental changes as individual organisms might. They are a kind of superorganism. Not everyone accepted (in whole or in part) Clements's views. Henry Gleason argues that the distribution of plants across the landscape results from the proclivities of individual members of plant species rather than the functioning of organized living units above the level of species. For Gleason, "every species of plant is a law unto itself, the distribution of which in space depends upon its individual peculiarities of migration and environmental requirements."[26]

Tansley brought forth the ecosystem concept "as an alternative to superorganicism."[27] His ecosystems include the living things and the nonliving components of the physical environment as well as interchanges among them. Ecosystems are physical systems "in the sense of physics" and contain living things but are not themselves alive.[28] He rejects the idea of communities as superorganisms but does not entirely abandon organismal ideas. According to Hagen, Tansley accepts the organismal metaphor but feels that Clements and others push it too far; communities may be like organisms in some respects but are not actually organisms. Indeed, Tansley writes that vegetation climaxes are "a relatively stable phase reached by successional change . . . [that] may be considered *quasi-organisms*" [emphasis in the original].[29] Tansley places the notion of equilibrium rather than organicism at the center of his ecosystem concept.

Tansley's ideas rest squarely on the presumption that equilibrium or movement toward it is the norm of nature. "Some systems develop gradually," writes Tansley, "steadily becoming more highly integrated and more delicately adjusted in equilibrium . . . ecosystems are of this kind." He goes on, "Normal autogenic succession is a progress towards greater integration and stability." "The 'climax,'" he feels, "represents the highest stage of integration and the nearest approach to perfect dynamic equilibrium that can be attained in a system."[30] Since Tansley first proposed the ecosystem concept, ecologists have kept the notion of equilibrium at its core. Writing in 1969, Eugene Odum concludes that ecological succession "is an orderly process . . . that is reasonably directional and, therefore predictable," and "it culminates in a stabilized ecosystem."[31]

Researchers find it difficult to identify orderly progressions toward stable climaxes in the field. Instead of confirming assertions regarding order in nature, the facts reveal quite the opposite. "Empirical studies have increasingly demonstrated either a lack of equilibrium or equilibrium conditions that are observed only at particular scales of time and space," write Norman Christensen and his colleagues.[32] Analyses by many ecologists over the last two decades point to "disequilibria, instability, and even chaotic fluctuations in biophysical environments."[33] Scientists' studies do not support one of the fundamental assumptions on which the ecosystem concept rests: that equilibrium is a basic condition of nature. For the University of Maryland's Mark Sagoff, this calls into question the entire ecosystem concept. "Ecological systems are the conceptual constructs of a theoretical ecology, the old equilibrium ecology, that is now defunct," he writes. "Just as the smile of the Cheshire Cat survived his demise, the idea of an ecological system [ecosystem] . . . has survived the demise of equilibrium theory . . . the ecosystem concept haunts the ecological literature as an apparition without substance."[34]

It is hard for many ecologists to set aside thoughts of equilibria despite the weight of the evidence against it, for, as Golley notes, it "has a special role in the development of the ecosystem concept . . . it reappears over and over through the history of the subject."[35] He asks a very pertinent question: "Recognizing that the environment is continually variable and that biological organisms continually are born, die, adapt, and evolve, what limits of variance in space, time, and response do we accept within the concept of equilibrium?"[36] The answer that he receives is vacuous; we are now told by those clinging to the concept that ecosystems "exhibit a suite of behaviors over all spatial and temporal scales, and the processes that generate these dynamics should be maintained."[37] So instead of abandoning the equilibrium idea, some ecologists simply reinvent it. Equilibrium no longer means equilibrium, so they no longer look for ecosystems to evolve toward a single predictable end point but allow that many different end points are possible.[38] These ecologists speak of an ecosystem's resiliency, persistence, and resistance to trajectory-altering disturbance events and admonish that "we must always remember that left alone, living systems are self-organizing, that is they will look after themselves."[39] But in the end it is the discredited equilibrium idea that lies behind these notions as well as the belief that ecosystems are self-organizing entities intent on maintaining their existence. In trying to keep the equilibrium idea attractive and alive via a massive facelift and multiple organ transplants, ecologists carry the ecosystem concept itself further into an intellectual landscape populated by loose images that inhibit intellectual advancement.

Sagoff points out that one area of intellectual enterprise vital to the growth of a science is the ability for practitioners to classify the objects they study.[40] The National Research Council (NRC) agrees: "Ecological classifications are extremely important for the management of biological resources on an ecosystem . . . basis."[41] Classifications, they say, "are needed for recognizing and mapping" ecosystems "and for communicating about their status, distributions, and trends."[42] Yet more than sixty years after Tansley proposed the ecosystem concept, after decades of dominance in graduate schools, after the publication of thousands of scholarly articles, monographs, and books, ecosystem ecologists still have not developed a recognized classification system for ecosystems.[43] The NRC finds that "there are currently no broadly accepted classification schemes for . . . ecological units above the level of species."[44] And it is but a pious hope that researchers will produce such classifications in the future.

How might we classify ecosystems? Despite claims by new paradigmists that ecosystems are natural landscape units, Christensen and his colleagues recognize that "nature has not provided us with a natural system of ecosystem classification," so we must look to our own devises.[45] Perhaps we could base a classification on the core characteristics of ecosystems? A reasonable idea, but the NRC points out that in the long years since the ecosystem concept was first proposed, scholars do not yet agree on what the core characteristics of ecosystems might be.[46] If ecosystems can be insect intestines, water drops, a ponderosa pine forest, or the entire planet, as ecologists tell us, there is little reason to believe that a universal set of core characteristics will ever be forthcoming from the scientific community. Essentially, we are left with idiosyncrasy, from which not even the most knowledgeable and energetic ecologist can fashion a universally acceptable classification system. Dutch ecologists Frans Klijn and Helias de Haes write that policy-relevant ecosystem classifications must be about "concrete recognizable ecosystems" rather than the abstractions of Tansley and Odum that "do not exist in reality, only in the minds of researchers and in their papers."[47] Yet the authors of the collection of papers that follow Klijn and de Haes do not offer anything substantial at all, just more author-specific, arbitrarily determined ecosystems classified in author-specific ways.[48] "No single classification system," says the NRC, "is likely to serve all management purposes."[49] This admission by the NRC again demonstrates that out on the landscape the ecosystem concept is an ad hoc notion.

If there is no classification system, then are no general rules or principles that enable ecologists to objectively determine where one ecosystem stops and another begins or to dispassionately assess landscape change in order to discern when one ecosystem is replaced by another or that give scholars a substantive foundation on which to place ideas such as sustainability, health, and integrity.

Ecosystem Sustainability, Health, Integrity, and Other Fancies of the Day

"Virtually everyone is for sustainability in much the same way that nearly everyone is for justice," observes Nels Johnson of the World Resources Institute.[50] Echoing what Secretary of the Interior Bruce Babbitt says of ecosystems themselves, Johnson finds that "sustainability is in the eye of the beholder." The phrase is so loose that some observers believe that it teeters on the brink of being a buzzword, " 'greenwash' that will fade away like so many previous rhetorical colorings."[51] Buzzword or not, researchers, politicians, and others use it over and over to convince the public they must make major changes in their relations with Mother Nature. What does the term mean among ecologists?

In a report done for the Ecological Society of America, Christensen and coauthors write that sustainability is the maintenance "of the ecological interactions and processes necessary to sustain ecosystem function and structure" over time frames covering generations.[52] They neither indicate what interactions or processes they are referring to, nor do they define ecosystem function and structure, so the reader is left to fend for him- or herself regarding the real meaning of sustainability. The Christensen group's use of the word *sustain* in their explanation of sustainability is a circular definition. Reviewers would wonder even more about circular reasoning on reading the section of the report headed "Ecosystem function depends on its structure, diversity, and integrity." Here the authors define biological diversity to include ecosystems, then go on to conclude that "biological diversity is central to the . . . sustainability of the earth's ecosystems."[53] As I read this, essentially the authors say (in part at least) that sustaining ecosystems is important for sustaining ecosystems.

Jerry Franklin of the University of Washington defines sustainability as "the potential for our land and water ecosystems to produce the same quantity and quality of goods and services in perpetuity."[54] He envisions a "broad range of goods and services," including "the regulation of stream flow and the minimization of erosional losses of nutrients and soil," and argues that sustainability means being able to provide conditions that would offer "habitat for the full array of organisms historically found" in an area and "continuing the capacity to provide the same quantity and quality of products for human consumption."[55] What is curious about his approach is that since the composition of landscapes changes through time, his basic desire for a constant output (or the potential of such output) of goods and services of the same kind and quality in perpetuity cannot be reached under any circumstance.

Franklin fundamentally adopts a nature-knows-best approach to sustainability, which, as Peters points out, is a vacant idea based on neither fact nor theory.[56] He sees "landscape dysfunction" associated with human-induced changes that alter a landscape's "ability to regulate aspects of the hydrologic cycle or provide habitat for forest dwelling organisms."[57] For the Pacific Northwest forests, his requirements for sustainability mean that more and bigger reserves—areas "where human disruptions of natural processes . . . are minimized"—will be needed, as will new land use and management practices on lands outside these reserves. He emphasizes focusing on ecosystems and implementing government planning at landscape and regional scales with particular concern for ensuring the preservation of biological diversity on private lands and also emphasizes that an appropriate mix of ecosystems is maintained throughout the larger region. How planners and managers could implement these ecosystem protection efforts without diminishing the production of goods and services for humans—another component of Franklin's view of sustainability—is a mystery.

In his book *Sustainable Forestry,* Chris Maser views sustainability as "the degree of overlap between what is ecologically possible and what is socially desired by the current generation . . . and the ability of both the ecosystem and the management system to adapt to changing environmental conditions over centuries." This requires "the current generation to protect both the social and ecological options for the future, because both will change in some related and probably unpredictable fashion."[58] For Maser, "the forest, like a human body, is a living thing" and presumably other ecosystems are as well.[59] He sees us in a state of ecological crisis, with forests in desperate need of "healing," but our response to the forest crisis is impeded by our dysfunctional economic theories, government agencies, and Congress. Americans must revise society's institutions in order to achieve sustainability by working together "within the ethos of the democratic social system and the biological capabilities of the ecosystem."[60] He writes of the need for humans to do only "what is ecologically possible." While high sounding, his approach rests on quicksand from the outset since there is no such thing as "the ecosystem." His idea of society adopting the "ecologically possible" as the standard of sustainability is odd because by definition everything that we are currently doing on the landscape that involves living things—from forestry to farming—is ecologically and biologically possible, otherwise we could not do it.

Since ecosystems are mental constructs, it is reasonable for us to ask, Just what are their biological capabilities in the first place? Where do they occur, and over what period of time? Leaving aside the thorny matters of place and time, assume that supporters of Maser's view present us with a laundry list of capabilities; the next questions are, How are we to measure and compare those individual capabilities and make judgments about exceeding some capability? If researchers determine, for example, that forest practices result in the population decline of five species but the increase in the population of another five, have the biological capabilities of the ecosystem been exceeded? How about the extinction of one species and the introduction of three? What if habitat availability for several aquatic species increases but the rate of soil formation declines? And so on. There is no apparent calculus by which researchers can integrate the individual measures of the presumed multiple biological capabilities of an ecosystem even if the phrase could be given coherent meaning.

When he looks to the future, Maser's idea of sustainability becomes particularly murky. He ties sustainability to "the ability of both the ecosystem and the management system to adapt to changing environmental conditions over centuries." But where is the substance in Maser's view? Recall from chapter 1 that scientists define ecosystems in terms of nonliving and living things occupying some fragment of the landscape plus the interrelations among them. The nonliving things on a given landscapes evolve over time largely in response to the forces that operate on the planet and sculpt its surface. They do not adapt to anything. Mountains are first uplifted and then eroded; it is not a matter of adaptation. Rivers do not adapt; rather, they are simply a collection of water molecules obeying the law of gravity. For the entire spectrum of abiotic things on the earth's surface that scientists normally include within their definitions of ecosystems, the idea of adaptation is inapplicable. On the other hand, ecologists and geographers know that biota come and go on the landscape in response to changing conditions. But they do so as individual species, not as entire plant and animal assemblages picking up and moving. It is these organisms—the living things—that adapt to changing conditions, not ecosystems, of which living things are only a part. Even if one assumes that ecosystems are discrete entities that adapt to change, how is such adaptability to be observed and measured, and by what standards will we judge the success or failure of the ecosystem at adaptation since scientists cannot even tell us what their core characteristics might be?

The concept of a sustainable ecosystem as routinely used by ecologists is an oxymoron. Members of that fraternity steadfastly refuse to acknowledge that their own professed understandings of ecosystems renders the idea of their sustainability illogical. Christensen et al., for example, urge making the maintenance of ecosystem sustainability the centerpiece of ecosystem management while telling us that "ecosystems are dynamic in space and time . . . [they] are constantly changing," "ecosystem processes operate over a wide range of spatial and temporal scales . . . thus, there is no single appropriate scale or time frame for management," and "boundaries defined for the study or management of one process are often inappropriate for the study of others."[61] How can sustainability have meaning when the entity to which ecologists would attach it is, in the considered view of those same ecologists, in constant flux in time and space? But I submit that the problem is even more fundamental. Recall from chapter 1 ecologist Simon Levin's conclusion that "what we call an ecosystem . . . is really just an arbitrary subdivision of a continuous gradation of local species assemblages"[62] or O'Neill's view that ecosystems are "merely . . . localized, transient experiment[s] in species interaction"[63] or Norton's claim that "ecological scientists . . . [treat ecosystems] . . . as temporary conveniences to be discarded and replaced at will"[64] or Donald Worster's finding that "we must conceive of ecosystems then, not as permanent entities engraved on the face of the earth but as shifting patterns in the endless flux, always new, always different."[65] There is no getting around it: Ecosystems are but heuristic devices, so there is nothing concrete on the landscape to sustain in the first place.

Politicians, political activists, and social scientists also take up the cry of sustainability, but in doing so they add no clarity to the concept as it appears in ecology. They speak of sustainable development and unite with those calling for sustainable ecosystems so that in the world of resource decision making and allocation, the two phrases are inseparable. Sustainable development became particularly fashionable when the World Commission on Environment and Development published its report in 1987.[66] Popularly known as the Brundtland Report after the name of the commission's chair, Norway's Prime Minister Gro Harlem Brundtland, it defines sustainable development as "development that meets the needs of the present generation without compromising the ability of future generations to meet their own needs."[67] President Bill Clinton's Council on Sustainable Development uses this definition in its plan for American sustainable development prepared for the United Nations Commission on Sustainable Development.[68]

Oxford University economist Wilfred Beckerman notes the barrenness of this definition. The criterion of needs is "totally useless," he writes. He says that "people at different points in time, or at different income levels, or with different cultural or national backgrounds will differ about what 'needs' they regard as important."[69] Geographers and others have made points such as this for decades, yet their findings are overlooked by the Brundtland Commission and those who adopt their definition. Tom Wilbanks, in his 1994 presidential address to the Association of American Geographers, calls on its members to "help fill the void of theory and intellectual substance that undermines sustainable development as a commitment for global and local action."[70] Wilbanks generally sees the concept as "a slogan (however oxymoronic) that has taken on a life of its own, becoming a screen behind which resources are being allocated and decisions made, regardless of whether the forcing term is understood or not."[71]

Now we are getting to the nub of the issue. Sustainable development has less to do with ecology than it has to do with power. It is fundamentally a "political concept."[72] Stephen Budiansky observes that "certainly a well managed cattle ranch is no less 'sustainable' than a well managed rain forest. In the end it is purely a value judgment which we will have."[73] New paradigmists seize the issues of sustainability and sustainable development as a means of using the political process to protect their goddess, Mother Earth. Sustainability is but one of the characteristics given ecosystems by ecologists. Two others, ecosystem health and ecosystem integrity, are widely used in the scientific and policy communities, and I examine these next.

The editors of the book *Ecosystem Health* find that there is no "clear conception of the term" and that it and ecological integrity "have never been defined well enough to make them useful" in policy documents, although they are widely used therein.[74] Others also struggle with the idea. "The definition of 'health' . . . for ecosystem has proved somewhat elusive," writes Peter Calow of Sheffield University.[75] Nonetheless, protecting ecosystem health is an imperative for some ecologists.[76] "Ecosystem health/integrity . . . stands as the central policy concept to guide ecologically understood environmental management," say Bryan Norton and Robert Ulanowicz.[77] But what is their understanding of ecosystem health? They argue that "the capacity for creativity constitutes the crux of what is normally referred to ecosystem health." And what is "creativity"? That is the ability of the ecosystem to solve problems that in turn requires, in Norton and Ulanowicz's view, the ecosystem to possess both "a requisite amount of ordered complexity" and an appropriate level of internal "incoherence." And what is "ordered complexity"? For them, an ecosystem must contain enough "apparatus" to respond to events via "a channelized sequence of reactions." And what is internal "incoherence"? That is the presence of "dysfunctional repertoires" that are not normally used by the ecosystem but that serve as reservoirs of "stochastic, disconnected, inefficient features that constitute the raw building blocks of effective innovation."[78] This sequence of fuzzy terms and ideas leads these ecologists to accept the proposition that "an ecological system is healthy and free from 'distress syndrome' if it is stable and sustainable; i.e., if it is active and maintains its organization and autonomy over time."[79] But they also say that "ecosystem health is a comprehensive measure . . . of resilience, organization, and vigor . . . concepts that are imbedded in the term 'sustainability'."[80] Confusion reigns, as they simultaneously assert that stability and sustainability are the fundamental requirements for ecosystem health and that ecosystem health is the way to measure sustainability.

Other ecologists advance different understandings of ecosystem health. Costanza finds that his colleagues variously think of ecosystem health in terms of homeostasis, absence of disease, diversity/complexity, stability/resilience, vigor, and balance among system components.[81] David Rapport of Guelph University observes that there "are a plethora of attempts to define ecosystem health . . . [that] range widely from very broad definitions which incorporate bio-physical, human and socio-economic components to definitions focusing primarily on the biophysical aspects to definitions which focus on a single indicator within the biophysical domain."[82] He goes on to note that there is no consistency in the way researchers measure the state of ecosystem health, some using a composite index, others multiple indicators, and still others a single indicator variable.

Those touting the concept of ecosystem health are pleading for an idea without scientific substance, an intellectual house of cards, so it is no wonder that many scientists fault the use of the idea.[83] "Health," writes fisheries scientist

Robert Lackey of the Environmental Protection Agency (EPA), "is one of those amoeba-like words that changes to fit the surrounding conditions." He concludes that "using the *health* concept in ecosystem management may not clarify but actually cloud the fundamental choices society must make" [emphasis in the original].[84] He shares Shrader-Frechette and McCoy's observation that while there is general agreement about what constitutes human health and that clinical norms exist to measure it, there is nothing comparable associated with ecosystems.[85] When considering human health, the doctor has a distinct patient. When considering ecosystem health, the ecologist must fabricate a patient. When considering human health, a living patient exists in the doctor's office. When considering ecosystem health, a nonliving patient exists in the ecologist's head. The thought that ecosystems are creative entities on the landscape trying to stay well or to sustain themselves in the face of problems that the larger world has foisted on them (or that have bubbled up from within like an appendix about to burst) collapses in the light of scientists' universal understanding that change is the only constant on the surface of the earth. As Golley observes, "When ecosystem scientists propose that ecosystems are self-regulated superorganisms with the purpose of maintaining stability," then criticisms of the ecosystem concept "are justified."[86] Clyde Goulden joins Golley in his criticism of the continued use of organismal metaphors such as "health" in ecology and reminds us that to do so is inconsistent with Tansley's overall concept of the ecosystem.[87] Although many ecologists who use the ecosystem health concept steadfastly deny it, when all is said and done those advancing the idea assume that ecosystems are real entities—living things that exist on the landscape.[88] This assumption is an inescapable and fatal flaw within the idea.[89]

Like so many other slogans associated with ecosystem management, ecologists cannot give ecological integrity cogent meaning because it is not a scientific term. John Lemons observes that "ambiguity of ecological integrity is a recognition that its definition, like many ecological concepts, is determined, in part, on the basis of value-laden judgments and not solely on so-called value-free or precisely defined scientific criteria."[90] Researchers routinely link ecological integrity with ecological health but disagree about the essence of the relationship. For Rapport, healthy ecosystems are "characterized by the systems integrity."[91] Kay turns the relationship around: "ecosystem health . . . is the first requisite for ecosystem integrity."[92] He sees ecological integrity as being about "our sense of the wholeness and well being" of the system.[93] Paul Angermeier and James Karr also invoke "wholeness," they believe that "integrity refers to a system's wholeness, including presence of all appropriate elements and all occurrence of all processes at appropriate rates."[94] *Appropriate* for them means whatever Mother Earth put someplace because pristine nature is the standard they use for judging integrity. "By definition," they say, "naturally evolved assemblages possess integrity but random assemblages do not." Any human activity that results in new species being introduced into an area lessens ecosystem integrity in their view. They propose cooperative efforts between the government and the private sector to maintain ecosystem integrity across the country but argue that if cooperative efforts fail, then government agencies adopt "safety-net measures analogous to the Endangered Species Act to prevent important or unique ecosystems and landscapes from being destroyed."[95] Of course, new paradigmists would determine just what constitutes "important or unique ecosystems" for public policy purposes.

Not all champions of ecosystem integrity subscribe to the belief that naturalness is the key to the concept. The University of Toronto's Henry Regier writes that "there is room for choice in the kinds of ecosystems with integrity that humans might prefer."[96] Others agree.[97] Just what integrity means for Regier, however, is unclear, for he writes that there is "no way to sketch the concept of ecosystem in a linear, closed way," arguing that "to do so would be contrary to its meaning." Instead, he (working with James Kay and Bruce Bandurski) offers a lengthy laundry list of partial ideas concerning the condition of an ecosystem in a "state of integrity." They include in their catalog such inoperational attributes as "contains some relatively large, longer-lived plant and animal organisms in locales where externally-driven factors—natural and cultural—do not reach extreme values with respect to preconditions necessary for life and do not fluctuate excessively with respect to amplitude and frequency."[98]

Many new paradigmists reject the human-values approach to ecosystem integrity favored by Regier because it does not give adequate reverence to Mother Earth. Philosophy professor Laura Westra writes of "the moral poverty of the anthropocentric position."[99] She finds the "attitudes, values, and lifestyles" of the United States and other Western nations to be rooted in environmental ignorance and moral shallowness. She proposes making the protection and restoration of ecosystem integrity the chief legal and normative goal of the nation. We must, in her view, protect the whole of nature, and "wild, undisturbed ecosystems" offer the ideal examples of ecosystems with intact integrity.[100] Because we are simply another component of nature, we must subordinate ourselves to the larger goal of defending Mother Earth via invoking her "principle of integrity" as "an imperative which must be obeyed *before* other human moral considerations are taken into account [emphasis in the original]."[101] According to Westra, integrity must be defined in nature's terms, for "if human 'good,' human 'rights and interests,' or even human 'justice' are to be the only arbiters of what defines moral action, then human definitions of these terms will retain ultimate power. And even enlightened humans may tend to define the 'good' or any other 'ultimates' in nonsustainable, unecological terms."[102] What nature's definition of ecological integrity might be remains a mystery as Westra returns to Aldo Leopold for guidance. Somehow we are to employ Leopold's idea of thinking like a mountain so as to make his admonition that "a thing is right when it tends to preserve the integrity, stability, and beauty of the biotic community [and] it is wrong when it tends otherwise" the basic ethical and statutory guide for human life in the United States.

Ecologists and others like Westra cannot give us a clear understanding of ecosystem integrity. It is just one more poorly defined empty term, like sustainability, stability, health, and diversity, that clouds policy debates. New paradigmists employ these notions in an effort to give the appearance that they and their ideas represent the ethical and scientific wisdom needed to guide the nation to eco-salvation and that their opponents are but advocates of ignorance, greed, and a base morality.

Is Nature Balanced?

Balance and harmony stand at the apex of nature's characteristics in the minds of many new paradigmists outside the scientific community. Vice President Al Gore writes that "the ecological perspective begins with a view of the whole, an understanding of how various parts of nature interact in patterns that tend toward balance and persist over time."[103] The first principle of the EPA's 1995 strategic plan calls for the EPA to protect ecosystems in order to reach their vision of a world wherein "the natural balance of all living things is not threatened."[104]

Writers have made claims for a balance or harmony in nature over the sweep of time; some are scientists, some are philosophers, some are poets, and some are romantics.[105] Ecological historian Frank Egerton traces the concept back to Herodotus, writing around 450 B.C.[106] Some ecologists and many in the public at large still cling to the notion. Harvard entomologist E. O. Wilson refers to the existence of the balance of nature in his popular book *The Diversity of Life,* and the National Audubon Society wants us to begin "thinking like an ecosystem" so as to grasp "the principles that we need to think about in preserving balance in our ecosystems."[107] A belief in natural harmony is closely associated with the redefining of equilibrium by ecologists that I noted earlier in this chapter and a similar repackaging of Clements's superorganicism as with Kay and Schneider's finding that ecosystems, like "all living systems go through cycles of birth, growth, death, and renewal."[108] Shrader-Frechette and McCoy see in the notion of stability a rewrapped balance of nature.[109] Yet despite such implicit support for a balance in nature in some corners of ecology, most researchers today reject the idea, in part at least because scholars have made such little progress with the idea over the past 2,000 years. In the first century B.C., Cicero observed, "Who cannot wonder at this harmony of things, at this symphony of nature which seems to will the well-being of the world," while in 1995

Kay and Schneider find that "ecosystem behavior is like . . . a symphony" whose integrity we must protect to ensure our well-being.[110]

Most scientists disavow the idea of balance in nature for the best of reasons; when they look for it in the field, they find disequilibrium, not balance.[111] Evidence from the landscape shows that selfishness rather than harmony far better characterizes the interaction among living things in nature because nature is neither equilibrial nor balanced.[112] "The balance of nature," write Pickett, Kolasa, and Jones, "is not a scientific theory or concept," it is only a metaphor, and a poor one at that, because it is tied to the generally abandoned notion of equilibrium in the environment.[113] Egerton describes the origin and perpetuation of the "balance of nature myth" over the long sweep of history and attributes its longevity to its vagueness, which prevented scientists from actually subjecting it to close scrutiny. He attributes the "abandonment of the balance of nature myth by ecologists" to multiple causes but finds scientists' rejection of a hypothesis that cannot be falsified to be "deadly to any mythology still lurking within science."[114] Although scientists reject them, the twin myths of a balance of nature and harmony of nature live on in classroom materials, in pronouncements by politicians, and in television specials, movies, cartoons, and other information sources. An Internet search using the key words "balance of nature" returned 1,022,981 hits in October 1997.[115] All of this is clear evidence in support of Egerton's claim that "a myth cannot be destroyed by facts."[116]

The perpetuation of ecological ignorance by activists and others is harmful. It will require a significant educational effort to overcome the popularly held but scientifically impoverished balance-harmony-equilibrium view of nature.[117] George Mason University ecologist Dan Botkin believes that continued public acceptance of the discredited "classic balance of nature" view of the world, in which "the idea that nature, undisturbed by human influences, is constant and that this constancy is desirable and good—and the best possible condition for all life," underlies many current environmental conflicts.[118] Public pressure on decision makers to have government impose its authority on segments of the population in order to reestablish harmony within nature or to allow nature to return to a balance presumably upset by clumsy humans is out of touch with current scientific understanding. Government attempts to impose harmony or balance policies are preordained to fail because nature is neither harmonious nor balanced.

Do Ecosystems Perform Services for Humans or Any Other Living Thing?

New paradigmists are fashioning another arrow for their quiver. Some ecologists and economists are now calling for ecosystem protection on the grounds that ecosystems provide services to humans. Ecosystem services, they say, are either not valued or are undervalued in normal economic and policy decision making. They see this situation as a major impediment to an appropriate relationship between humans and nature and one that threatens human well-being, if not the very survival of our species. Some advocates advise us that the "economies of the earth would grind to a halt without the services of ecological life-support systems."[119] Their observation is true but pointless, as it is akin to saying that if there were no life on earth, there would be no economic activity. In this section I assess two efforts by researchers to place a price tag on nature's services.

University of Maryland ecologist Robert Costanza and twelve colleagues from ecology and economics at several other institutions estimate the value of the planet's ecosystem services in a study that emerged from a workshop sponsored by the National Science Foundation in 1996.[120] These scholars write that failure to capture these services in policy decisions "may ultimately compromise the sustainability of humans in the biosphere."[121] They want to change this situation so that "the natural capital stock that produces these services [is given] adequate weight in the decision process."[122]

For their analysis the researchers use seventeen major categories of ecosystem services: atmospheric gas regulation, climate regulation, disturbance regulation, water regulation, water supply, erosion control and sediment retention, soil formation, nutrient cycling, water treatment, pollination, biological control, refugia, food production, raw materials, genetic, recreation, and cultural. They divide the world into sixteen major ecosystem types (open ocean, coastal ocean, coral reefs, continental shelf, tropical forests, temperate and boreal forests, grasslands and rangelands, wetlands, tidal marshes and mangroves, swamps and floodplains, lakes and rivers, deserts, ice and rock, tundra, cropland, and urban) and estimate the area of each category. Costanza and his coauthors put a per hectare value on each service applicable for each of the ecosystems, multiply by the number of hectares in a given ecosystem, and sum the results, which they express in 1994 U.S. dollars. For the planet they conclude that the value of ecosystem services ranges between $16 trillion and $54 trillion per year with the average being $33 trillion annually. The authors compare this with a global gross national product of about 18 trillion U.S. dollars per year and believe that their estimate of ecosystem services is conservative. While acknowledging that their numbers are just a "crude initial estimate," they argue that their findings "should help to modify systems of national accounting" and be used in "project appraisal, where ecosystem services lost must be weighed against the benefits of a specific project."[123] They envision government land use and environment regulations as well as permitting actions that factor in ecosystem services and believe that their work can help establish ecosystem-use taxes.[124]

The Costanza group's claim that ecosystems provide services to humans (or any other living thing for that matter) is erroneous, owing to the falsity of two of their underlying assumptions. First, they assume that ecosystems are real, an illusion that I dispel in earlier chapters. For their effort to have meaning, they demand that the planet be completely covered by tangible organized entities that exist above the level of species. They call these entities ecosystems and see them as laboring at least in their own behalf, if not for ours. If their ecosystems are real, then why is there no consistency in their geographic portrayal? Why do Costanza et al. see sixteen categories of ecosystems, for example, while Robert Bailey partitions the earth into nine kinds of ecoregions and forty-six subcategories?[125] If their ecosystems were real, why do Costanza et al. find it necessary to consult eleven different sources in deciding on the number and location of their ecosystems? Surely, if ecosystems were real, scholars would have long ago located and mapped them and declared the matter closed, except perhaps for some cartographic fine-tuning.

Second, the benefits that Costanza et al. see as services provided humans by ecosystems (as organized, knowable, and discrete entities on the surface of the earth) are in essence the serendipitous by-products of individual living organisms seeking to survive and nonliving things following fundamental laws of physics and chemistry. For example, they claim that pollination is a benefit that humans derive from the actions of ecosystems; yet ecosystems do not pollinate anything. Individual food-seeking insects, birds, and other biota that move from flower to flower in normal pursuit of their species' lifestyle are responsible for pollination. Costanza and his colleagues make a huge leap when they attribute to ecosystems the fruits of insect labor.

They make similar errors in asserting that ecosystems provide such services as climate control and water management. Researchers have long known that plants can affect climate: They both absorb and reflect incoming solar radiation, they alter the mix of atmospheric gases, and they can modify ground-level circulation of the atmosphere. But they do so as individual plants, not as participants in some union of life forms intent on altering the atmosphere. Likewise, ecosystems do not manage or regulate water. The fate of individual water molecules falling as rain, for example, is determined by physical forces, such as gravity not the edicts of ecosystems. So it is that some of these molecules are bound up in the soil, others percolate downward to gather in aquifers, and still others eventually reach lakes and oceans. The presence or absence of vegetation, vegetation composition, soil makeup, topography, and other factors influence these happenings but not in any orchestrated way demonstrating the existence of a higher-level entity called an ecosystem directing the behavior of the water molecules to provide a service for humans or for itself. Scientists do not point to evidence suggesting that the internal geological forces of the earth that produced the Sierra Nevada Mountains of California did so for the purpose of controlling rainfall in California and Nevada. As Calow argues, "Ecosystems are not superorganisms in which parts are actively controlled for the sake of the well being of the whole."[126]

A second attempt to value the services of nature is that of David Pimentel of Cornell University and eight colleagues who estimate the worldwide value of biodiversity at $3 trillion.[127] They see a world wherein "pollution of ecosystems and the depletion of basic resources have reached dangerous levels." They believe that their effort serves "as a foundation to develop strategies and policies to preserve biological diversity and maintain ecosystem integrity" and to do so "before it is too late for meaningful action."[128] They share with Costanza and his coauthors a view that we are in dire ecological straits and a desire to influence policymaking in the direction of ecosystem protection by valuing services they attribute to nature.

Pimentel et al. focus their evaluation on the activities of biota rather than ecosystems, although they refer to ecosystems throughout their analysis and clearly link the two. They value twenty-one biological activities: waste disposal, soil formation, nitrogen fixation, bioremediation of chemicals, crop breeding, livestock breeding, biotechnology, biocontrol of crop and forest pests, host plant resistance for crops and forests, perennial grains, pollination, fishing, hunting, seafood, other wild foods, wood products, ecotourism, pharmaceuticals from plants, and forest sequestration of carbon. Like the Costanza group, these researchers combine work done by others and make extrapolations to cover the entire planet.

But unlike the Costanza group, who openly acknowledge that their estimates are crude and who report many limitations of their study, Pimentel et al. make no such admission. They believe that their point estimate of $2,928,000,000 is a meaningful value for "the vital services that are provided by all biota (biodiversity), including their genes and biomass, to humans and to the environment."[129] Scholars normally reject point estimates in favor of a range of values when the calculations depend heavily on very crude estimates and are little more than guesswork, as is the case here.

I illustrate the weakness of this work with two examples. First, Pimentel's group places a worldwide value on the decomposition of organic wastes emanating from human activity at $760 billion annually "based on the information that it costs $0.04–0.044/kg to collect and dispose of organic wastes that are produced in the Village of Cayuga Heights, N.Y. or in the city of Madison, Wisconsin."[130] I scarcely need to note that the cost of collecting and disposing of organic wastes in Cayuga Heights or Madison are unlikely to be typical of international conditions. Nonetheless, Pimentel et al. simply cut the value in half and use $0.02 per kilogram as the value of waste disposal. The authors provide no reason for dividing $0.04 kilograms by two instead of by ten or twenty or some other number. If they had divided that figure by ten or twenty, then the value of biodiversity in the waste disposal category falls from $760 billion to $152 billion and $76 billion, respectively, thereby making a major change in the overall benefit that the authors attribute to biodiversity. Second, despite the earth's wide range of climate, vegetation, parent rock, topography, and soil-living biota—all factors in soil formation—Pimentel's group purports to have a single meaningful number for the amount of topsoil created by earthworms and their subsurface brethren that applies worldwide: one ton per hectare per year.

Advocates' claims that evaluations of so-called ecosystem services can result in better decision making are inaccurate. One must ask how analyses based on ill-defined concepts such as ecosystems, ecosystem health, and sustainability can inform rather than confuse or mislead decision makers. Even if the Costanza et al. estimate of the value of the world's ecosystem services is accurate, of what use is it to policymakers? How might they factor the knowledge that a single asset—ecosystem services—is more valuable than the gross income of all humans on the planet into the formation of laws and regulations? If the value of ecosystem services exceeds the totality of human income, how will people pay for it? Duke University economist V. Kerry Smith concludes that the Costanza et al. "analysis seems to combine bad economics with bad ecological science," while Sagoff sees within their analysis a substantial confusion of ideas.[131]

But what else can we expect? There is little hope for objective analysis when the subjects of the analyst's attention—ecosystems—are nothing more than arbitrary constructs in the researcher's mind. There is little likelihood for meaningful work products when researchers understand that these constructs shift in time and space and that analysts' attempts to pin them down for evaluative purposes are necessarily idiosyncratic. One's selection of services to evaluate—that is, extract from the myriad of biotic and physical activities occurring on the planet only a handful of items for investigation—is also individual. Researchers employ methods to generate values for ecosystem service that are awash in subjectivity and yield widely varying results. Scholars' attempts to evaluate ecosystem services add confusion, not clarity, to resource decision making. In the end ecosystems no more provide services to humans than do they possess the attribute of health. We should reject efforts to infuse the decision-making process with nebulous ideas and mindless numbers. As Smith observes, "The Costanza et al. estimates are seriously flawed . . . [and] their results should not be used in any form."[132] That does not mean that we Americans should be careless in our relations with the environment, ignore environmental impacts in our decision making, or be oblivious to the benefits we derive from our natural endowment. Happily, as I illustrate in chapter 4, we do none of those things.

The attempts of new paradigmists to value ecosystem services points the discussion toward the often heard contention that healthy economies require healthy ecosystems. I look at that claim in the next section.

Healthy Economies and Healthy Ecosystems

Believers in ecosystem management routinely intone that healthy economies require healthy ecosystems. The White House Interagency Ecosystem Management Task Force concludes that "the link between a healthy economy and a healthy environment has highlighted the need to actively maintain our natural infrastructure."[133] Paul and Anne Ehrlich believe that "the economy is a wholly owned subsidiary of natural ecosystems" and see human activities as an enormous threat to those ecosystems.[134] Lester Brown repeats this theme annually in his *State of the World* books.[135] Other new paradigmists are more aggressive. "Maintaining healthy ecosystems . . . is vital to ensuring a high quality of life for future generations of Americans," writes Vice President Gore.[136] Norton and Ulanowicz also move beyond a concern with simple economic well-being. "We are arguing," they say, that "public values—aesthetic, economic and moral—all depend on protecting the processes that support the health of larger-scaled ecological systems."[137] This is heady stuff.

The new paradigmist claim that healthy economies require healthy ecosystems is not self-evident, but they must assume that it is since they rarely present a substantive argument to support their assertion. Their claim is not a statement of fact. The idea cannot be directly demonstrated by its proponents or falsified by its critics because its key terms, *healthy ecosystems* and *healthy economies,* are not clearly defined by those touting the relationship. As I point out earlier in this chapter, the idea of a healthy ecosystem has no objective meaning under any circumstance, so the claim of a relationship between ecosystems and the economy is properly assigned to political speech rather than to scientific discourse. Economist Ron Johnson notes that "the desire by proponents to link ecosystem management and ecosystem health is understandable" because it helps build political support for the "new world order" that many supporters of ecosystem management desire.[138] Their new world order calls for governments to protect Mother Earth, and it shows up here with proponents equating healthy ecosystems to natural ecosystems. Once again they presume that an earth unaffected by human works is the ideal.

When one takes a broad look at the asserted relationship between economic health (I think of economic health as one yielding surpluses that translate into high standards of living) and ecosystem health (for the sake of argument, I assume that a healthy ecosystem is one little impacted by human action), the supposed dependency vanishes. Over the long sweep of history, there are few landscapes that one might consider less healthy (by the deviation from natural condition standard) than the wet rice landscapes of South China and Japan. Farmers here displaced natural vegetation communities centuries ago and have grown rice in paddies for hundreds of years, yet the well-being of these people for the most part has hardly lagged behind that of the remainder of the world's population over much of that time. Conversely, in the United States today, Alaska's lands and waters are reasonably reckoned by scholars to be the closest we have to an area unaffected by humans, yet native Alaskans in traditional settlements are hardly among the best-off groups in the United States.

The evidence from the landscape leads an objective observer to conclude that new paradigmists have cause and effect backward, in that a healthy ecosystem depends on a healthy economy, not the other way around.[139] Well-off peoples—not those struggling to secure adequate food, clothing, and shelter—take the lead in devoting societal wealth to environmental protection. In world terms it is the wealthier industrial and postindustrial nations of the North that initiate and implement policies to reduce air and water pollution and that set aside large tracts of land as nature parks and wilderness.

The issue for us is not over whether Americans want clean air or clean water or national parks or to be good stewards of the environment; we answered those questions affirmatively long ago. The issue is whether new paradigmists will succeed with their effort to bamboozle a large segment of the population into believing that a societal objective of improving economic well-being is dependent on the fiction of protecting or restoring ecosystem health. Improving the quality of life—including a healthier economy—for Americans is in no way dependent on reconstructing landscapes in an image of some mythic vision of a North America never occupied by Europeans and their descendants.

Where Does This Leave Us?

Recall a few words from the new paradigm supporters I presented in the introduction to this book. Interior Secretary Babbitt tells us that we must live in "harmony with the environment" and that we must find "the missing link in nature's self-regulating cycle of life" to restore "our forests to presettlement equilibrium." The White House Interagency Ecosystem Management Task Force says that ecosystem management is about "sustaining or restoring natural systems." The Bureau of Land Management tells Americans that its goal is to devise a management strategy that "conserves, restores, and maintains ecological integrity." The Fish and Wildlife Service wants to perpetuate "dynamic, healthy ecosystems." For the earth, Vice President Gore says that "all its [environment] parts exist in a delicate balance of interdependency" that we must "preserve and nurture" through a "wrenching transformation of society." What do we know about the key ideas in these statements? We know that nature possesses neither harmony nor balance nor equilibrium.

How about the ecosystem concept? We know that Sir Arthur Tansley coined the term in 1935 to as a way of advancing the debate among fellow ecologists of the day. He fashioned a heuristic device to aid researchers in organizing knowledge to better understand the workings of the world. He did not offer ecosystems as a geographic unit of governance.

We know that during this same period, researchers have gone through a series of paradigms (e.g., static equilibrium to dynamic equilibrium to nonequilibrium). We know that scholars employ a lengthy list of ideas in conjunction with ecosystems (e.g., stability, integrity, health, sustainability, and diversity) and that these thoughts remain ambiguous, nebulous, and controversial among ecologists and other scientists. We know that scientists have not produced agreed-on definitions of these key ideas.

We know that in the sixty-plus years since Tansley surfaced his concept, ecologists have produced little if any useful theory regarding ecosystems for public policy purposes. They cannot agree on such seemingly fundamental questions as What are the core characteristics of ecosystems? or offer generic guidelines on such equally elementary matters as how to determine when one ecosystem is replaced by another.

What about ecosystems themselves? When ecologists apply the ecosystem concept to the landscape, we know that they delimit ecosystems arbitrarily and without recourse to theory or agreed-on methods or well-established protocols. We know that the boundaries of ecosystems are guesswork and rarely represent real features on the landscape. We know that the landscape is in constant flux so that the ecosystems depicted by researchers constantly change in space and time in poorly understood ways, turning ideas such as ecosystem stability and sustainability into oxymorons. We know that ecosystems are geographic fictions that exist in the heads of their creators, not on the surface of the earth. On this point in particular, many ecologists may strongly disagree, but until they produce a single definitive ecosystem map for the United States, develop an agreed-on classification system for ecosystems, and are able to generically explain (or, more convincingly, be able to predict) when, where, why, and how one ecosystem ceases to exist and a new one takes its place on the same landscape, I will stand by the statement.

The weight of the evidence plainly demonstrates that the assertions of the new paradigmists of a firm scientific basis for their efforts to force Americans into a "wrenching transformation society" so as to save us from environmental disaster are false. Even ecologists who strongly support ecosystem management find that "a scientifically defensible and comprehensive view of ecosystem management has yet to be articulated."[140]

Ecology and the work of ecologists can help inform particular decisions at particular times in particular places. But the discipline cannot provide a rational basis for radical new directions in public policy in any way related to the ecosystem concept because of the depth of confusion and uncertainty that continue to surround the idea more than half a century after Tansley introduced it.

Despite the nebulousness of ecosystems as landscape features, Congress includes the concept in numerous federal statutes. In the next chapter, I investigate ecosystem protection in federal law.

Notes

1. R. V. O'Neill, D. L. DeAngelis, J. B. Waide, and T. F. H. Allen, *A Hierarchical Concept of Ecosystems* (Princeton, N.J.: Princeton University Press, 1986), 1.
2. Steward T. A. Pickett, Jurek Kolasa, and Clive G. Jones, *Ecological Understanding: The Nature of Theory and the Theory of Nature* (San Diego: Academic Press, 1994), 177.
3. Jack Ward Thomas, "Foreword," in *Biodiversity in Managed Landscapes,* ed. Robert C. Szaro and David W. Johnston (New York: Oxford University Press, 1996), ix.
4. Robert H. Peters, *A Critique for Ecology* (Cambridge: Cambridge University Press, 1991).
5. Peters, *A Critique for Ecology,* 1.
6. Robert P. McIntosh, "Pluralism in Ecology," in *Annual Review of Ecology and Systematics* (Palo Alto, Calif.: Annual Reviews, 1987), 327.
7. Claudia Pahl-Wostl, *The Dynamic Nature of Ecosystems: Chaos and Order Intertwined* (Chichester: John Wiley & Sons, 1995), xi, 1; see also Robert P. McIntosh, *The Background of Ecology: Concept and Theory* (Cambridge: Cambridge University Press, 1985).
8. Joel B. Hagen, *An Entangled Bank: The Origins of Ecosystem Ecology* (New Brunswick, N.J.: Rutgers University Press, 1992), 196.
9. Robert M. May, "Foreword," in Pickett et al., *Ecological Understanding,* x.
10. Peters, *A Critique for Ecology,* 95.
11. Peters, *A Critique for Ecology,* 96.
12. Julie Ann Miller, "Biosciences and Ecological Integrity," *BioScience* 41, no. 4 (April 1991): 206–10.
13. K. S. Shrader-Freschette and E. D. McCoy, *Method in Ecology: Strategies for Conservation* (Cambridge: Cambridge University Press, 1993), 46.
14. Lionel Johnson, "The Far-from-Equilibrium Ecology Hinterlands," in *Complex Ecology: The Part-Whole Relation in Ecosystems,* ed. Bernard C. Patten and Sven E. Jorgensen (Englewood Cliffs, N.J.: Prentice Hall, 1995), 52.
15. L. Scott Mills, Michael E. Soulé, and Daniel F. Doak, "The Keystone-Species Concept in Ecology and Conservation," *BioScience* 43, no. 4 (April 1993): 219–24.

16. Mary E. Power, David Tilman, James A. Estes, Bruce A. Menge, William J. Bond, L. Scott Milles, Gretchen Daily, Juan Carlos Castilla, Jane Lubchenco, and Robert T. Paine, "Challenges in the Quest for Keystones," *BioScience* 46, no. 8 (September 1996): 618.

17. McIntosh, *The Background of Ecology.*

18. National Research Council, *A Biological Survey for the Nation* (Washington, D.C.: National Academy Press, 1993), 76; Ronald Amundson and Hans Jenny, "On a State Factor Model of Ecosystems," *BioScience* 47, no. 8 (September 1997): 536–43.

19. James J. Kay and Eric Schneider, "Embracing Complexity: The Challenge of the Ecosystem Approach," in *Prospective on Ecological Integrity,* ed. Laura Westra and John Lemons (Dordrecht: Kluwer, 1995), 55.

20. Pickett et al., *Ecological Understanding.* This movement reflects broader changes in the philosophy of science.

21. Bryan G. Norton and Robert E. Ulanowicz, "Scale and Biodiversity Policy: A Hierarchical Approach," in *Ecosystem Management,* ed. Fred B. Samson and Fritz L. Knopf (New York: Springer, 1996), 424.

22. Helen Ingram, H. Brinton Milward, and Wendy Laird, "Scientists and Agenda Setting: Advocacy and Global Warming," in *Risk and Society: The Interaction of Science, Technology and Public Policy,* ed. Marvin Waterstone (Dordrecht: Kluwer Academic Press, 1992), 38.

23. For detailed accounts prepared by ecologists, see Frank Golley, *A History of the Ecosystem Concept in Ecology* (New Haven, Conn.: Yale University Press, 1993), and Hagen, *An Entangled Bank.* A briefer account by an ecologist can be found in Robert E. Ricklefs, *Ecology,* 3rd ed. (New York: W. H. Freeman, 1990), 173–88; for a journalist's view, see Stephen Budiansky, *Nature's Keepers: The New Science of Nature Management* (New York: The Free Press, 1995).

24. Golley, *A History of the Ecosystem Concept,* 35.

25. Hagen, *An Entangled Bank,* 22.

26. Henry Allan Gleason, "The Individualistic Concept of the Plant Association," *Bulletin of the Torrey Botanical Club* 53 (1926), as quoted in Hagen, *An Entangled Bank,* 29.

27. Golley, *A History of the Ecosystem Concept,* 201.

28. A. G. Tansley, "The Use and Abuse of Vegetational Concepts and Terms," *Ecology* 16, no. 3 (1935): 306.

29. Tansley, "The Use and Abuse of Vegetational Concepts and Terms," 299.

30. Tansley, "The Use and Abuse of Vegetational Concepts and Terms," 300.

31. Eugene P. Odum, "The Strategy of Ecosystem Development," *Science* 164, no. 3877 (April 18, 1969): 262.

32. Norman Christensen et al., "Report of the Committee on the Scientific Basis for Ecosystem Management," *Ecological Applications* 6, no. 3 (1996): 674.

33. Karl S. Zimmerer, "Human Geography and the 'New Ecology': The Prospect and Promise of Integration," *Annals of the Association of American Geographers* 84, no. 1 (March 1994): 108.

34. Mark Sagoff, "Muddle or Muddle Through? Takings Jurisprudence Meets the Endangered Species Act," *William and Mary Law Review* 38, no. 3 (March 1997): 893.

35. Golley, *A History of the Ecosystem Concept,* 195.

36. Golley, *A History of the Ecosystem Concept,* 195.

37. Christensen et al., "Report of the Committee on the Scientific Basis for Ecosystem Management," 675.

38. C. S. Holling, "What Barriers? What Bridges?" in *Barriers and Bridges to the Renewal of Ecosystems and Institutions,* ed. Lance H. Gunderson, C. S. Holling, and Stephen S. Light (New York: Columbia University Press, 1995), 3–36.

39. Kay and Schneider, "Embracing Complexity: The Challenge of the Ecosystem Approach," 55.

40. Sagoff, "Muddle or Muddle Through?"

41. National Research Council, *A Biological Survey,* 75.

42. National Research Council, *A Biological Survey,* 75.

43. Sagoff, "Muddle or Muddle Through?"

44. National Research Council, *A Biological Survey,* 75.

45. Christensen et al., "Report of the Committee on the Scientific Basis for Ecosystem Management," 670.

46. National Research Council, *A Biological Survey,* 75.

47. Helias A. Udo de Haes and Frans Klijn, "Environmental Policy and Ecosystem Classification," in *Ecosystem Classification for Environmental Management,* ed. Frans Klijn (Dordrecht: Kluwer Academic Publishers, 1994), 3.

48. Klijn, ed., *Ecosystem Classification for Environmental Management.*

49. National Research Council, *A Biological Survey,* 75.

50. Nels Johnson, "Introduction," in *Defining Sustainable Forestry,* ed. Gregory H. Aplet, Nels Johnson, Jeffrey T. Olson, and V. Alaric Sample (Washington, D.C.: Island Press, 1993), 11.

51. Dahpne Gail Fautin, "Preface," in *Annual Review of Ecology and Systematics,* vol. 26, ed. Dahpne Gail Fautin (Palo Alto, Calif.: Annual Reviews, 1995), v.

52. Christensen et al., "Report of the Committee on the Scientific Basis for Ecosystem Management," 668.

53. Christensen et al., "Report of the Committee on the Scientific Basis for Ecosystem Management," 671.

54. Jerry F. Franklin, "The Fundamentals of Ecosystem Management with Applications in the Pacific Northwest," in Aplet et al., eds., *Defining Sustainable Forestry,* 127.

55. Franklin, "The Fundamentals of Ecosystem Management," 127.

56. Peters, *A Critique for Ecology.*

57. Franklin, "The Fundamentals of Ecosystem Management," 131.

58. Chris Maser, *Sustainable Forestry: Philosophy, Science, and Economics* (Delray Beach, Fla.: St. Lucie Press, 1994), 309.

59. Maser, *Sustainable Forestry,* 229.

60. Maser, *Sustainable Forestry,* 308.

61. Christensen et al., "Report of the Committee on the Scientific Basis for Ecosystem Management," 666.

62. Simon Levin, "The Problem of Pattern and Scale in Ecology," *Ecology* 73, no. 6 (December 1992): 1960.

63. R. V. O'Neill, "Get Ready to Rewrite Your Lecture Notes," *Ecology* 77, no. 2 (March 1996): 660.

64. Bryan G. Norton, "A New Paradigm for Environmental Management," in *Ecosystem Health: New Goals for Environmental Management,* ed. Robert Costanza, Bryan G. Norton, and Benjamin D. Haskell (Washington, D.C.: Island Press, 1992), 35.

65. Donald Worster, *Nature's Economy—a History of Ecological Ideas,* 2nd ed. (Cambridge: Cambridge University Press, 1994), 412.

66. World Commission on Environment and Development, *Our Common Future* (New York: Oxford University Press, 1987).

67. World Commission on Environment and Development, *Our Common Future,* 43.

68. White House, Office of the Press Secretary, "On Earth Summit Anniversary President Creates Council on Sustainable Development," June 14, 1993, downloaded from The White House Virtual Library, at http://library.whitehouse.gov, October 10, 1997.

69. Wilfred Beckerman, *Green-Colored Glasses: Environmentalism Reconsidered* (Washington, D.C.: Cato Institute, 1996), 146.

70. Thomas J. Wilbanks, "Sustainable Development in Geographical Perspective," *Annals of the Association of American Geographers* 84, no. 4 (December 1994): 550.

71. Wilbanks, "Sustainable Development," 541.

72. Wilbanks, "Sustainable Development," 544.

73. Stephen Budiansky, *Nature's Keepers: The New Science of Nature Management* (New York: The Free Press, 1995), 98.

74. Benjamin D. Haskell, Bryan G. Norton, and Robert Costanza, "Introduction," in Costanza et al., eds., *Ecosystem Health,* 4.

75. Peter Calow, "Ecosystem Health—a Critical Analysis of Concepts," in *Evaluating and Monitoring the Health of Large Scale Ecosystems,* ed. David J. Rapport, Connie L. Gaudet, and Peter Calow (New York: Springer, 1995), 33.

76. See David Rapport, Robert Costanza, Paul R. Epstein, Connie Gaudet, and Richard Levins, eds., *Ecosystem Health* (London: Blackwell Science, 1998). There is now an International Congress on Ecosystem Health that annually gathers scientists to discuss questions related to the issue.

77. Norton and Ulanowicz, "Scale and Biodiversity Policy," 429.

78. Norton and Ulanowicz, "Scale and Biodiversity Policy," 430.

79. Costanza et al., eds., *Ecosystem Health,* 9.

80. Robert Costanza, "Toward an Operational Definition of Ecosystem Health," in Costanza et al., eds., *Ecosystem Health,* 240.

81. Costanza, "Toward an Operational Definition of Ecosystem Health," 239.

82. David J. Rapport, "Ecosystem Health: An Emerging Integrative Science," in Rapport et al, eds., *Evaluating and Monitoring the Health of Large Scale Ecosystems,* 6.

83. Rapport, "Ecosystem Health: An Emerging Integrative Science," identifies four lines of criticism: the inappropriateness of the health metaphor, lack of objective basis for measurement, inability to define norms of health, and use of the medical model.

84. Robert T. Lackey, "Pacific Salmon, Ecological Health, and Public Policy," *Ecological Health* 2, no. 1 (March 1996): 65–66.

85. Shrader-Freschette and McCoy, *Method in Ecology;* Robert T. Lackey, "Ecosystem Management: Paradigms and Prattle, People and Prizes," *Renewable Resources Journal* 16, no. 1 (1998): 8–13.

86. Golley, *A History of the Ecosystem Concept,* 204.

87. Clyde E. Goulden, "Ecological Comprehensiveness," *Science* 264 (April 29, 1994): 726–27.

88. Rapport et al., eds., *Evaluating and Monitoring the Health of Large Scale Ecosystems.* This book contains the proceedings of the NATO Advanced Research Workshop held in Montreal, October 10–15, 1993. In the preface, the editors write that "all who have been involved in this Workshop reject the concepts of ecosystems as superorganisms and as homeostatic systems."

89. David Ehrenfeld offers a broader view of ecosystem health. Conceding that it is not a scientific concept and that evaluations of the ecosystem health are subjective, he still desires ecologists to use the term on occasion so as not to lose what he feels is the intuitive value of the word; see "Ecosystem Health," *Orion* (winter 1993): 12–15.

90. John Lemons, "The Conservation of Biology: Scientific Uncertainty and the Burden of Proof," in *Scientific Uncertainly and Environmental Problem Solving,* ed. John Lemons (Cambridge: Blackwell Science, 1996): 216.

91. David J. Rapport, "What Is Clinical Ecology?" in Costanza et al., eds., *Ecosystem Health,* 145.

92. James J. Kay, "On the Nature of Ecological Integrity: Some Closing Comments," in *Ecological Integrity and the Management of Ecosystems,* ed. Stephen Woodley, James Kay, and George Francis (Delray Beach, Fla.: St. Lucie Press, 1993), 205.

93. Kay, "On the Nature of Ecological Integrity," 201.

94. Paul L. Angermeier and James R. Karr, "Biological Integrity versus Biological Diversity as Policy Directives," *BioScience* 44, no. 10 (November 1994): 692. For an approach to mapping class of ecological integrity on the basis of the deviation from Küchler's ideas of potential natural vegetation (chapter 2), see Robert V. O'Neill, Carolyn T. Hunsaker, K. Bruce Jones, Kurt H. Riitters, James D. Wickham, Parul M. Schwartz, Iris A. Goodman, Barbara L. Jackson, and William S. Biallargeon, "Monitoring Environmental Quality at the Landscape Scale," *BioScience* 47, no. 8 (September 1997): 513–19.

95. Angermeier and Karr, "Biological Integrity versus Biological Diversity," 696.

96. Henry A. Regier, "The Notion of Natural and Cultural Integrity," in Woodley et al., eds., *Ecological Integrity and the Management of Ecosystems,* 16.

97. James W. Crossley, "Managing Ecosystems Integrity: Theoretical Considerations for Resource and Environmental Managers," *Society and Resources* 9, no. 5 (1996): 465–81; Kay, "On the Nature of Ecological Integrity"; and Anthony W. King, "Considerations of Scale and Hierarchy," in Woodley et al., eds., *Ecological Integrity and the Management of Ecosystems,* 19–46.

98. Regier, "The Notion of Natural and Cultural Integrity," 14.

99. Laura Westra, *An Environmental Proposal for Ethics: The Principle of Integrity* (Lanham, Md.: Rowman & Littlefield, 1994), 5.

100. Westra, *An Environmental Proposal for Ethics,* 59.

101. Westra, *An Environmental Proposal for Ethics,* 6.

102. Westra, *An Environmental Proposal for Ethics,* 4.

103. Al Gore, *Earth in the Balance: Ecology and the Human Spirit* (New York: Penguin, 1993), 2.

104. U.S. Environmental Protection Agency, *New Generation of Environmental Protection: Agency-Wide Strategic Plan Summary,* downloaded from the EPA's Web site at http://www.epa.gov, February 2, 1995 (copy on file with the author). The reference to natural balance does not appear in the 1997 version of the plan.

105. Clarence J. Glacken, "Changing Ideas of the Habitable World," in *Man's Role in Changing the Face of the Earth,* ed. William L. Thomas Jr. (Chicago: University of Chicago Press, 1956), 70–92; Nash, *The Rights of Nature;* and Robert McIntosh, *The Background of Ecology: Concept and Theory* (Cambridge: Cambridge University Press, 1985); see also Frank N. Egerton, "The History and Present Entanglements of Some General Ecological Perspectives," in *Humans as Components of Ecosystems,* ed. Mark J. McDonnell and Steward T. A. Pickett (New York: Springer-Verlag, 1993), 9–23, and Frank N. Egerton, "Changing Concepts in the Balance of Nature," *Quarterly Review of Biology* 48 (1973): 322–50.

106. Egerton, "Changing Concepts."

107. National Audubon Society, *Almanac of the Environment,* ed. Valerie Harm (New York: G. P. Putnam, 1994).

108. Kay and Schneider, "Embracing Complexity: The Challenge of the Ecosystem Approach," 53.

109. Shrader-Freschette and McCoy, *Method in Ecology,* 34.

110. Cicero, as quoted in Daniel P. Botkin, *Our Natural History: The Lessons of Lewis and Clark* (New York: G. P. Putnam, 1995), 12; Kay and Schneider, "Embracing Complexity: The Challenge of the Ecosystem Approach," 55.

111. Shrader-Freschette and McCoy, *Method in Ecology;* Pickett et al., *Ecological Understanding;* Hagen, *An Entangled Bank;* Daniel Botkin, *Discordant Harmonies: A New Ecology for the Twenty-First Century* (New York: Oxford University Press, 1990); and R. V. O'Neill, D. L. DeAngelis, J. B. Wade, and T. F. H. Allen, *A Hierarchical Concept of Ecosystems* (Princeton, N.J.: Princeton University Press, 1986).

112. Michael E. Soulé, "The Social Siege of Nature," in *Reinventing Nature? Responses to Postmodern Deconstruction,* ed. Michael E. Soulé and Gary Lease (Washington, D.C.: Island Press, 1995): 137–70.

113. Pickett et al., *Ecological Understanding,* 181.

114. Egerton, "The History and Present Entanglements," 15.

115. The search was done on October 28, 1997, using America Online's search engine.

116. Egerton, "The History of Present Entanglements," 14.

117. Steward T. A. Pickett, V. Thomas Parker, and Peggy L. Fiedler, "The New Paradigm in Ecology: Implications for Conservation Biology above the Species Level," in *Conservation Biology,* ed. Peggy L. Fiedler and Subodh K. Jain (New York: Chapman & Hall, 1992), 66–88.

118. Botkin, *Our Natural History,* 12.

119. Robert Costanza, Ralph d' Arge, Rudolf de Groot, Stephen Farber, Monica Grasso, Bruce Hannon, Karin Limburg, Shahid Naeem, Robert V. O'Neill, Jose Paruelo, Robert G. Raskin, Paul Sutton, and Marjan van den Belt, "The Value of the World's Ecosystem Services and Natural Capital," *Nature* 387, no. 6230 (May 15, 1997), 253–60. *Nature* posts major articles and other materials on its Web site (http://www.nature.com). The Costanza et al. article was downloaded on May 20, 1997, and page references in the notes are to those pages; the quote is from p. 2. Several of my comments on the Constanza et al. work are taken from Allan K. Fitzsimmons, "Clearing Ecosystem Misunderstandings," *Regulation* 40, no. 4 (fall 1997): 3–4.

120. Costanza et al., "The Value of the World's Ecosystem Services."

121. Costanza et al., "The Value of the World's Ecosystem Services," 2.

122. Costanza et al., "The Value of the World's Ecosystem Services," 10.

123. Costanza et al., "The Value of the World's Ecosystem Services," 10.

124. Wade Roush, "Putting a Price Tag on Nature's Bounty," *Science* 276 (May 16, 1997): 1029.

125. Robert Bailey, *Ecosystem Geography* ((New York: Springer, 1996).

126. Calow, "Ecosystem Health," 38.

127. David Pimentel, Christa Wilson, Christine McCullum, Rachel Huang, Paulette Dwen, Jessica Flack, Quynh Tran, Tamara Saltman, and Barbara Cliff, "Economic and Environmental Benefits of Biodiversity," *BioScience* 47, no. 11 (December 1997): 747–57.

128. Pimentel et al., "Economic and Environmental Benefits of Biodiversity," 747, 755.

129. Pimentel et al., "Economic and Environmental Benefits of Biodiversity," 747.

130. Pimentel et al., "Economic and Environmental Benefits of Biodiversity," 748.

131. V. Kerry Smith, "Mispriced Planet," *Regulation* 40 (summer 1997): 17. See also Mark Sagoff, "Can We Put a Price on Nature's Services?" Report from the Institute for Philosophy & Public Policy, University of Maryland, downloaded from http://www.puaf.umd.edu/ippp/nature.htm, February 10, 1998.

132. Smith, "Mispriced Planet," 16.

133. White House Interagency Ecosystem Management Task Force, *The Ecosystem Approach: Healthy Ecosystems and Sustainable Economies. Vol. II: Implementation Issues* (Washington, D.C.: The White House, November 1995), 1.

134. Paul R. and Anne H. Ehrlich, *Betrayal of Science and Reason* (Washington, D.C.: Island Press, 1996), 186.

135. Lester Brown (and various coauthors by year), *The State of the World* (Washington, D.C.: World Resources Institute, annual).

136. Al Gore, *Creating a Government That Works Better & Costs Less: Reinventing Environmental Management* (Washington, D.C.: Office of the Vice President, September 1993), 11.

137. Norton and Ulanowicz, "Scale and Biodiversity Policy," 429; see also Westra, *An Environmental Proposal for Ethics,* and Robert Goodland and Herman Daly, "Environmental Sustainability: Universal and Non-Negotiable," *Ecological Applications* 6, no. 4 (November 1966): 1002–17.

138. Ronald Johnson, "Ecosystem Management and Reinventing Government," in *Breaking the Policy Gridlock,* ed. Terry L. Anderson (Stanford, Calif.: Stanford University Press, 1997), 34.

139. Beckerman, *Green-Colored Glasses;* Julian Simon, *Ultimate Resource 2* (Princeton, N.J.: Princeton University Press, 1996).

140. Christensen et al., "Report of the Committee on the Scientific Basis for Ecosystem Management," 676.

Chapter 7

The Law and Ecosystem Protection

> A strategy common to the environmental movement everywhere is to revise the ground rules under which society operates. This objective is attained . . . through written law . . . and treaties or comparable agreements at the international level.
>
> —Lynton Caldwell, *Between Two Worlds: Science, the Environmental Movement, and Policy Choice,* 97

Congress refers to ecosystems in over 100 federal laws. Regulators cite them in nearly 150 sections of the *Code of Federal Regulations* that they use to administer those laws. The law offers new paradigmists opportunities to redirect society's ground rules, and Vice President Al Gore urges all federal "agencies to interpret their existing authorities as broadly as possible to implement the ecosystem management policy and process."[1] In this chapter, I explore ecosystem protection in federal law and international agreements. I look at the U.S. Forest Service's effort to shift to an ecosystem approach for all agency programs and their plan (along with the Bureau of Land Management) to make ecosystem protection the standard for federal land management in the Pacific Northwest's Interior Columbia Basin. I go on to explore the World Heritage Convention, the Man and the Biosphere Programme, and the UN Convention on Biological Diversity and address the issue of national sovereignty as it relates to these international pacts.

Legal scholars agree that protecting ecosystems is not a fundamental component of federal environmental or natural resource law. The White House Interagency Ecosystem Management Task Force concludes that "no single federal statute contains an explicit, overarching national mandate to take an ecosystem approach to management, and Congress has never declared that a particular federal agency has the ecosystem approach as its sole, or even, primary mission."[2] University of Utah law professor Robert Keiter, a strong advocate of ecosystem management, concurs, saying that "the existing legal system governing public lands and resources does not expressly endorse the concept of ecosystem management" while Richard Haeuber observes that "there is no existing or proposed statutory basis for EM" [ecosystem management].[3] What the White House task force, Keiter, and others do see in federal law, however, are several statutes that can be read to mandate or permit ecosystem management under particular circumstances.[4] Existing law offers opportunities for new paradigmists to advance their agenda, but thus far Congress refuses to make ecosystem protection the centerpiece of federal environmental or land use policy.[5]

In the 1990s Congress rejected multiple opportunities to embrace the generic protection of ecosystems. On November 19, 1993, President Bill Clinton formally transmitted the UN Convention on Biological Diversity to the U.S. Senate for its approval.[6] The convention calls for government protection of ecosystems for their own sake as a component of biological diversity.[7] The Senate refuses to consent. Earlier, the 102nd Congress did not enact The National Biological Diversity Conservation and Environmental Research Act, which highlights the failure of existing law to emphasize ecosystem management, seeks to protect and manage ecosystems, and attempts to make all activities of all federal agencies consistent with a national goal of protecting biological diversity, which it defines as including ecosystems.[8] During the 104th Congress the Senate took no action on the Ecosystem Management Act of 1995, which would make it "the policy of the Federal Government to carry out ecosystem management with respect to public lands."[9] Congress not only declined to enact legislation creating the National Biological Service within the Department of the Interior—whose purpose it was to gather data to support ecosystem management—but also killed the agency when it was subsequently established administratively by Secretary of the Interior Bruce Babbitt.[10] Many members of Congress oppose a 1997 agency proposal to make ecosystem protection the fundamental goal of federal land managers in the Pacific Northwest.[11] Congress has yet to board the new paradigm bandwagon.

New paradigmists face another difficulty with respect to the law. Boston College of Law professor Zygmunt Plater writes that much environmental law enacted in the last three decades accepts as ecological truth the presumption that "nature has developed a richly diverse and interacting natural equilibrium" that may be "jostled out of balance" by human activities.[12] Law professors Fred Bosselman and A. Dan Tarlock agree that "in the rush to implement Leopold's dictum to 'think like a mountain' in the heady days of the rise of environmentalism," lawyers unquestioningly incorporated the equilibrium paradigm into federal law.[13] They point out, however, that the balance-of-nature view of the world has not withstood close scrutiny by scientists so that "current environmental law . . . rests on a simple ecological paradigm which the science has now rejected."[14] The scientific basis for laws that ecosystem management proponents enlist to support their cause no longer exists.

During what Bosselman and Tarlock call "the heady days of the rise of environmentalism," members of Congress not only missed the debate within ecology calling into question the balance, harmony, nature-knows-best view of the world but also indicated a poor grasp of the subject in general. For example, in the Wild Free-Roaming Horses and Burros Act of 1971, Congress stands the scientific understanding of exotic species on its head. Scientists consider exotic species those whose occurrence in an area is due to human actions as opposed to arrival via natural process. Classic examples of exotic species are the horses and burros that Europeans brought to North America. Congress, however, defines these exotic species (wild horses and burros) "as an integral part of the natural system," then goes on to direct that their management should be done "in a manner that is designed to achieve and maintain a thriving natural ecological balance on the public lands."[15] One should not be surprised at such language. If members and their staff read any books on the environment, they were likely to be Rachel Carson's 1962 popular *Silent Spring* and Eugene Odum's college text *Fundamentals of Ecology,* both of which championed balance and equilibria. In policy debates, environmental activists accentuated the balance of nature and ecosystems as a means of securing scientific justification for what Bosselman and Tarlock call their "neo-pagan theories that nature . . . was divine."[16] Congress responded. Plater concludes that Congress passed thirty-four major environmental laws in the three years after enactment of the National Environmental Policy Act (NEPA) of 1969 and another twenty during the Carter administration.[17] Laws such as NEPA, the Endangered Species Act (ESA), and Section 404 of the Clean Water Act (CWA) are rooted in the belief that nature, undisturbed by humans, is a place of harmony, balance, and equilibrium.[18]

With respect to science and law, advocates of ecosystem protection try to have it both ways. Even though the old equilibrium model in ecology is essentially contravened by the nonequilibrium model, they claim that both perspectives offer scientific legitimacy for their policy prescriptions.[19] After describing the equilibrium-to-nonequilibrium paradigm shift in ecology that I examine in chapter 6, University of Georgia ecology professor Judy Meyer writes that "if we follow the non-equilibrium paradigm . . . we need to preserve the ecosystem . . . and the process that has given rise to that interacting set of species, so the assemblage can continue to change in response to environmental change."[20] Conservation biologist and activist Reed Noss agrees. He says that "the new ecological paradigm, described in Professor Meyer's article, suggests that . . . conservation must be extended to entire landscapes and regional ecosystems."[21] Many ecologists share Noss and Meyers's perspective.[22] If, however, the underlying science changes course 180 degrees and one's policy agenda does not, then it is difficult to reasonably claim a close linkage between the two.

Federal Laws and Ecosystem Management

Lacking any overarching legal authority to protect ecosystems, researchers must sift through numerous statutes to find a legal basis to do so. The White House Interagency Ecosystem Management Task Force finds congressional mandates or permission to protect ecosystems in many places in the law. "The federal government," it writes, "currently has significant statutory authority available to take an ecosystem approach to federal activities." It casts a wide net and finds statutory support for ecosystem management in "the authorities to protect the environment, and the authorities to ensure sustainable economies."[23] Thus, the task force hauls in over two dozen laws from the obvious, such as NEPA and ESA, to the obscure (from an ecosystem protection perspective), such as the Intermodal Surface Transportation Efficiency Act.[24]

The White House group categorizes these authorizes as "laws that apply to all ecosystems," "laws that govern management of federal lands," and "laws that apply to congressionally designated ecosystems."[25] In their view, federal laws that apply to all ecosystems include NEPA; ESA; CWA; Federal Insecticide, Fungicide, and Rodenticide Act; Clean Air Act; Comprehensive Environmental Response, Compensation, and Liability Act; Migratory Bird Treaty Act; and Fish and Wildlife Coordination Act. They err because geographically none of the laws they refer to apply to all ecosystems. The geographic reach of NEPA, for example, extends only to ecosystems impacted by a particular federal action, the ESA touches only ecosystems associated with a threatened or endangered species, and Section 404 of CWA reaches only wetland ecosystems, which the FWS estimates cover about 5 percent of the conterminous states.[26] Despite careful searching, analysts can point to no federal law covering all ecosystems, witness the congressional finding in the proposed National Biological Diversity Conservation and Environmental Research Act that "existing laws and programs relevant to the loss of biological diversity" [defined in the bill to include ecosystems] contain "gaps in geographic . . . coverage."[27]

Legal scholars regularly plumb NEPA for its relevance to the management and protection of ecosystems. Writing for American Bar Association natural resource specialists, Rebecca Thompson notes that some reviewers and litigants try to invoke its purpose of declaring "a national policy which will encourage productive and enjoyable harmony between man and his environment; . . . promote efforts which will prevent or eliminate damage to the environment; . . . [and] enrich the understanding of ecological systems" and implementing regulations requiring the consideration of "effects on natural resources and on components, structures, and functioning of affected ecosystems" in the defense of ecosystems.[28] James McElfish of the Environmental Law Institute argues for an expansive reading of NEPA that goes "far beyond . . . simple environmental impact assessment requirements."[29] "Reading NEPA," he says, "makes clear that it applies to policies, not just projects, that it can affect the shape of legislation, that it can also shape programs. And that it is not limited to federal actions that affect U.S. lands and waters."[30] Dinah Bear, general counsel for the President's Council on Environmental Quality (the executive branch organization that guides NEPA's implementation) also sees NEPA as "broad and prescient."[31] She says that the act "is this country's charter for environmental policy" and reaches "every creature we know about today and that we are going to discover." Most observers and the courts, however, find NEPA far less sweeping.[32]

After nearly three decades of litigation, legal experts generally concur that NEPA imposes procedural but not substantive requirements on federal agencies. Law professor Gary Meyers writes that "the Supreme Court has considered NEPA on 12 different occasions . . . [and] has been consistent on one point: NEPA is essentially a procedural statute."[33] The courts are unable to find any concrete land or resource management requirements in NEPA's flowery language calling for harmony between people and nature. NEPA only requires agencies to consider environmental factors in their decision making; it does not compel them to act on their findings. Congress, in enacting NEPA, in no way requires federal agencies to protect the health, integrity, or sustainability of ecosystems on or off federal lands. It does not endorse ecosystem management in NEPA or provide a means whereby federal land management agencies can reach beyond their boundaries to influence actions on adjacent lands in the name of ecosystem management.[34]

Supporters of the new paradigm usually link NEPA with other statutes, hoping to produce a legal synergy for ecosystem protection. The Forest Service reports that between 1970 and 1989 it averaged less than five significant NEPA-related lawsuits per year but by 1995 were facing eleven major new NEPA suits annually as advocates of the new paradigm sought to bend the law to suit their policy agenda of protecting ecosystems over large regions.[35] Litigants often assert that the Forest Service fails to comply with provisions of NEPA when doing planning under the National Forest Management Act (NFMA).[36] "The nature of litigation," say government land managers, has "shifted from individual development projects (like timber sales or grazing allotments) to land use decisions and management over large geographic areas."[37] New paradigmists' use of NEPA with other statutes can lead to courts requiring protection of ecosystems covering large expanses of land. In the case of the endangered northern spotted owl—*Seattle Audubon Society v. Lyons*—federal district court judge William Dywer invoked sections of NEPA, NFMA, and ESA in concluding that "given the current condition of the forest, there is no way the agencies could comply with the environmental laws without planning on an ecosystem basis."[38] His decision covers land in Washington, Oregon, and California.

Secretary of the Interior Babbitt describes the ESA as "an extraordinary piece of legislation because it allows the Federal Government to preserve, maintain, and foster the recovery of endangered species wherever they occur, without regard to geography, location, or land ownership. Here is a law of great reach and power."[39] Most students of the act would agree with the secretary's characterization, and new paradigmists look to its provisions for support for ecosystem management.[40] They do not have to look far because Congress speaks directly to a substantive requirement to protect ecosystems in the act. It says that one of the act's purposes is "to provide a means whereby the ecosystems upon which endangered species and threatened species depend may be conserved."[41] Researchers' efforts to find such ecosystems have set off firestorms across the country, as with the multiyear, multiscientist, multimillion-dollar project to identify lands essential to preservation of the northern spotted owl that yielded a multiplicity of maps showing different ecosystems supposedly critical for the owl's survival.[42]

Congress makes it clear, however, that the ESA does not protect ecosystems for ecosystems' sake. The act's implementing provisions require that only habitat "critical" for the survival of a listed threatened or endangered species be protected.[43] Congress does not contemplate that the ESA will yield a comprehensive set of ecosystems blanketing the nation and does not require any coherent ecological or geographic nexus among ecosystems devised pursuant to the statute. Consequently, these ecosystems are simply a collection of disjoint areas that researchers devise using totally different criteria from one ecosystem to the next.

Champions of ecosystem management note the infirmities of the ESA. "With respect to ecosystem protection," writes environmental attorney Jason Patlis, "there is no sweeping mandate in the ESA, nor are there any sweeping judicial decisions."[44] He notes that "although ecosystem protection has been explicitly mentioned several times in the legislative history of the ESA," what he calls the "legislative consciousness" did not find its way "into regulatory obligation or judicial interpretation." Alyson Flournoy of the University of Florida Law School says that "the shortcomings of our statutory framework are understandable, in that the ESA and our public land management laws were not designed with ecosystem preservation . . . in mind,"[45] and Tulane Law School's Oliver Houck believes that "one of the more rational conclusions to emerge from America's experience with the Endangered Species Act is that we need to manage ecosystems and protect biological diversity on a scale larger than individual species on the brink of doom."[46] He wants to "eliminate the middleman of endangered species" and proceed straight to the protection of large ecosystems. As we see in chapter 9, new paradigmists propose an Endangered Ecosystem Act to overlay the ESA.

From start (listing of a species as endangered or threatened) to finish (delimiting "critical habitat," e.g., ecosystems), Congress provides for only a highly subjective approach to ecosystem protection in the act. Air Force Judge Advocate John Kunich concludes that "the ESA has no objective, measurable standards for determining when a species is 'endangered,' or 'threatened' or even whether it qualifies as a 'species' or other taxon protectable under the Act. Likewise, 'critical habitat' is vaguely defined."[47] In her analysis of the Fish and Wildlife Service's implementation of the act, wildlife biologist Andrea Easter-Pilcher finds that it uses "a plethora of undefined or loosely defined biological terms, nonbiological terms, phrases, and lengthy descriptions" and goes on the say that "the listings lacked a large amount of biological information regarding species selected to be listed as endangered or threatened."[48] "There is," she writes, "little evidence of objective standards against which decisions for selecting species to list as endangered or threatened can be measured." Using Mark Sagoff's standard that "a decision may be rational if it can be tested or 'verified' against criteria or data determined independently," she concludes that the lack of standards in the listing process means that "ultimate decisions may be less than rational."[49] It is out of this highly subjective, sometimes less than rational, and often data poor process that analysts devise ecosystems to protect under provisions of the ESA.[50] Once these ecosystem are designated, Congress declares that protection of their habitat value to a listed species is paramount when determining land use regardless of who owns the land or the costs involved. Both NEPA and the ESA apply to private and public lands alike, but other laws speak directly to the management of federal lands.

Law and the Federal Lands

The White House task force and others conclude that laws governing the management of federal lands provide an opening for the new paradigm without either endorsing it or requiring it. Keiter, who writes extensively on law and ecosystem management on federal lands, concludes that "it is possible . . . only to derive a fragmentary and incomplete ecosystem management imperative from the law."[51] Congress, for example, does not explicitly direct the National Park Service to protect ecosystems in our national parks and monuments, but Keiter finds a backdoor to ecosystem management in existing law. He cites the "unimpaired clause" in the National Park Service's Organic Act's declaring the purpose of the parks "to conserve the scenery and the natural and historic objects and wildlife therein and to provide for the enjoyment of the same in such manner and by such means as will leave them unimpaired for the enjoyment of future generations" as constituting "a clear substantive standard requiring the Park Service to protect the natural integrity of the landscape."[52]

There are difficulties with Keiter's interpretation of this language. Congress uses the word *conserve,* not *preserve,* in the Organic Act. It understood the distinction between the two terms when it passed the act in 1916, for conservationist Gifford Pinchot and preservationist John Muir had been outspoken and influential in the development of federal natural resource policy for many years. It concluded that conservation was more in keeping with its intent that the national parks be places of public use and enjoyment rather than nature reserves placed off limits to human use. More specifically, neither scientists nor land managers can give substantive meaning to the idea of "unimpaired" and, as I show in chapter 6, they do no better with the notion of "integrity." Writing for the Conservation Foundation in 1969, ecologist Frank Darling and geographer Noel Eichhorn labeled the idea of maintaining parks unimpaired for future generations as "rhetoric."[53] The National Park Service tells its managers that "over the years, legislative and administrative actions have been taken that have brought some measure of change to . . . components of our national parks . . . yet they are not necessarily deemed to have impaired resources for the enjoyment of future generations."[54] It goes on to instruct park officials that "whether an individual action is or is not an 'impairment' is a management decision" based on their evaluation of "such factors as the spatial and temporal extent of the impacts, the resources being impacted . . . the relation of the impacted resources to other park resources, and the cumulative as well as individual effects." The National Park Service gives individual managers a great deal of discretion in judging impairment.

New paradigmists seize on managerial discretion in a far broader context in their effort to justify ecosystem-based management of the federal lands. When Congress enacts laws that guide how agencies are to manage lands in their charge, it often provides little specific guidance about what it means in these statutes. It implicitly grants the agencies administrative flexibility in the interpretation of the law. The courts defer to the land-managing agencies' reading of their management statutes.[55] The Supreme Court, for example, in *Clark v. the Community for Creative Non Violence,* writes that judges do not have "the authority or . . . the competence to judge how much protection of park lands is wise and how that level of conservation is to be attained."[56] While Vice President Gore issues a top-down directive for "agencies to interpret their existing authorities as broadly as possible to implement the ecosystem management policy and process," new paradigmists within the agencies seek to change agency culture so as to impact legal interpretations from the bottom up. For example, the activist group the Association of Forest Service Employees for Environmental Ethics (AFSEEE) believes that "the land is public trust, to be passed on with reverence from generation to generation." Consequently, they proclaim that "our mission is to forge a socially responsible value system for the Forest

Service."[57] For the AFSEEE that means increasing protection of nature by placing more public land off limits to most human uses and greatly restricting human activity where our presence is permitted. They argue that "the Forest Service and other public agencies must follow the footsteps of Aldo Leopold . . . and become leaders in the quest for a new resource ethic." Their response to President Clinton's call for an ecosystem-based management plan for federal lands in the Pacific Northwest is a proposal that the interagency planning team directing the project characterizes as not meeting the "need to support the economic and/or social needs of people, cultures and communities, and to support predictable and sustainable levels of goods and services from National Forest System and BLM-administered lands."[58]

Congress assigns to each of the four principal federal land management agencies—Bureau of Land Management, Forest Service, Fish and Wildlife Service, and National Park Service—a general direction for land management that they can be modify on a unit-by-unit basis. For example, they generally forbid timber harvesting, mining, and oil and gas production in units of the national park system, but legislation creating the Big Cypress National Preserve in Florida provides that oil can be produced therefrom. Congress tells the National Park Service and the Fish and Wildlife Service to manage lands chiefly for conservation purposes so that nonrecreational economic use of the land is normally forbidden. Conversely, they direct the Bureau of Land Management (BLM) and the Forest Service (FS) to manage lands to accommodate many different uses so as to improve human well-being. Except for the BLM and FS lands that Congress chooses to place in the national wilderness system, it expects that the agencies will provide for multiple uses and sustained yields from the public lands.

Three principal statutes cover BLM and FS lands: the Multiple-Use and Sustained Yield Act (MUSYA), the National Forest Management Act (NFMA), and the Federal Land Policy and Management Act (FLPMA).[59] MUSYA (enacted in 1960) directs the Forest Service to "develop and administer the renewable surface resources of the national forests for multiple use and sustained yield of the several products and services obtained therefrom."[60] It goes on to define multiple use as "the management of all the various renewable surface resources . . . so that they are utilized in the combination that will best meet the needs of the American people; making judicious use of the land for some or all of these resources." Congress recognizes in the definition "that some land will be used for less than all resources," and multiple use must be accomplished via "harmonious and coordinated management . . . without impairment of the productivity of the land."[61] "Sustained yield," says Congress, "means the achievement and maintenance in perpetuity of a high-level annual or regular periodic output of the various renewable resources of the national forests without impairment of the productivity of the land."[62] In FLPMA (passed in 1976), Congress borrows heavily from MUSYA in defining multiple use and sustained yield. It makes no substantive change in the definition of sustained yield. It adds consideration of the needs of future generations in its definition of multiple use. Additionally, in FLPMA Congress alters the MUSYA language regarding "impairment" in two ways. MUSYA refers only to "impairment," while FLPMA qualifies this by referring to "permanent impairment." MUSYA speaks to "impairment of the productivity of the land," while FLPMA expands the coverage to include "the quality of the environment," although it offers no guidance on what that may be.[63] In these acts, Congress consistently makes clear that no one of the several potential multiple uses of the public lands should take precedence over all others across the landscape.

Analysts see in these and related statutes some promising means to advance the new paradigm. Congress places significant planning requirements on the FS in NFMA. One such requirement pertains to biological diversity that professors Keiter and Houck read as offering support for ecosystem management.[64] Keiter notes other ecosystem-management-friendly aspects of NFMA, including requirements for interdisciplinary planning and management, protection for riparian zones and steep slopes, and public participation requirements. He points out, however, that Congress also directs the FS to plan at the geographic level of an individual forest, and this frustrates the ecosystem approach. Nonetheless, he concludes that "the NFMA reflects clear congressional concern over safeguarding the ecological integrity of national forests, which should afford the Forest Service a firm statutory basis for its current ecosystem management initiatives."[65]

With FLPMA, Congress gives the BLM "an organic mandate that can be construed to impose some ecosystem management responsibilities," in Keiter's view.[66] He sees glimmers of ecosystem management in the act's policy admonition to BLM to "protect the quality of scientific, scenic, historical, ecological, environmental, air and atmospheric, water resource, and archeological values" as well as its definition of multiple use that calls for the agency to take into account the long-term needs of future generations and to manage lands so as not to permanently impair the productivity of the land and the quality of the environment. Lewis and Clark College law professor Michael Blumm takes a more negative view of the ecosystem protection possibilities in statutory definitions of multiple use and sustained yield. He argues that because of the discretion granted land managers this language is used to subsidize what he calls the "Lords of Yesterday": miners, ranchers, loggers, and others who he believes cause "environmental desecration."[67] Absent new statutes making ecosystem protection the law of the land or sweeping Supreme Court reinterpretations of existing law, if land-managing agencies are to implement the new paradigm, they must utilize the very agency discretion that Blumm and others fault to accomplish the ends that he and other new paradigmists desire. At the end of the 1990s, the executive branch is producing creative reinterpretations of existing law and arbitrarily redefining the concept of sustained yield in a manner inconsistent with decades-long understanding of the term.

It is more difficult for observers to find a rationale for the new paradigm in MUSYA. The courts hold that the act requires the FS only to consider various potential uses of the land prior to making land management decisions so that even if language in MUSYA could be read to enumerate protection of ecosystem health or integrity as one of several potential multiple uses of a tract, it does not take precedence over other legitimate land uses.[68] Enter the White House Interagency Ecosystem Management Task Force, which writes that "the multiple-use and sustained yield directives" in MUSYA, NFMA, and FLPMA "provide an opportunity for promoting sustainable economies and communities," which it claims are dependent on healthy ecosystems, which in turn requires us to sustain and restore natural ecosystems.[69] In other words, if the agencies are to successfully comply with congressional multiple-use and sustained-yield requirements, land managers must first make sure that healthy natural ecosystems are widespread across the landscape.[70] Presto! The statutory bugaboos of multiple use and sustained yield that many new paradigmists believe significantly hinder their agenda now become friends of the ecosystem through artful reinvention.

Much more brazen is the BLM's and FS's attempt to overturn clear congressional direction on the meaning of sustainability as it applies to the management of federal lands. As part of the Clinton administration's effort to make ecosystem management the centerpiece of national environmental policy, the president directed that the FS and the BLM prepare an ecosystem-based management plan for federal lands in the Pacific Northwest. This project became known as the Interior Columbia Basin Ecosystem Management Project (ICBEMP), which I review more closely in the following section. In moving forward with their planning, BLM and FS officials ignore congressional intent by significantly changing the congressional meaning of sustainability in the draft environmental impact statements for the project. They offer two definitions of sustainability: "Meeting the needs of the present without compromising the abilities of future generations to meet their needs; emphasizing and maintaining underlying ecological processes that ensure long-term productivity of goods, services, and values without impairing productivity of the land;" and, "in commodity production, [sustainability] refers to the yield of a natural resource that can be produced continually at a given intensity of management."[71] Their incorporation of language calling for "emphasizing and maintaining underlying ecological processes" has no obvious basis in law, and the

"over time" means and is vague about the central ideas of composition, function, and structure. According to the FS, "Ecosystem composition includes the plant and animal species that make up an ecosystem," but that is not illuminating.[81] Scientists have yet to completely inventory the living things for any part of the earth's landscape, and such inventories are a practical impossibility for the lands of the national forest system, so what is to be included within the "composition of an ecosystem?"[82] The agency does not tell us, but it does suggest that the questionable notion of keystone species does not offer an answer in noting that "there is diminishing scientific support for focusing on individual species as indicators of the welfare of a group of associated species."[83]

The authors write that "ecosystem function includes processes and the relationships among processes, such as nutrient cycling, in a system."[84] What other processes in addition to nutrient cycling does the FS contemplate as components of ecosystem function? Hydrologic flux and storage? Biological productivity and decomposition?[85] What other features may land managers include in the notion of ecosystem function in addition to processes and their interrelationships? How are assessments of these processes and their interrelationships to be made? Do all ecosystems have the same functions? Are all functions of equal importance in all ecosystems and over the same time period? If not, how will the land managers assess the differences for the purposes of land management decision making? No one answers these questions.

Ecosystem structure, the FS says, "includes the distribution and pattern of ecosystem elements such as forest openings and riparian corridors at a landscape scale, and the amount and arrangement of special habitat features such as seeps, snags, and down woody material at smaller scales."[86] The idea of ecosystem structure in this language is unclear. It is also circular since each of the so-called ecosystem elements that the agency identifies can themselves be thought of as ecosystems, according to the rule's authors, who tell us that "ecosystems exist at multiple spatial scales and are infinite in number . . . the span of ecosystems can range from the microscopic world of life occurring on a fallen tree trunk to the range of a migratory bird that travels annually from the tropics to the arctic."[87] The FS offers no guidance on how it will address the issue of geographically nested ecosystems in its efforts to manage ecosystem structure or composition or function; rather, it simply directs that planners "should address those ecosystems of most relevance to forest plan decisionmaking."[88]

In the end, the agency concedes that "the terms 'sustainable,' 'restoration,' 'maintenance,' or 'deteriorated ecosystem' are all subject to varying and evolving interpretations" and acknowledges that they have "no expectation that there will ever be precise and universally accepted understanding or measure of what sustainable ecosystems are and the actions appropriate to maintain or restore them."[89] Further, it points out that "there is nothing in the proposed rule that establishes a concrete standard regarding ecosystem sustainability" so that land managers would determine the "conditions indicative of sustainability" on a plan-by-plan basis.[90]

To summarize, the FS seeks to oversee the national forest system in order to sustain undefined conditions on undefined landscape units that exist in limitless numbers in undefined locations and that are dynamic and constantly changing over time and space in unclear ways. Moreover, there are no standards by which land managers are to measure the undefined landscape conditions that the rule is intended to guarantee. This is an unintelligible basis for the management of the national forest system, but it is indicative of new paradigm thinking, for now "the *principal goal* of managing the National Forest System is to *maintain or restore ecosystems*" while "meeting people's needs and desires" must be done "*within the capacity of natural systems*" [emphasis added]. There is no suggestion of balance here, as each statement subordinates human activities to ecosystem protection. Resources for the Future forestry scholar Roger Sedjo concludes that ecosystem-based management of the national forests is rigid and arbitrary and "makes one objective dominant and essentially impervious to trade-offs."[91]

With the ICBEMP, FS and BLM managers try to apply the new paradigm to specific federal lands in the Pacific Northwest. The ICBEMP covers 144 million acres, mostly in Oregon, Idaho, and Washington. The FS and BLM oversee 72 million of these acres using 74 separate land and natural resource management plans that they would change to conform with ICBEMP direction. Government planners explain their ideas for ecosystem management in the region in a series of documents—including draft environmental impact statements (DEISs)—covering several thousand pages. However, on none of these pages do the planners describe a statutory basis for the shift to ecosystem-based land management. They cite a presidential directive, memos from agency heads, and court decisions related to the ESA but fail to tie their action to existing law in meaningful ways.[92] At a May 1997 joint hearing on the proposals, Congresswoman Helen Chenoweth of Idaho concludes that "it is my strong feeling that this project has far exceeded the statutory bounds."[93]

The agencies say that change in land management is required by two distinct needs: "There is a need to restore and maintain forest, rangeland, and aquatic and riparian ecosystem health and integrity and to identify desired ranges of future landscape conditions for landscape structure, composition, succession, and disturbances; for hydrologic processes and functions; and for aquatic habitat structure and complexity," and "there is a need to contribute to the vitality and resiliency of human communities and to provide for human uses and values of natural resources *consistent with maintaining healthy, diverse ecosystems*" [emphasis added].[94] "In essence," the agencies write, "ecosystems must be healthy, diverse, and productive to meet the needs of society today as well as those needs of future generations." The management priorities flowing from these needs are restoring deteriorated ecosystems, maintaining ecosystem health and diversity, and meeting people's needs "within the limitations of ecological integrity, health, and diversity."[95]

Planners evaluate seven alternatives in the DEISs. The authors declare that the "underlying philosophy" of the first two alternatives "is one of multiple-use . . . to produce goods and services in helping meet the needs of the American people . . . current plans emphasize sustained yields . . . while maintaining site productivity and environmental quality."[96] These alternatives carry forward traditional understandings of multiple use and sustained yield and would continue to require officials to place enhancement of human well-being at the center of land management. Agency managers, however, reject these alternatives out of hand because they are inconsistent with the new paradigm even though they concede that under present policies "many ecological conditions and trends have improved in the past two decades" and that "the current conditions of rangelands appears to be the best since the turn of the century."[97]

The remaining alternatives reflect the new paradigmists' assumption that the perfect landscape results from entirely natural processes; Mother Nature is the ideal model for land managers. According to agency planners, exemplary regional landscapes in the Interior Columbia River Basin developed prior to the arrival of Europeans, and landscape change resulting from European settlement represents a retreat from the ideal. They use terms such as "loss of ecological integrity" and "decline in landscape health" to describe that retreat. The assumption requires that managers seek to emulate nature and move the landscape toward a condition that might exist had Europeans not occupied the region. Senator Slade Gorton of Washington observes that the project represents the view that "nature is the ideal situation and that everything human beings do to change the state of nature is bad."[98]

The agencies' selection of the preferred alternative in the DEISs illustrates this point. With it, managers seek to create a landscape "designed to resemble natural vegetative mosaics historically created by disturbance processes."[99] Specifically, the officials call for restoration as the management emphasis for the overwhelming majority of the 72 million acres directly impacted by the project.[100] "Restoration," they say, "generally refers to the process of compensating for disturbances on an ecosystem so that the system can resume acting, or continuing to act, as if those disturbances were absent. Ecological restoration includes well-laid plans and is targeted toward a specific historical ecosystem model."[101] Tom Mills, director of the FS's Pacific Northwest Experiment Station, tells Congress that "nature has been managing these systems for a long time . . . and that the thought is if we could design management regimes that . . .

manage these systems as nature did, we would more likely have sustainable systems."[102] Project scientists acknowledge that a new set of values, not scientific findings, stand behind the assumption that an unfettered nature produces ideal landscapes. They find that "the change to ecosystem management incorporates a struggle about changing values" and note that "many of the scientific concepts elevated to the status of principles are in fact judgments reflecting the values of the scientists who define the principles."[103]

The planners' belief in the inherent wisdom of nature permeates the project and preordains analytic outcomes. The Science Integration Team (SIT), whose work underpins much of the analysis and evaluations in the DEISs, believes that "the purpose of ecosystem management is to maintain the integrity of ecosystems over time and space."[104] "Integrity," claims the SIT, "is the quality or state of being complete, a sense of wholeness," which it equates with naturalness so that the closer the landscape is to conditions that might be present today had Europeans not settled in the region, the greater its ecological integrity.[105] Consider this: "Measures [of integrity] were developed by the Science Integration Team using direct and indirect variables to indicate how much various elements have departed from historical conditions. For purposes of this analysis, 'high departure' signifies that an area is significantly different than the condition expected for its biophysical environment, and roughly indicates 'low integrity,' "[106] and "the current composite integrity . . . was rated by alternative as having high, moderate, or low trends compared to historical integrity."[107] Using these assumptions, researchers' conclusions, such as "forest ecosystem[s] . . . with the highest integrity ratings were those that are largely unroaded" and "range ecosystems with the highest overall integrity ratings were those upland shrublands that are less developed, less roaded, and more remote," are not surprising.[108]

The agencies' adoption of the nature-knows-best philosophy—and the lengths to which they must go to make it seem scientific—is exemplified by their using what they term the Historic Range of Variation (HRV) to evaluate the alternatives. Researchers define the HRV as "the natural fluctuations of components of healthy ecosystems over time [that] in this EIS, refers to the range of conditions and processes that are likely to have occurred prior to settlement of the project area by people of European descent (approximately the mid-1800s)."[109] Using the HRV idea, they model the landscape as it might appear 100 years from now had Mother Nature been free from the effects of Europeans and their progeny and then compare that picture against their vision of regional landscapes after 100 years of management under each of the alternatives. They use the midpoint of the HRV to generate their yardstick and rank alternatives according to their degree of fidelity to the HRV-generated picture.

Agency researchers show little appreciation of the overall weakness of long-term modeling. Even the most enthusiastic government modelers tremble at making predictions of conditions a century or more hence for the purposes of guiding today's policymaking. For example, the Department of Energy (DOE) makes no effort to present data past the year 2030 in the *National Energy Strategy* (NES) because it understands the inability to make worthwhile long-term appraisals of all the factors that influence energy production and consumption.[110] It notes that even a forty-year time horizon is "unusually distant." The number of elements influencing the evolution of landscapes is far greater than those associated with energy, but where DOE and its analysts fear to tread, personnel from the FS and BLM rush in.

ICBEMP researchers' "simulations [of the HRV landscapes] used the historical (circa 1850s) map as a starting input layer," and they make modeling runs out 400 years.[111] In chapter 4, I show that reconstruction of past landscapes is a difficult proposition, so ICBEMP analysts can produce hundreds of different vegetation maps from 1850 for the project area. Each map is nothing more than its author's conjecture about the generalized distributions of a handful of the region's thousands of plant species.[112] Different initial conditions result in different model trajectories and end points, and by extending models farther and farther into the future, analysts magnify the impact of even the smallest differences in model starting points. At the point in time (the year 2100) where they use the modeling runs to compare HRV landscapes against those of the DEIS alternatives, they stretch the model 250 years from its 1850 starting point while pictures of landscapes emanating from the alternatives require 100-year projections. There is little reason anyone should have confidence in the ability of such efforts to intelligently inform public policymaking.

The difficulty with people using the HRV notion as a guide for land use management decisions, however, reaches beyond unreliable model results to the heart of the concept itself. Geographers and other observers understand that landscapes contain countless components that constantly change across time and space at different rates and in response to different processes.[113] They realize that change in one landscape component may or may not have any spatial or temporal relationship with change in another and that even when relationships exist, their strength varies with time and place. They know that detailed data regarding historic landscape change is very limited. Thus, the government officials' belief that landscapes have a historic range of variation that these officials can calculate for public policy purposes is suspect from the outset.

However, for the sake of argument, suppose that scientists know everything about every living thing that occupied the Interior Columbia River Basin from 1500 to 1850, the pre-European period used to determine the HRV.[114] Suppose that they know, for example, when every rabbit was born, what it ate every day hour by hour, where it moved, when it died, and so on. Suppose that they know, hour by hour, the status of every conceivable physical measure applicable to every square foot of land and water within the basin. What possible calculus could scholars employ to combine these data into a single meaningful HRV with a high, a low, and a midpoint? How will experts meld numbers portraying the comings and goings of rabbits over 350 years with those on soil erosion? How will they combine rainfall data with variations in wolf populations? Would the midpoint of the HRV be the average number of rabbits, plus the average number of acres where Douglas fir were the principal tree species, plus the average amount of rainfall, and so on? How does one sum the average of rabbits with the average of rainfall? Of what possible value to anyone is the resulting number, even if some worthwhile calculus should appear to allow its calculation? At a time when ecologists, geographers, and other students of the environment tell us that the landscape is in constant flux and that change is often unpredictable and certainly multidirectional and without balance or equilibrium, the agencies' notion of the existence of a meaningful midpoint in landscape evolution or ecosystem fluctuation is remarkably unscientific.

American law does not offer a substantial legal underpinning for ecosystem management, and new paradigmist efforts to turn it to that purpose are intriguing to watch as they result in some inventive regulatory efforts. However, what of international pacts of which this nation is a part? Can advocates of ecosystem management fare better using such instruments? I address these questions in the next section.

Using International Agreements to Implement Ecosystem Management

Various international agreements relate to the protection of ecosystems. In 1992 the United Nations held the Conference on Environment and Development in Rio de Janeiro (the Rio Earth Summit). In Principle 7 of the 1992 *Rio Declaration on the Environment and Development,* nations agree that "states shall cooperate in a spirit of global partnership to conserve, protect and restore the health and integrity of the Earth's ecosystem."[115] Governments also approved *Agenda 21* at Rio. In this wide ranging blueprint—whose importance the United Nations underscores with its establishment of the UN Commission on Sustainable Development to monitor the document's implementation—states pledge themselves to set the world on the path to a more appropriate relationship with the environment. According to *Agenda 21,* "Urgent and decisive action is needed to conserve and maintain . . . ecosystems . . . [and] effective national action and international cooperation is required for the in situ protection of ecosystems."[116] New paradigmists and others strongly support the UN Convention on Biological Diversity that also emerged at Rio. Geographically, these accords extend to all the territory of signatory nations because there are no particular restrictions on the ecosystems of interest. Framers of other international mechanisms, however, target ecosystems of special interest, for example, the UN Convention Concerning the Protection of the World Cultural and Natural Heritage, the Man and the Biosphere Programme, and pacts treating specific species.[117] Agreements of this kind have a limited spatial reach. By 1996 program officials, in conjunction with the affected national governments, had designed 506 World Heritage sites (twenty in the United States) and 337 Biosphere Reserves (forty-seven in the United States).[118]

New paradigm supporters link international agreements with domestic efforts at ecosystem protection whenever possible. In 1995 backers of ecosystem protection began a concerted effort to employ the World Heritage machinery to prevent development outside Yellowstone National Park, which is both a World Heritage site and a Biosphere Reserve. The specific proposal arousing their interest was for the New World gold mine outside Yellowstone but near its northeast corner. Fourteen environmental organizations—including the National Parks and Conservation Association, Yellowstone Coalition, Sierra Club, Wilderness Society, Friends of the Earth, National Audubon Society, National Wildlife Federation, and Natural Resources Defense Council—wrote the chairman of UN Educational, Scientific and Cultural Organization's (UNESCO) World Heritage Committee calling for that body to investigate the "extremely serious threats presented to the park and its larger ecosystem by the proposed New World gold mine" to determine whether Yellowstone National Park should be placed on the "List of World Heritage in Danger."[119] When one reads the letter, it quickly becomes apparent the mine issue is simply a rallying point to invoke the weight of an international agreement to limit human use of the park and surrounding area. The environmentalists see threats to Yellowstone in everything from park tourism to many economic uses on millions of acres of multiple-use national forestlands outside the park. Responding to the environmentalists' letter, Bernd von Droste, director of the World Heritage Centre, wrote Assistant Secretary of the Interior (DOI) George Frampton to express concern about the issues they raised.[120]

The Clinton administration quickly joined the effort to involve the United Nations. Frampton replied to Droste that "the Committee should be informed that the property [Yellowstone National Park] . . . is in danger" and invited the Committee, along with the World Conservation Union, to inspect the area.[121] UN representatives visited the region in September 1995 and subsequently voted to place Yellowstone on the endangered list at their next meeting in December 1995. In August 1996 the mine owners, faced with an onslaught of negative publicity and open hostility from new paradigmists and others in the administration, abandoned the project. In an agreement reached with President

Clinton in Jackson Hole, Wyoming, on August 12, they gave up their mining claim at the site in return for unidentified public lands valued at $65 million. In addition, they agreed to establish a $22 million fund to clean up damage to the immediate area during earlier mining operations in which they had no part.[122]

The Yellowstone and New World mine project illustrates two important points about international agreements. First, they are of significant propaganda value, as new paradigmists effectively use global designations to intimidate politicians, business interests, and private landowners into forgoing development or other actions that may harm Mother Nature. Few people are willing to stand up to accusations that they are participating in the destruction of something so precious as to be deemed vital to all peoples of the world. On the World Heritage Committee's finding that Yellowstone was endangered, the environmentalists said that "today, the world has affirmed what we have said repeatedly: that Yellowstone is in jeopardy," so "it is our hope that the Committee's findings will help America redouble its efforts to protect this global treasure" because the decision is "a powerful expression of world opinion and concern about the integrity of the park."[123] For ecosystem management advocates, it is a question of good versus evil, and the UN proclamation of an endangered Yellowstone National Park provides additional ammunition for what in their minds are the forces of good. The committee's action aided opponents of the mine by increasing public pressure for action prior to the completion of scientific analyses whose outcome new paradigmists were less able to control.

Ecosystem advocates in the administration opposed the mine long before results of the technical assessments required by NEPA were in. Frampton tells Droste in his letter of June 27, 1995, that Interior Secretary Babbitt said that "placing a giant mine just across the boundary from Yellowstone is a bad idea, pure and simple."[124] Five months later, in a December 1, 1995, letter, Frampton informs Droste about the preparation of a draft environmental impact statement for the project in fulfillment of NEPA.[125] "The purpose of the document," he writes, "is to provide decision makers with relevant information to assist in . . . . making final permit and other decisions." The draft document "will be issued in early 1996" for public comment, so until then "there will not be a factual basis" on which to make decisions. He tells Droste that until professionals analyze public comments, "it will not be possible to evaluate whether all of the studies and analyses that need to be part of such an assessment have been included in the analysis" and that analysts do not expect to issue a final environmental impact statement before the fall 1996 at the earliest. Babbitt characterizes the president's August 12 announcement as bringing "a sense of relief to every American who cares about our oldest and most beloved national park." He says the agreement ends a "great environmental threat" to Yellowstone as well as "wilderness, a Wild and Scenic River . . . threatened and endangered wildlife, watersheds, and groundwater."[126] This is lofty rhetoric, but it is premature, as according to his own assistant secretary, the scientific work intended under U.S. law to provide a sound basis for decision making was not yet completed.

A second lesson of the New World mine is that international agreements such as the World Heritage Convention and the Man and the Biosphere Programme do not put UN bureaucrats in charge of land use in the United States as some new paradigmists may hope and as feared by some people concerned with national sovereignty. At the December 1995 meeting, where it lists Yellowstone as endangered, the World Heritage Committee writes, "During the discussion it was noted that whether the State Party should grant a permit to the mining company or not is entirely a domestic decision of the State Party. It was further stated that there is no wording in the Convention or Operational Guidelines which could lead to an interference in sovereignty."[127] The Department of the Interior's reaction to the listing of Yellowstone as endangered emphasizes that this matter is "a domestic issue" and that the action by the United Nations "does not supersede any U.S. law."[128] The environmentalists who pushed for the UN declaration of endangerment and agree with this interpretation say that "the Committee's decision to add Yellowstone to the list of World Heritage In Danger in no way affects jurisdictional or management responsibility for the park."[129]

Public and congressional concern about these programs is nonetheless justified. In his analysis of the New World mine issue, Cornell professor of government Jeremy Rabkin observes the following:

> In the Yellowstone dispute, environmentalists ridiculed protests over international intervention as "black helicopter arguments"—paranoid ravings about sinister U.N. military squads poised to take control of American national parks. Though some people do seem given to paranoid ravings, one may take the threat to sovereignty quite seriously without being a raving paranoiac.[130]

Government officials muddy the sovereignty waters. Yellowstone superintendent Mike Finley told reporters that as a signatory to the World Heritage treaty, the "U.S. has a statutory responsibility to ensure that Yellowstone, a designated World Heritage Site, is preserved and protected."[131] Frampton fuels people's suspicion by telling Droste that "the United States must assume full responsibility for assuring the integrity of World Heritage values is not compromised by adverse environmental actions taken either internal or external to World Heritage Site boundaries."[132] Here Frampton suggests that while it is the United States that will take action, it will be in part to protect values established by the international community. What might those values be? In "The Vision from

Seville for the 21st Century"—a UNESCO sponsored conference on Biosphere Reserves—participants declare that the reserves "can become theatres for reconciling people and nature" and will be a "means for the people who live and work within and around them to attain a balanced relationship with the natural world."[133] Participants at Seville see an expanding system to embrace more and more ecosystems. They envision spatial expansion of individual reserves so as to influence decision making over larger areas. They believe that the reserves should become an important means of implementing *Agenda 21* and the UN Convention on Biological Diversity. This is an aggressive plan for a program that, as Rabkin writes, "has never actually been presented to participating governments as a formal treaty to be formally ratified [so] it is a drifting cloud of fairy dust" rather than an operation rooted in treaty obligations.[134]

Many elected state and local officials, business interests, landowners, and others look to Congress to rectify what they view as harm caused by international designations.[135] They do not view World Heritage or Biosphere Reserve designations as benign international testimonials. Instead, they see them as costly and intrusive and speak in favor of legislative proposals such as the American Land Sovereignty Protection Act (H.R. 901). With this bill Congress would amend existing U.S. law—particularly the National Historic Preservation Act Amendments of 1980—that guides how the U.S. government proposes international designations for places in the United States.[136] In H.R. 901, Congress finds that these amendments allow international programs to excessively intrude on specific constitutional responsibilities of Congress, improperly impact nonfederal lands, and inappropriately impinge on the sovereignty of the individual states. With H.R. 901, Congress amends the National Historic Preservation Act accordingly.[137] The secretary of the interior "may not nominate any lands owned by the United States for inclusion on the World Heritage list" unless he or she finds that "commercially viable uses" of the land in question, plus land within ten miles of the site's boundary, "will not be adversely affected by inclusion of the lands on the World Heritage List."[138] Further nominations must be specifically authorized by law, just as are the individual units of the national park system. The act prohibits any federal official from nominating lands as Biosphere Reserves and terminates existing designations unless they are individually authorized by law, consist entirely of federal lands, and have a management plan "that specifically ensures that the use of intermixed or adjacent non-Federal property is not limited or restricted as a result of that designation." More generally, Congress prohibits any federal official from setting aside any federal land pursuant to international agreements or treaties without express congressional authorization and prohibits seeking international designation for private lands without the written permission of the landowner.[139] On October 8, 1997, the House of Representatives passed the American Land Sovereignty Protection Act by a vote of 236 to 191.[140] An administration witness threatened a presidential veto should Congress enact the bill.[141]

The UN Convention on Biological Diversity (CBD) is a new paradigmist proposal that is much more far reaching than any other international agreement related to ecosystem management. The CBD "presupposes a new contract between people and nature," says Elizabeth Dowdeswell, executive director of the UN Environmental Programme. Adopting discredited ecological ideas of the past, she tells attendees of the June 20, 1994, session of the Intergovernmental Committee on the Convention of Biological Diversity that "every element in nature, however small, is part of the whole and contributes in its own fashion to the harmony of the whole and to its overall balance . . . every species has a right to survival because its existence is linked to that of the entire community of life on earth."[142] The new contract is part of the struggle to meet what William Snape, legal advisor to Defenders of Wildlife, sees as "perhaps the ultimate challenge posed by environmental law . . . redefining national sovereignty for the twenty-first century."[143]

The CBD emerged from the Earth Summit in Rio de Janeiro; its objectives are "the conservation of biological diversity, the sustainable use of its components and the fair and equitable sharing of the benefits . . . of genetic resources."[144] The framers of the convention define biological diversity "as the variability among living organisms from all sources including, *inter alia* terrestrial, marine and other aquatic ecosystems and the ecological complexes of which they are part: this includes diversity within species, between species and of ecosystems."[145]

The CBD preserves biological diversity by obliging governments to protect ecosystems using all the vague and imprecise language of the new paradigm I discuss in earlier chapters. The preamble asserts "that the fundamental requirement for the conservation of biological diversity is the in-situ conservation of ecosystems and natural habitats and the maintenance and recovery of viable populations of species in their natural surroundings." The convention accomplishes this through several mechanisms. "In accordance with its particular conditions and capabilities," Article 6 directs nations "to develop national strategies, plans, and programmes for the conservation and sustainable use of biological diversity . . . which shall reflect . . . the measures set out in this Convention." Countries must "integrate" the conservation and the sustainable use of biological diversity into "sectoral and cross-sectoral plans, programmes and policies." Article 7 lays the groundwork for coast-to-coast application of the CBD with its requirement that nations identify those components of biological diversity of significance by applying the categories of Annex I. According to Annex I, any ecosystem or habitat exhibiting one of the following characteristics is a legitimate focus for the actions under the convention: "containing high diversity, large numbers of endemic or threatened species or wilderness; required by migratory species; of social, economic, cultural or scientific importance; or, which are representative, unique, or associated with evolutionary or other biological processes." Every part of the United States falls under at least one of these criteria.

In Article 8 the CBD obliges "each Contracting Party, as far as possible and as appropriate," to do a number of things, including the following: "establish a system of protected areas"; "regulate or manage biological resources important for the conservation of biological diversity whether within or outside protected areas"; "promote environmentally sound and sustainable development of areas adjacent to protected areas with a view to furthering protection of these areas"; "rehabilitate and restore degraded ecosystems . . . through the development and implementation of plans or other management strategies"; and, "where a significant adverse effect on biological diversity has been determined pursuant to Article 7, regulate or manage the relevant processes and categories of activities." The CBD carries with it sweeping requirements for national-level land use planning and management to protect ecosystems.

Many observers wonder about the relationship of the CBD with U.S. law and issues of sovereignty. Stanford University law professor John Barton writes that "the core of the treaty lies in its conservation commitments, but these commitments have received relatively little publicity. The commitments are sensitive because they infringe on traditional national sovereignty."[146] Prior to Senate debate on the convention, thirty-five U.S. Senators (all Republicans) write George Mitchell, the Senate majority leader, that "it appears that the treaty may have implications for U.S. domestic law and environmental policies."[147] Convention proponents ensure members of Congress and others that there is no need for worry, that existing U.S. law is quite sufficient to fulfill all obligations that come with its ratification. In his letter transmitting the convention to the Senate, President Clinton writes that "existing programs and authorities are considered sufficient to enable any activities necessary to effectively implement our responsibilities under the Convention."[148] The Senate Foreign Relations Committee notes in its favorable report on ratification that "the Administration has determined that existing law provides sufficient statutory authority for the United States to meet its obligations under the Convention; as a result no new implementing legislation will be necessary."[149] This same point is made in a letter from Assistant Secretary of State Wendy Sherman to Senator Dole and thirty-four other senators. Sherman tells Dole that "no implementing legislation is required—the US meets and surpasses all treaty provisions."[150] But does existing U.S. law really offer a firm basis to implement the convention?[151] No.

Analysts cannot locate in federal law statutes providing for national land use or cross-sectoral planning mechanisms. In its 1996 report to the United Nations on the status of our national implementation of *Agenda 21,* the administration finds that "there are no mandatory national or regional land use planning policies in the United States"; thus, there is no basis to meet the CBD's national planning requirements as they pertain to land use planning.[152] We know that the White House and legal scholars agree that no federal laws require ecosystem protection on federal land (to say nothing of the 70 percent of the nation in non-federal ownership) and that Congress refuses to pass any such statutes. We know that in the 1990s Congress specifically rejected legislation making protection of biological diversity a principal component for U.S. environmental policy.[153] So what then is the statutory basis for fulfillment of Article 8—which Snape calls "the heart of the CBD"—requirements that government "regulate or manage the relevant processes and categories of activities" that are determined to have a "significant adverse impact on biological diversity?"[154] What legal authority exists to generally "rehabilitate and restore degraded ecosystems," as stipulated in Article 8? None. The article requires nations to establish a system of protected areas with surrounding protective buffer zones. Americans place over 5,000 natural areas under federal or state protection through designation as parks, wildlife refuges, wilderness areas, wild and scenic rivers, and so on. Although some may wish for them, federal law does not provide for buffer zones around such places, so the United States does not have a statutory basis "to regulate or manage biological resources . . . outside protected areas" as required by Article 8.

Overall, the conclusion by the administration that our existing statutes are sufficient to meet our obligations if the Senate ratifies the CBD is hard to justify. At a minimum it requires that attorneys, jurists, and others engage in expansive reinterpretations of existing law and that government agencies significantly broaden the scope of implementing regulations to account for the convention's requirements. In either case, Senate ratification opens the door to an extensive period of litigation in which new meaning is sought in old laws and in which judges seek to find the substance in ecosystem jargon that has eluded ecologists and others for decades. Constitutional and environmental law authority Mark Pollot finds that

> this treaty will result in more litigation than all other treaties entered into by the United States combined. This is true . . . [because of] the effects of the treaty on existing state, federal, and local laws . . . statutory and regulatory changes that will flow from implementation of the Convention over time . . . [and] is inevitable as the Convention calls for massive intrusion into not only private activities, but also the functioning of state and local governments.[155]

Thus, the CBD offers advocates of ecosystem management a significant opportunity to impose their policy agenda on the nation via judicial rather than legislative processes.

Legal scholars agree that neither present U.S. law nor international agreements offer firm legal footing for implementing ecosystem management in the United States. Consequently, many new paradigmists advocate changes in the law to make it more consistent with their efforts to, as Caldwell puts it, "revise the ground rules under which society operates" in order to implement their agenda for America. These and other proposals to implement the new paradigm are the subjects of the next chapter.

Notes

1. Al Gore, *Reinventing Environmental Management* (Washington, D.C.: Office of the Vice President, 1993), 14.

2. White House Interagency Ecosystem Management Task Force, *The Ecosystem Approach: Health Ecosystems and Sustainable Economies. Vol. II: Implementation Issues,* 69; Richard Haeuber, "Setting the Environmental Policy Agenda: The Case of Ecosystem Management," *Natural Resources Journal* 36, no. 1 (winter 1996): 3.

3. Robert B. Keiter, "Beyond the Boundary Line: Constructing a Law of Ecosystem Management," *University of Colorado Law Review* 65, no. 2 (1994): 295.

4. See generally Bruce M. Pounder, "Reforming Livestock Grazing on the Public Domain: Ecosystem Management-Based Standards and Guidelines Blaze a New Path for Range Management," *Environmental Law* 27, no. 2 (summer 1997): 513–610; Rebecca W. Thompson, "Ecosystem Management: Great Idea, But What Is It, Will It Work, and Who Will Pay?" *Natural Resources and the Environment* 9, no. 3 (winter 1995): 42–45, 70–72.

5. Robert B. Keiter, "Toward Legitimizing Ecosystem Management on the Public Domain," *Ecological Applications* 6, no. 3 (1996): 727–30.

6. William J. Clinton, "Message from the President of the United States Transmitting the Convention on Biological Diversity, with Annexes, Done at Rio de Janeiro June 5, 1992, and Signed by the United States in New York on June 4, 1993," Senate, 103rd Cong., 1st sess., treaty doc. 103–20.

7. United Nations, *Convention on Biological Diversity,* 1992; see Articles 2, 6, 8, and 14.

8. U.S. Congress, Senate Subcommittee on Environmental Protection of the Committee on Environment and Public Works, *Hearing on S. 58, the National Biological Diversity Conservation and Environmental Research Act,* 102nd Cong., 1st sess., July 26, 1991.

9. Section 216 of S. 93, the Ecosystem Management Act of 1995, introduced by Senator Hatfield on January 4, 1995, and referred to the Committee on Energy and Natural Resources, which took no action.

10. H.R. 1845, The National Biological Survey Act of 1993.

11. U.S. Congress, Senate/House Subcommittee on Forests and Public Land Management of the Senate Committee on Energy and Natural Resources and the Subcommittee on Forests and Forest Health of the House Committee on Resources, *Joint Hearing on the*

Interior Columbia Basin Ecosystem Management Plan, 105th Cong., 1st sess., May 15, 1997. (hereafter *Joint Hearing*)

12. Zygmunt J.B. Plater, "From the Beginning, a Fundamental Shift of Paradigms: A Theory and Short History of Environmental Law," *Loyola of Los Angeles Law Review* 27, no. 3 (April 1994): 982.

13. Fred P. Bosselman and A. Dan Tarlock, "The Influence of Ecological Science on American Law: An Introduction," *Chicago-Kent Law Review* 69, no. 4 (1994): 868.

14. Bosselman and Tarlock, "The Influence of Ecological Science on American Law," 847.

15. 16 U.S.C. §§ 1331, 1333(a) (1994).

16. Bosselman and Tarlock, "The Influence of Ecological Science on American Law," 863.

17. Plater, "From the Beginning, a Fundamental Shift of Paradigms."

18. A. Dan Tarlock, "The Nonequilibrium Paradigm in Ecology and the Partial Unraveling of Environmental Law," *Loyola of Los Angeles Law Review* 27, no. 3 (April 1994): 1121–44.

19. Some ecologists argue the nonequilibrium paradigm includes a consideration of equilibrium conditions in some circumstances. See Peggy L. Fiedler, Peter S. White, and Robert A. Leidy," The Paradigm Shift in Ecology and Its Implications for Conservation," in *The Ecological Basis of Conservation,* ed. S. T. A. Pickett, R. S. Ostfeld, M. Shachak, and G. E. Likens (New York: Chapman & Hall, 1997), 83–92.

20. Judy L. Meyer, "The Dance of Nature: New Concepts in Ecology," *Chicago-Kent Law Review* 69, no. 4 (1994): 881. I first discovered this quote in Mark Sagoff, "Muddle or Muddle Through? Takings Jurisprudence Meets the Endangered Species Act," *William and Mary Law Review* 38, no. 3 (March 1997): 825–993.

21. Reed F. Noss, "Some Principles of Conservation Biology, as They Apply to Environmental Law," *Chicago-Kent Law Review* 69, no. 4 (1994): 894.

22. See generally Pickett et al. eds., *The Ecological Basis of Conservation.*

23. White House Interagency Ecosystem Management Task Force, *The Ecosystem Approach. Vol. II: Implementation Issues,* 69.

24. White House Interagency Ecosystem Management Task Force, *The Ecosystem Approach. Vol. II: Implementation Issues,* 97. 23 U.S.C. §§ 134, 135 (1994).

25. White House Interagency Ecosystem Management Task Force, *The Ecosystem Approach. Vol. II: Implementation Issues,* 79.

26. T. E. Dahl and C. E. Johnson, *Status and Trends of Wetlands in the Conterminous United States, Mid-1970s to Mid-1980s* (Washington, D.C.: U.S. Department of the Interior, Fish and Wildlife Service, 1991).

27. U.S. Congress, Senate Subcommittee on Environmental Protection of the Committee on Environment and Public Works, *Hearing on S. 58, the National Biological Diversity Conservation and Environmental Research Act,* 92, sec. 2 (9).

28. Thompson, "Ecosystem Management," 44; 42 U.S.C. § 4321 (1988); 40 C.F.R. 1508.8.

29. James McElfish, "Back to the Future," *The Environmental Forum* (September/October 1995): 15.

30. McElfish, "Back to the Future," 16.

31. Dinah Bear, "The Promise of NEPA," in *Biological Diversity and the Law,* ed. William Snape III (Washington, D.C.: Island Press, 1996), 178.

32. Thompson, "Ecosystem Management."

33. Gary D. Meyers, "Diving Common Law Standards for Environmental Protection: Application of the Public Trust Doctrine in the Context of Reforming NEPA and the Commonwealth Environmental Protection Act," *Environment and Planning Law Journal* (August 1994): 291; Gary D. Meyers, "Old Growth Forests, the Owl, and Yew: Environmental Ethics versus Traditional Dispute Resolution under the Endangered Species Act and Other Public Lands and Resources Laws," *Boston College Environmental Affairs Law Review* 18, no. 4 (summer 1991): 623–68.

34. Robert B. Keiter, "NEPA and the Emerging Concept of Ecosystem Management on the Public Lands," *Land and Water Law Review* 25, no. 1 (1990): 43–60.

35. U.S. Department of Agriculture, Forest Service, and Department of the Interior, Bureau of Land Management, *Upper Columbia River Basin Draft Environmental Impact Statement. Vol. 2: Appendices* (Boise, Idaho: U.S. Department of Agriculture, May 1997), app. B.

36. 16 U.S.C. §§ 1600–14 (1994).

37. U.S. Department of Agriculture, Forest Service, and Department of the Interior, Bureau of Land Management, *Upper Columbia River Basin Draft Environmental Impact Statement. Vol. 2: Appendices,* 21.

38. As quoted in Keiter, "Toward Legitimizing Ecosystem Management," 728.

39. Bruce Babbitt, testimony before Congress on H.R. 1845, The National Biological Survey Act of 1993, House of Representatives, before the Subcommittee of Technology, Environment and Aviation, and the Subcommittee on Investigations and Oversight of the Committee on Science, Space, and Technology, 103rd Cong., 1st sess., September 14, 1993, Committee print 56, 41.

40. Jason Patlis, "Biodiversity, Ecosystems, and Endangered Species," in Snape, ed., *Biological Diversity and the Law,* 43–58; Charles C. Mann and Mark L. Plummer, *Noah's Choice: The Future of Endangered Species* (New York: Alfred Knopf, 1995); Alston Chase, *In a Dark Wood* (Boston: Houghton Mifflin, 1995); Jon Welner, "Natural Communities Conservation Planning: An Ecosystem Approach to Protecting Endangered Species," *Stanford Law Review* 47 (January 1995): 319–61; "The Endangered Species Blueprint," *NWI Resource* 5, no. 1 (special issue, fall 1994); Joseph L. Sax, "Nature and Habitat Conservation and Protection in the United States," *Ecology Law Quarterly* 20 (1993): 47–56.

41. 16 U.S.C. § 1531(b) (1994).

42. FEMAT, Department of the Interior, "Recovery Plan for the Northern Spotted Owl: Draft" Washington, D.C., April 1992.

43. 16 U.S.C. § 1536(a)(2) (1994). This a difficult task, and so far researchers have determined "critical habitat" for less than half the species listed as threatened or endangered, according to data downloaded from U.S. Fish and Wildlife Service, Division of Endangered Species, "Endangered Species Home Page," at http://www.fws.gov/r9endspp/boxscore.html, May 23, 1997.

44. Patlis, "Biodiversity, Ecosystems, and Endangered Species," 43.

45. Alyson C. Flournoy, "Coping with Complexity," *Loyola of Los Angeles Law Review* 27, no. 3 (April 1994): 814.

46. Oliver A. Houck, "On the Law of Biodiversity and Ecosystem Management," *Minnesota Law Review* 81, no. 4 (April 1997): 869.

47. John Charles Kunich, "The Fallacy of Deathbed Conservation under the Endangered Species Act," *Environmental Law* 24, no. 2 (1994): 567.

48. Andrea Easter-Pilcher, "Implementing the Endangered Species Act," *BioScience* 46 (May 1996): 357.

49. Easter-Pilcher, "Implementing the Endangered Species Act," 362; Mark Sagoff, "Where Ickes Went Right or Reason and Rationality in Environmental Law," *Ecology Law Quarterly* 14 (1987): 265–82.

50. Robert Gordon, "When the 'Best Available Data' Is B.A.D.: The Data Error Plague," *NWI Resource* 4, nos. 2–3 (summer 1993): 3–7.

51. Keiter, "Beyond the Boundary Line," 304.

52. 16 U.S.C. § 1 (1994).; Keiter, "Beyond the Boundary Line," 304; Robert B. Keiter, "Preserving Nature in the National Parks: Law, Policy, and Science in a Dynamic Environment," *Denver University Law Review* 74, no. 3 (1997): 649–95.

53. F. Fraser Darling and Noel D. Eichhorn, *Man and Nature in the National Parks,* 2nd ed. (Washington, D.C.: Conservation Foundation, 1969), 28.

54. Department of the Interior, National Park Service, *Management Policies* (Washington, D.C.: 1988), chap. 1 at 3.

55. Julie Weis, "Eliminating the National Forest Management Act's Diversity Requirement as a Substantive Standard," *Environmental Law* 27, no. 2 (summer 1997): 641–62; Pounder, "Reforming Livestock Grazing on the Public Domain;" Walter Kuhlman, "Wildlife's Burden," in Snape, ed., 189–201.

56. 468 U.S. 288, 299 (1984), as quoted in Keiter, "Preserving Nature in the National Parks," 676.

57. From the AFSEEE's Web page at http://www.afseee.org/front.html, downloaded December 11,1997.

58. U.S. Department of Agriculture, Forest Service, and Department of the Interior, Bureau of Land Management, *Upper Columbia River Basin Draft Environmental Impact Statement: Vol. 1,* chap. 3 at 1.

59. 16 U.S.C. §§ 528–31 (1994); 16 U.S.C. §§ 1600–14 (1994) ; and 43 U.S.C. §§ 1701–84 (1994).

60. 16 U.S.C. § 529 (1988).

61. 16 U.S.C. § 531(a) (1988).

62. 16 U.S.C. § 531(b) (1988).

63. 43 U.S.C. § 1702(c) (1988).

64. 16 U.S.C. § 1604(g)(3)(B) (1994); Keiter, "Beyond the Boundary Line;" Houck, "On the Law of Biodiversity and Ecosystem Management."

65. Keiter, "Beyond the Boundary Line," 311.

66. Keiter, "Beyond the Boundary Line," 311.

67. Michael Blumm, "Public Choice Theory and the Public Lands: Why 'Multiple Use' Failed," *Harvard Environmental Law Review* 18, no. 2 (1994): 406.

68. Scott W. Hardt, "Federal Land Management in the Twenty-First Century: From

Wise Use to Wise Stewardship," *Harvard Environmental Law Review* 18, no. 2 (1994): 345–403.

69. White House Interagency Ecosystem Management Task Force, *The Ecosystem Approach. Vol. II: Implementation Issues,* 83.

70. The General Accounting Office came to a similar conclusion in *Ecosystem Management: Additional Actions Needed to Adequately Test a Promising Approach,* GAO/RCED-94–111 (Washington, D.C.: U.S. General Accounting Office, August 1994).

71. U.S. Department of Agriculture, Forest Service, and Department of the Interior, Bureau of Land Management, *Upper Columbia River Basin Draft Environmental Impact Statement: Vol. 1,* chap. 5 at 51 (hereafter cited as *UCRB DEIS*). For purposes of preparing the draft environmental impact statements, the Interior Columbia River Basin was divided into parts geographically, each with its own planning team. The Eastside team addressed the area from the crest of the Cascade Mountains eastward to the borders of the states of Oregon and Washington, and the Interior Columbia Basin team considered the remainder of the project area. The two teams drew on common resources by which each was supported by the same scientific assessment team and issued documents that were identical in nearly all respects, including their respective draft environmental impact statements. Hereafter, the *Eastside Draft Environmental Impact Statement* is cited as *Eastside DEIS.*

72. *UCRB DEIS,* vol. 1, chap. 4 at 173.

73. *UCRB DEIS,* vol. 1, chap. 4 at 173.

74. U.S. Forest Service, "National Forest System Land Planning and Resource Management Planning Proposed Rule," *Federal Register* 60, no. 71 (April 13, 1995): 18886–932 (hereafter cited as FS Proposed Rule), quote at 18922; the analysis of this rule draws heavily on Allan Fitzsimmons, "Sound Policy or Smoke and Mirrors: Does Ecosystem Management Make Sense?" *Water Resources Bulletin* 32, no. 2 (April 1996): 217–27. Kennedy and Quigley argue not only that the FS is compelled to adopt ecosystem management in word and spirit but also that the agency should expand it to include "human socio-economic systems." See James J. Kennedy and Thomas M. Quigley, "Evolution of USDA Forest Service Organizational Culture and Adaptation Issues in Embracing an Ecosystem Management Paradigm," *Landscape and Urban Planning* 40 (1998): 113–22.

75. FS Proposed Rule, 18919, 18922.

76. FS Proposed Rule, 18889.

77. FS Proposed Rule, 18892.

78. Jay E. Anderson presents an attempt to measure naturalness in "A Conceptual Framework for Evaluating and Quantifying Naturalness," *Conservation Biology* 5, no. 3 (1993): 347–52.

79. FS Proposed Rule, 18920.

80. FS Proposed Rule, 18921.

81. FS Proposed Rule, 18893.

82. Carol Yoon, "Counting Creatures Great and Small," *Science* 260 (April 30, 1993): 620–22.

83. FS Proposed Rule, 18896; Mary E. Power, David Tilman, James A. Estes, Bruce

A. Menge, William J. Bond, L. Scott Milles, Gretchen Daily, Juan Carlos Castilla, Jane Lubchenco, and Robert T. Paine, "Challenges in the Quest for Keystones," *BioScience* 46, no. 8 (September 1996): 618; and L. Scott Mills, Michael E. Soulé, and Daniel F. Doak, "The Keystone-Species Concept in Ecology and Conservation," *BioScience* 43, no. 4 (April 1993): 219–24.

84. FS Proposed Rule, 18893.

85. Curtis J. Richardson, "Ecological Functions and Human Values in Wetlands: A Framework for Assessing Forestry Impacts," *Wetlands* 14, no. 1 (1993): 1–9.

86. FS Proposed Rule, 18893.

87. FS Proposed Rule, 18897.

88. FS Proposed Rule, 18897.

89. FS Proposed Rule, 18894.

90. FS Proposed Rule, 18892.

91. Roger A. Sedjo, "Ecosystem Management: An Uncharted Path for Public Forests," *Resources* 10 (fall 1995): 20.

92. *UCRB DEIS,* vol. 1, chap. 1 at 8, and vol. 2, appendix B; *Eastside DEIS,* vol. 1, chap. 1 at 8, and vol. 2, appendix 1.

93. Joint Hearing, 48.

94. *UCRB DEIS,* vol. 1, chap. 1 at 6; *Eastside DEIS,* vol. 1, chap. 1 at 5.

95. *UCRB DEIS,* vol. 1, chap. 1 at 9; *Eastside DEIS,* vol. 1, chap. 1 at 10.

96. *UCRB DEIS,* vol. 1, chap. 3 at 6; *Eastside DEIS,* vol. 1, chap. 3 at 6.

97. *UCRB DEIS,* vol. 1, chap. 2 at 2.

98. Joint Hearing, 33.

99. Joint Hearing, 27.

100. *UCRB DEIS,* vol. 1, chap 3 at 28.

101. *UCRB DEIS,* vol. 1, chap. 1, facing p. 1; *Eastside DEIS,* vol. 1, chap. 1, facing p. 1.

102. Joint Hearing, 34.

103. U.S. Department of Agriculture, Forest Service, Pacific Northwest Research Station, *A Framework for Ecosystem Management in the Interior Columbia Basin,* ed. Richard W. Haynes, Russell T. Graham, and Thomas M. Quigley (Portland, Ore.: U.S. Department of Agriculture, 1996), 10 (hereafter cited as *Framework*).

104. *Framework,* abstract.

105. *Framework,* 45.

106. *UCRB DEIS,* vol. 1, chap. 2 at 221; *Eastside DEIS,* vol. 1, chap. 2 at 227.

107. *UCRB DEIS,* vol. 1, chap. 4 at 211; *Eastside DEIS,* vol. 1, chap. 4 at 211.

108. *UCRB DEIS,* vol. 1, chap. 2 at 222; *Eastside DEIS,* vol. 1, chap. 2 at 228.

109. *UCRB DEIS,* vol. 1, chap. 5 at 41; *Eastside DEIS,* glossary at 11.

110. Department of Energy, *National Energy Strategy* (Washington, D.C.: U.S. Government Printing Office, 1991); Department of Energy, *National Energy Strategy—Technical Annex* 2 (Washington, D.C.: U.S. Government Printing Office, 1991), available from the National Technical Information Center as document DOE/S-0086P.

111. Quigley et al., *Evaluation,* 47.

112. See generally A. W. Küchler and I. S. Zonneveld, *Vegetation Mapping* (Dordrecht: Kluwer Academic Publishers, 1988).

113. John Leighly, "Some Comments on Contemporary Geographic Method," *Annals of the Association of American Geographers* 27, no. 3 (September 1937): 125–41; Monica Turner, Robert O'Neill, Robert Gardner, and Bruce Milne, "Effects of Changing Spatial Scale on the Analysis of Landscape Pattern," *Landscape Ecology* 3 (1989): 153–62. See also Michael Conroy and Barry Noon, "Mapping of Species Richness for Conservation of Biological Diversity: Conceptual and Methodologic Issues," *Ecological Applications* 6, no. 3 (1996): 763–73, and Ling Bian and Stephen Walsh, "Scale Dependencies of Vegetation and Topography in a Mountainous Environment of Montana," *Professional Geographer* 45, no. 1 (February 1993): 1–11.

114. This analysis draws on Allan K. Fitzsimmons, "Why a Policy of Federal Management and Protection of Ecosystems Is a Bad Idea," *Landscape and Urban Planning* 40 (1998): 195–202.

115. United Nations, *Report of the United Nations Conference on Environment and Development, Rio Declaration on Environment and Development,* Rio de Janeiro, June 3–14, 1992, A/Conf.151/26, vol. I, August 12,1992, downloaded from gopher://gopher.un.org/00/conf/unced/English/riodecl.txt, December 26, 1997.

116. United Nations, *Report of the United Nations Conference on Environment and Development, Agenda 21,* chap. 15, para. 15.3, Rio de Janeiro June 3–14, 1992, A/Conf. 151/26, vol. II, August 12, 1992, downloaded from gopher://gopher.un.org/00/conf/unced/English/a21-txt, December 26, 1997.

117. A listing of international instruments and institutions related to ecosystem management can be found in White House Interagency Ecosystem Management Task Force, *The Ecosystem Approach. Vol. II: Implementation Issues,* 125–27. The UN Convention Concerning the Protection of the World Cultural and Natural Heritage was suggested by the United States in 1972. The United States was the first signatory. It became operational in 1975 after twenty members of the UN Educational, Scientific, and Cultural Organization (UNESCO) signed on. The Man and the Biosphere Programme originated with a UNESCO task force in 1974, and the Biosphere Reserve network began in 1976. The program operates without a specific treaty. The United States withdrew from UNESCO in 1984 but remains active in the World Heritage and Biosphere Reserve programs.

118. UNESCO Man and the Biosphere (MAB) Programme, "Conservation and Management of Terrestrial Ecosystem Resources," downloaded from http://www.unesco.org/mab/home/mabis.htm, December 27, 1997. The U.S. National Committee for the Man and the Biosphere Program, *U.S. MAB Bulletin* 21, no 3. (December 1997): 1.

119. *U.S. Newswire,* "Letter from the National Parks and Conservation Association to the World Heritage Committee on Threats to Yellowstone," February 28, 1995.

120. Bernd von Droste, director, World Heritage Centre, letter to George Frampton, March 6, 1995 (copy on file with the author).

121. George T. Frampton, assistant secretary for fish, wildlife, and parks, U.S. Department of the Interior, letter to Dr. Bernd von Droste, director, World Heritage Centre, June 27, 1995 (copy on file with the author).

122. Paul Bedard, "Clinton, Mining Firm Sign Yellowstone Deal," *Washington Times,* August 13, 1996, A4.

123. National Parks and Conservation Association, "Yellowstone Declared 'In Dan-

ger'—'Ascertained and Potential' Threats to World's First National Park Cited by International Committee," *NPCA News,* December 6, 1995, downloaded from http://www.npca.org/news/hertg.html, January 2, 1998.

124. Frampton, letter to von Droste, June 27, 1995.

125. George T. Frampton, assistant secretary for fish, wildlife, and parks, U.S. Department of the Interior, letter to Dr. Bernd von Droste, director, World Heritage Centre, December 1, 1995 (copy on file with the author).

126. Bruce Babbitt, "Statement by Secretary of the Interior Bruce Babbitt on Announcement by President Clinton That Will Close Mine on Outskirts of Yellowstone National Park," press release, Office of the Secretary, Department of the Interior, Washington, D.C., November 12, 1996.

127. UNESCO, World Heritage Committee, "Report," Nineteenth Session, Berlin, Germany, December 4–9, 1995, chap. 7, 20, downloaded from http://www.unesco.org/whc/archive/repcom95.htm, December 26, 1997.

128. Department of the Interior, Office of the Secretary, "Statement by the Department of the Interior on Designation of Yellowstone National Park as a World Heritage Site in Danger," press release, Office of the Secretary, Department of the Interior, Washington, D.C., December 5, 1996.

129. National Parks and Conservation Association, "Yellowstone Declared 'In Danger'."

130. Jeremy Rabkin, "The Yellowstone Affair: Environmental Protection, International Treaties, and National Sovereignty," Competitive Enterprise Institute, Washington, D.C., May 1997, 7.

131. As quoted in Chris Tollefeson, "Park Service Defends Decision to Invite Panel," *Casper Star Tribune,* September 9, 1995, A1.

132. Frampton, letter to von Droste, June 27, 1995.

133. International Conference on Biosphere Reserves, "The Seville Strategy for Biosphere Reserves—the Vision from Seville for the 21st Century," Seville, Spain, March 1995, downloaded from http://www.unesco.org/mab/home/stry-1.htm, December 26, 1997. The model biosphere reserve has a natural area at its core that is surrounded by a buffer zone that is itself surround by a transition zone. Permissible human uses of the land increases as you move away from the core.

134. Rabkin, "The Yellowstone Affair," 9.

135. See generally the record of hearings held on May 5, 1997, and June 10, 1997, printed as "Hearings on the American Land Sovereignty Protection Act," House of Representatives, before the Committee on Resources, 105th Cong., 1st sess., June 10, 1997, Committee print 105–26.

136. 16 U.S.C. § 470.

137. U.S. Congress, House, "American Land Sovereignty Protection Act," H.R. 901, 105th Cong., 1st sess., downloaded from ftp://ftp.loc.gov/pub/thomas/c105/h901.eh.txt, December 27, 1997.

138. U.S. Congress, House, "American Land Sovereignty Protection Act"; section 3 addresses World Heritage sites, section 4 deals with Biosphere Reserves, and section 5 deals with international designations in general.

139. The act provides exceptions to lands to be set aside under agreements made in con-

junction with the North American Wetlands Conservation Act and the Fish and Wildlife Improvement Act of 1978.

140. "Detailed Legislative History of H.R. 901," downloaded from the congressional Web site at http://thomas.loc.gov, December 27, 1997.

141. Denis P. Galvin, testimony before Congress on H.R. 901, The American Land Sovereignty Protection Act of 1997, House of Representatives, before the Committee on Resources, 105th Cong., 1st sess., June 10, 1997, Committee print 105–26, 61.

142. As quoted in Mark Pollot, "The Unconventional Convention," in *Technical Review of the Convention on Biological Diversity* (Washington, D.C.: National Wilderness Institute and the Alexis de Tocqueville Institution, 1994), 15.

143. William J. Snape III, "International Protection: Beyond Human Boundaries," in Snape, ed., *Biological Diversity and the Law,* 88. See also David R. Downes, "Global Trade, Local Economies, and the Biodiversity Convention," in Snape, ed., *Biological Diversity and the Law,* 202–16.

144. United Nations, Convention on Biological Diversity, Art. 1.

145. United Nations, Convention on Biological Diversity, Art. 2.

146. John H. Barton, "Biodiversity at Rio," *BioScience* 42, no. 10 (November 1992): 773.

147. Letter to George Mitchell, majority leader, U.S. Senate, signed by Senators Bennett, Bond, Brown, Burns, Coats, Cochran, Coverdell, Craig, D'Amato, Danforth, Dole, Domenici, Faircloth, Gorton, Gramm, Grassley, Hatch, Helms, Hutchison, Kempthorne, Lott, Mack, McCain, McConnell, Murkowski, Nickles, Packwood, Pressler, Simpson, Smith, Spector, Stevens, Thurmond, Wallop, and Warner, August 5, 1994.

148. William J. Clinton, "Message from the President of the United States Transmitting the Convention on Biological Diversity."

149. U.S. Congress, Senate, *Report on the Convention on Biological Diversity,* Committee on Foreign Relations, 103rd Cong., 2nd sess., July 11, 1994, Executive Rep. 103–30, 16.

150. Wendy R. Sherman, assistant secretary of state for legislative affairs, letter to Senator Robert Dole transmitting the administration's response to questions raised by Senator Dole and others in their August 5, 1994, letter to Senator Mitchell, August 8, 1994 (copy on file with the author).

151. U.S. Congress, Senate, *Report on the Convention on Biological Diversity,* at 17, lists the Endangered Species Act; Marine Mammal Protection Act; Marine Protection, Research and Sanctuaries Act, and Coastal Zone Management Act as allowing the United States to meet the convention's biological diversity protection requirements. This interpretation is false on its face, as the CBD reaches all land and water, while the combined geographic reach of the cited laws is confined to coastal areas and critical habitat for threatened or endangered species.

152. United Nations, *Implementation of Agenda 21: Review of Progress Made since the United Nations Conference on Environment and Development,* U.S. Country Profile, U.N. Commission on Sustainable Development, 1997. The profile was prepared by the U.S. Interagency Group for UNCSD, Department of State, December 1996, downloaded from http://www.un.org/dpcsd/earthsummit/USA-cp.htm, December 26, 1997.

153. Several bills to protect biological diversity came before Congress in the 1990s,

including H.R. 1845, the National Biological Survey Act of 1993, 103rd Cong.; H.R. 2082, the Biological Diversity Conservation Research Act, 102nd Cong.; H.R. 585, the National Biological Diversity Conservation and Environmental Research Act, 102nd Cong.; and S. 58, the National Biological Diversity Conservation and Environmental Research Act, 102nd Cong. None were passed.

154. Snape, "International Protection," 82.

155. Pollot, "The Unconventional Convention," 27.

Chapter 8

Ecosystem Protection Proposals

> Determine which ecosystem types are most endangered (i.e., those that have declined or been degraded since European settlement). . . . [then] Pass legislation that will set recovery goals for endangered and threatened ecosystems and standards for representing all ecosystem types in protected areas, regardless of current endangerment. Specifically we propose a Native Ecosystem Act.
>
> —Reed Noss and Allen Cooperrider, *Saving Nature's Legacy,* 335

New paradigmists pursue an active agenda for making ecosystem protection the de facto basis for land use management as well as requiring it through changes in the law. I begin this chapter by looking into an aggressive grassroots effort on behalf of ecosystems and continue by examining proposals designed to make ecosystem management the law of the land. These include a hoped-for constitutional amendment, general legislative recommendations, and specific bills introduced in Congress.

The Wildlands Project

The Wildlands Project ties together many threads of the new paradigm I discuss in earlier chapters: false claims of environmental calamity, biocentric reverence for nature, and weak science, all commingled in a political effort to make the protection of Mother Nature the basis for land use management throughout North America. Dave Foreman, Reed Noss (the project's science director), Michael Soulé (University of California at Santa Cruz), David Johns (Portland State University), and a handful of others devised the North American Wilderness Recovery Program (the official name for what is popularly known as the Wildlands Project) in 1991–92 and officially unveiled their plan in a special issue of the magazine *Wild Earth* in December 1992.[1]

The sponsors believe that Americans threaten the environment. In their "Mission Statement," they write that "the environment of North America is at risk and an audacious plan is needed for its survival and recovery."[2] We must get about "healing the land" to restore its "vital flows"; "we must allow the recovery of whole ecosystems and landscape in every region." Doing so, they tell us, "is fundamental to the integrity of nature." Buzzwords abound:

> Our vision is simple, we live for the day when Grizzlies in Chihuahua have an unbroken connection to Grizzlies in Alaska; when Gray Wolf populations are continuous from Durango to Labrador, when vast unbroken forests and flowering plains again thrive and support pre-Columbian populations of plants and animals; when humans dwell with respect, harmony, and affection for the land . . . our vision is continental: from Panama and the Caribbean to Alaska and Greenland.

Can it really be that they want to return large, carnivorous predators such as grizzly bears, gray wolves, and mountain lions to much if not all of their ranges of 500 years ago and take vast amounts of rich agricultural land out of production, all to the detriment of human well-being? Yes, because that is what nature-worshiping biocentrism requires. As Noss writes, "The native ecosystem and the collective needs of non-human species must take precedence of the needs and desires of humans."[3]

For these new paradigmists, Mother Nature stands as a goddess defiled. They long for a pre-Columbian North America when, in their worldview, Earth Mother lived in peace with her worshipers and all was harmonious; when the landscape reflected an "Earth wisdom" that began to vanish from the continent with the arrival of Europeans.[4] Foreman writes that the plan "is a bold attempt to grope our way back to October 1492" and seek a new path, one that does not lead to "gold, empire, and death" but one that "leads to beauty, abundance, wholeness, and wildness."[5] He and his coauthors want to rescue Mother Nature from the predations of modern society by setting aside "vast interconnected areas of true wilderness . . . as the home for unfettered life, free from industrial human intervention."

The project is breathtaking in its geographic scope and its impacts on humans. Writing in the journal *Science,* Charles Mann and Mark Plummer remark that the plan "is the most ambitious proposal for land use management since the Louisiana Purchase of 1803."[6] Noss indicates what proponents have in mind for land use in the United States. He says that "at least half of the land in the conterminous states should be encompassed in core reserves and inner corridor zones."[7] Project planners see these cores as wilderness where traces of humans cannot be found: "extensive areas of native vegetation . . . off-limits to human exploitation" and "vast landscapes without roads, dams, motorized vehicles, powerlines, overflights, or artifacts of civilization."[8] They would place much of the remaining 50 percent of the lower forty-eight states in buffer zones wherein "only human activity compatible with protection of the core reserves and corridors would be allowed" and where "ecological health" would be restored.[9] Thus, in order to achieve their vision of a proper relationship between people and Mother Nature, advocates ultimately require the resettling of North America. The new paradigmists would have Americans abandon towns, small communities, and rural homes throughout the country. We would have to cease farming large portions of our most productive agricultural land, end forestry on millions of acres, stop going to national parks such as Yellowstone (because they would now be core wilderness areas and therefore generally unavailable for public use), and otherwise stop using at least one-half of the contiguous forty-eight states.

What ecosystems do the creators of the Wildlands Project want to protect and restore? They do not say, but they take elaborate steps to make their nonsaying sound scientific. One finds the ecological heart of the project in Noss's article "The Wildlands Project: Land Conservation Strategy."[10] But his analysis, even with its impressive bibliography of over 100 references to the scientific literature, fails to get around the deficiencies with the ecosystem concept that I illustrate in several chapters. Noss, for example, says that "representation of all ecosystems and environmental gradients is the first step toward maintaining the full spectrum of biodiversity in a region."[11] But, as I point out in chapter 1, any region has an infinite number of arbitrarily defined ecosystems, so his admonition to represent all of them in protected areas is operationally impossible. He argues that "all native ecosystem types and seral stages across their natural range of variation" must be protected.[12] In chapter 6, I illustrate that ecologists do not have an accepted typology for ecosystems (a point that Noss concedes; see the next section), so Noss's "all native ecosystem types" is oratory, not science. Lack of a classification system also renders his concept of "natural range of variation" meaningless since ecologists have no way of objectively and verifiably determining when and where one ecosystem fades into another in space and time. Confusion mounts as Noss fails to come to grips with the geography of ecosystems even though he writes that " regionalization is a central issue in the Wildlands Project."[13] He cites James Omernik's work (see chapter 2, map 2.2) as an example of ecosystem regionalization. In the end, however, Noss neither endorses any particular national pattern of ecosystems nor identifies a specific set of variables to use in their construction. He tells supporters to find roadless areas and make them the geographic cores of their land management proposals.[14] Roadlessness is not an ecological characteristic and bears no connection to any scientific definition of an ecosystem that I am aware of (see chapter 2). Noss suggests several sources of geographic data that activists may use to graft additional land on to the roadless core, but in the end the resulting ecosystems that they want to protect are just more arbitrarily determined uniform regions (see chapter 2).

To estimate the land use results of the Wildlands Project, the reader should refer to Omernik's map and consider the human implications of placing 50 percent of each of the ecosystems he depicts into wilderness status and portions of the other 50 percent into buffer zones. Promoters of the Wildlands Project would, for example, take half of our most productive farmland in Indiana, Illinois, Iowa, and Kansas out of production in order to try and re-create "vast unbroken . . . flowing plains" supporting thriving "pre-Columbian populations of plants and animals." Under their plan, no longer would these lands yield wheat, corn, soybeans, dairy products, hogs, or beef to feed people in the United States and abroad; instead, they would become wilderness cathedrals dedicated to the worship of Mother Earth. Omernik depicts narrow coastal ecosystems along our Pacific, Atlantic, and Gulf coasts that Americans occupy in substantial numbers. In order to meet Foreman and his colleagues' 50 percent wilderness requirement, many of these people must relocate as well as raze their settlements, roads, and other signs of human habitation in the process. Mann and Plummer conclude that execution of the Wildlands Project would make the United States look like "an archipelago of human inhabited islands surrounded by a sea of natural areas."[15] It is easy for one to dismiss the project as so outrageous to warrant no further consideration by reasonable people, but doing so is a mistake.

As radical as this proposal sounds, widely known figures such as Paul Ehrlich of Stanford University and E. O. Wilson of Harvard University endorse its principles, and others are working directly on its implementation.[16] Grassroots activists across the country are quickly beginning the long process of transforming the project from vision to land management reality; these groups include the Alliance for the Wild Rockies, Northwest Ecosystem Alliance, Sky Island Alliance, Siskiyou Regional Education Project, Coast Range Association, Northern Appalachian Restoration Project, Southern Appalachian Biodiversity Project, Southern Appalachian Forest Coalition, and California Wilderness Coalition. They busily identify areas for protection and use existing land use management planning procedures to argue their case, particularly when public lands are involved.

One such organization is the Southern Rockies Ecosystem Project (SREP), a 1992 coalition of over a dozen regional and local environmental groups. Its members want the government to adopt ecosystem protection plans in a region extending from southern Wyoming through Colorado to northern New Mexico.[17] By using cores, buffer zones, and corridors, they want to "Represent all native ecosystem types and age classes across their range of natural variation;

Maintain viable populations of all native species in natural patterns of abundance and distribution; Maintain ecological and evolutionary processes. . . . [and] Provide for the resilience of species and ecosystems in the face of long-term change."[18] They desire to "implement a wildlands recovery proposal which will reverse the decline of wildlife and wilderness and will recover whole ecosystems and landscapes." The group recruits volunteers, trains students, does geographic information systems (GIS) mapping, prepares proposals for public land managers, meets with local officials, and generally pushes their nature-first agenda.

A second coalition actively pursuing implementation of the Wildlands Project is the Southern Appalachian Forest Coalition (SAFC), put together in 1995 by the Sierra Club, Wilderness Society, and regional and local environmental groups.[19] The goal of these activists is the "protection and restoration of native biodiversity and ecological integrity" with a system of core areas, corridors, and buffer zones in a region that runs from Virginia to Alabama.[20] Citing work by Noss and Cooperrider as providing scientific justification, they seek to

> represent, in a systems of protected areas, all native ecosystem types and seral stages across their natural range of vegetation. Maintain viable populations of all native species in natural patterns of abundance and distribution. Maintain ecological process. . . . [and] Design and manage the system to be responsive to short-term and long-term environmental change and to maintain the evolutionary potential of the region.[21]

Like their counterparts in the SREP, they emphasize using the U.S. Forest Service's forest plan revision process to inculcate their agenda into land management decision making and have already roughed out several million acres in the Blue Ridge from southern Virginia to northern Georgia wherein nature protection is to dominate land use management.[22] Since biocentrists can best protect Mother Nature by keeping people away from her, prevention of road building in unroaded areas is a high, immediate policy priority for the group. Thus, the SAFC writes in a January 1998 policy bulletin for its members and others that

> the Administration is deciding this week how strong the interim roadless protection policy they have been considering will be. They must hear from us. . . . Key phrases to note in your message are. . . . The policy should prohibit all road building including temporary road building and road reconstruction. The policy should prohibit all logging. . . . All inventoried roadless areas, regardless of size, should be included.[23]

On January 22, 1998, the Forest Service announced an "interim policy" to suspend road construction in a formula to paralyze the judiciary as well as the economy of the nation. Envision judges trying to sort out competing claims to the benefits of living resources, as when one litigant wants to hunt deer but another wants to look at them. Picture judges deciding between members of the Wildlands Project who want to turn farmlands into "vast unbroken . . . flowing plains" supporting thriving "pre-Columbian populations of plants and animals" while others want to produce food. What happens when the habitat needs of one species conflict with those of another? How do you decide which ecosystems to protect? Scientists cannot offer significant guidance to the courts in such matters.

And what of the conflict between his amendment and constitutional protections for private property? Schlickeisen sees Fifth Amendment protections of property rights, coupled with what he deems its ecologically unenlightened interpretation by the Supreme Court (see chapter 9), as the foremost barriers to establishing the new paradigm in law. He solves these problems by effectively sweeping away Fifth Amendment protections with his suggested amendment. Under that amendment, no longer would a landowner really own a forest or a field or a family really control management of the yard surrounding their home because Schlickeisen's amendment declares that "the living natural resources in the United States are the common property of all people . . . all persons . . . have an inalienable, enforceable right to the benefits of those resources for themselves." His amendment eviscerates private property rights via mandatory communal ownership, planning, and decision making.

Legislative Proposals

Advocates of ecosystem management offer a variety of proposals requiring legislation for their implementation. The common attribute of these offerings is more government control, more land off limits to human use, and more restrictions on human use of the land in those areas where it is permitted at all. For example, Foreman, in a series of proposals separate from the Wildlands Project, calls for 75 million acres (more land than is in the states of Florida and Georgia combined) of new congressionally designated wilderness in roadless areas in the western states.[32] The government would restore grizzlies and wolves to several western states and manage people accordingly. Humans would be forbidden entry onto 1.5 million acres of the Los Padres National Forest near Los Angeles to protect the California condor. Congress would require the removal of dams and the elimination of grazing from public lands to protect ecosystems. For the eastern two-thirds of the nation, he offers several land management proposals needing congressional action. He wants government to create a 10- to 20-million acre Great Plains National Park and return grizzlies and gray wolves to it. In the Ohio

River Valley he seeks a large wilderness recovery area complete with eastern panthers and wolves while eastern panthers would be returned to a new 4-million-acre wilderness area in southern Appalachia. For New York he wants to reintroduce wolves and panthers into a 1.5-million-acre wilderness and likewise return major predators to a new 10-million-acre national park in Maine covering 50 percent of the state. Everglades National Park and the adjacent Big Cypress Nation Reserve presently cover 2.2 million acres; Foreman would expand their combined territory to 5 million of the state's 35 million acres. The great majority of eastern lands that Foreman targets are privately owned.

David Brower thinks that Congress should establish a National Ecosystem Reserve System, in which they should place all our biosphere reserves plus "inadequately protected" federal, state, local, and private land. To facilitate matters, Congress should "change the name and mission of the Bureau of Land Management to the National Land Service . . . [and] give it a roll in protecting *all* U.S. land, not just public land" [emphasis in the original].[33] He concedes that these actions are "hard to do" but believes that they are "essential to sustainability."

Noss and Cooperrider ask Congress to enact a Native Ecosystem Act.[34] In trying to single out ecosystems for protection via new federal legislation, they take a hierarchical view of ecosystem classification and assert that ecosystems at all levels of this spatial ordering must be protected because all can become endangered and all are required for a proper human relationship with nature. Consistent with the ecocentric view of the paradigm, Noss and Cooperrider judge endangerment by deviation from conditions that might otherwise exist had not Europeans settled in North America. The law should not only reach presently so-called threatened or endangered species in their view but also include the entire spectrum of "all native ecosystem types and seral stages across their natural range of variation."[35] They say that this could require as much as 99 percent of the land in a few regions but estimate that fulfilling the intent of the law could be accomplished by including between 25 and 75 percent of a region in "core reserves and inner buffer zones" in most cases.[36] Congress would set goals for the recovery and restoration of degraded ecosystems and for those in decline.

With their proposal, Noss and Cooperrider—like Foreman and Brower—are seeking to advance biocentrism, not scientifically based land management. As I note in earlier chapters, there simply is no scientific agreement on any of the key ingredients needed to implement legislation such as Noss and Cooperrider suggest (see chapters 1, 2, and 6). Researchers cannot provide Congress with as much as a definitive list or snapshot map of the ecosystems that new paradigmists want the law to protect, let alone determine their "natural range of variation" in space and time for policy purposes. Ecosystem degradation, health, buzzwords of the new paradigm, congressional acceptance of their proposals inevitably leads to biocentrism because it is biocentrists who will give legal and regulatory substance to the terms.

Thus far I have looked at general ideas for fresh legislation. In the next section I examine specific bills introduced into Congress that would incorporate the new paradigm into federal law.

Ecosystem Management Bills Introduced in Congress

Keiter observes that one approach Congress can take to making ecosystem protection the prime goal of land management is to simply require it on a place-by-place basis. The Northern Rockies Ecosystem Protection Act of 1997 is an example of this approach.[52] Congressman Christopher Shays of Connecticut introduced the bill on May 23, 1997; by early 1998, fifty-seven members of the House had signed on as cosponsors. With this bill these legislators would codify principles of the Wildlands Project within what they call the Northern Rockies Bioregion, which extends into Idaho, Montana, Oregon, Washington, and Wyoming. Like Muir and others before them, however, they hide their biocentrism in anthropocentric garb, fully aware that the American people are not yet ready for legislators to openly write nature worship into federal law.

The authors begin by lauding a comparatively undefiled Mother Nature within the region. "Many areas of undeveloped National Forest System lands . . . possess outstanding natural characteristics which give them high values . . . and will, if properly preserved, be an enduring resource of wilderness, wild land area, and biodiversity for the benefit of the American people."[53] But people threaten her. "Roadbuilding, timber harvesting, mining, oil and gas exploration . . . and other activities has [sic] severe effects on the wildlife populations, their habitat, the water quality, the ancient forests and the greater ecosystems of the Northern Rockies Bioregion." Should people do further harm, they "would cause a loss to the Nation of an entire wild land region."[54] Thus, congressional supporters of the bill believe that they have uncovered an "urgent need for an ecological reserve system . . . which includes core ecosystem reserve areas and biological connecting corridors necessary to ensure wildlife movements and genetic interchange between the core reserves."[55] They see an urgency for congressional action to "promote, perpetuate, and preserve the wilderness character of the land," "protect the ecological integrity and contiguity of major wild land ecosystems," "protect and maintain biological and native species diversity," and ensure that federal agencies engage in "holistic ecosystem management and protection of ecosystems."[56] Backers of the bill want a vast area of natural landscapes unblemished by humans.

Proponents of the bill propose several actions to accomplish their goals. In the bill they designate six core ecosystems based largely on existing national park and wilderness areas: the Greater Glacier–Northern Continental Divide Ecosystem, Greater Yellowstone Ecosystem, Greater Salmon-Selway Ecosystem, Greater Cabinet-Yaak-Selkirk Ecosystem, Islands in the Sky Wilderness, and Blackfeet Wilderness. The bill calls for expansion of these cores by adding 13 million new acres of wilderness and establishes corridors covering an additional 7.4 million acres to link the core areas. In addition to the cores, corridors, and new wilderness areas, the proposal directs the secretary of the interior to study the feasibility of creating two new units of the national park system.[57] Members further include the new paradigmist philosophy by minting a fresh protective category of federal lands, the National Wildland Restoration and Recovery System, in which they place eight areas cumulatively covering over 1 million acres.[58] The bill justifies the new system with a congressional finding that "unwise resource extraction and development" have damaged the land. Healing requires government managers to "restore such lands, as much as possible, . . . to their natural condition as existed prior to their entry and development."[59] I understand why it is in the interest of the eco-church's "Cardinal Green" and other new paradigmists to require government land managers to try and re-create romanticized pre-European landscapes, but it is not clear how such actions would serve the broader public interest.

Hardt and Blumm think that Congress should amend existing federal land management statutes to put ecosystem management in place; the Forest Biodiversity Act of 1997 (H.R. 1861) would do that.[60] Representatives Maurice Hinchey of New York and Christopher Shays of Connecticut introduced H.R. 1861 on June 11, 1997, and garnered seventeen cosponsors by February 1998. Hinchey and others want to amend the Forest and Rangeland Renewable Resources Planning Act of 1974, the Federal Land Policy and Management Act of 1976, the National Wildlife Refuge System Administration Act of 1996, the National Indian Forest Resources Management Act, and Title 10 of the U.S. Code covering forests on military lands. In general the sponsors write that "human beings depend on native biological resources" so that "alteration of native biodiversity has serious consequences for human welfare."[61] In particular they believe that "alteration of biological diversity in Federal forests adversely affects the functions of ecosystems and critical ecosystem processes."[62]

The authors introduce the new paradigmist approach into the statutes that they amend by first giving the cabinet secretary in question a new management directive. In revising the FLPMA, for example, H.R. 1861 changes FLPMA to read that "regardless of any other provision of this Act, in each stand and each watershed throughout the forested area, the Secretary shall provide for the conservation or restoration of native biodiversity except during the extraction stage of authorized . . . development . . . in which events the Secretary shall conserve native biodiversity to the extent possible."[63] And what is native biodiversity? It is straight out of the new paradigmist mantra. "The term 'native biodiversity'," say backers of the act, "means the full range of variety and variability within and among living organisms and the ecological complexes in which they would have occurred in the absence of human impact."[64] Thus, Mother Nature, acting without human interference, produces the ideal landscape. Leaving nothing to chance, they write that native biodiversity

> encompasses diversity within a species (genetic diversity), among species (species diversity), within a community of species (within-community diversity), between communities (between community diversity), within a total area such as a watershed (total area diversity), along a plane from ground to sky (vertical diversity), and along the plane of the earth-surface (horizontal diversity). Vertical and horizontal diversity apply to all the other aspects of diversity.[65]

They go on to explain that

> the terms "conserve" and "conservation" refer to protective measures for maintaining existing native biological diversity and active and passive measures for restoring diversity through management efforts, in order to protect, restore, and enhance as much of the variety of species and communities as possible in abundances and distributions that provide for their continued existence and normal functioning, including the viability of populations throughout their natural geographic distributions.[66]

Overall, the ecological thinking of subscribers to this kind of language is hopelessly addled. The addition of the notions of horizontal and vertical diversity to the already Gordian idea of biodiversity is especially troublesome because it indicates that several members of the House of Representatives are willing to incorporate grossly uninformed language into law. I demonstrate in chapter 4 that scientists are unable to give officials a calculus to measure biological diversity as it is. What, then, would backers of H.R. 1861 expect federal land managers to do if they had to implement a congressional directive to simultaneously protect and restore horizontal and vertical diversity within all the multitude diversities identified in the bill? What, for example, is meant by "diversity between communities along a plane from ground to sky"? And just where does the sky begin or end for public policy purposes? Statutory text of this kind invites litigation, which the bill openly encourages.

Sponsors of the Forest Biodiversity Act of 1997 ensure that new paradigmists can use the courts to paralyze human activities anywhere the amended laws apply. They do so by adding language to provide the widest possible latitude for individuals or organizations to bring suit. Their "purpose . . . is to foster the widest possible enforcement" of the amendments because they find "that all people of the United States are injured by actions on the lands" to which the amendment applies.[67] The bill declares that "any citizen harmed by a violation of the Act [the amended statute] may . . . [bring] an action for declaratory judgment, temporary restraining order, injunction, civil penalty, and other remedies against any alleged violator including the United States, in any district court of the United States."[68] Efforts to improve human well-being would thus be made hostage to new paradigmists, who could invoke this language to make endless claims of injury and violations of procedural requirements as they currently do using the ESA.

The ESA is arguably the most powerful federal statute in the new paradigmist arsenal; "a law of great reach and power" according to Interior Secretary Bruce Babbitt.[69] Consider, for example, the small (staff of sixteen) Southwest Center for Biological Diversity in Tucson.[70] *Washington Post* staff writer Tom Kenworthy says that this group alone filed eighty-five (mostly successful) lawsuits over four years that use the ESA to force federal agencies to protect ecosystems.[71] He quotes the group's executive director, Kieran Suckling, as saying that "social change comes from social stress and our job is to intensify that stress until large-scale change results." Along with other new paradigmist organizations, such as the Forest Guardians based in Santa Fe, New Mexico, they employ the ESA to eliminate forestry and ranching over vast areas, such as those envisioned by Noss and others, so as to defend Mother Nature. Kenworthy notes that these environmentalists offer "no apologies for their assault on a way of life." Indeed, Suckling opines that "yes, we are destroying a way of life that goes back 100 years. But it is a way of life that is one of the most destructive in the country. . . . Ranching is one of the most nihilistic life styles this planet has ever seen. It should end. Good riddance." And good riddance, too, to farming, home building, mining, road construction, camping, and multitudinous other human infringements on Mother Nature that new paradigmists stop by using ESA provisions that prevent or limit human activity in ecosystems defined by the habitat needs of threatened or endangered species. Little wonder they strongly support congressional efforts to expand and strengthen the act.[72]

Representative George Miller of California introduced the Endangered Species Recovery Act of 1997 (ESRA; H.R. 2351) on July 31, 1997.[73] The bill, which amends the ESA, attracted eighty-nine cosponsors by February 1998 as well as enthusiastic backing from environmental organizations.[74] The authors begin the bill by announcing that Americans want to protect the "natural environmental legacy of this Nation."[75] Next they say, "It is only through the protection of all species of plants and animals and the ecosystems on which they depend that we will conserve a world for our children with the spiritual, medicinal, agricultural, and economic benefits that plants and animals offer. Moreover, we have a moral responsibility not to drive other species to extinction. . . . We are rapidly proceeding in a manner that will deny a world of abundant, varied species to future generations."[76] Representative Miller and the others tell Americans that undisturbed nature is good by definition, that plants and animals have spiritual value, and that we are on the way to destroying both unless we protect the ecosystems on which "all species of plants and animals" depend (which I take to mean all ecosystems). The bill signals an unmistakable move toward biocentrism, for the ESA presently contains no language declaring the existence of either a spiritual value inherent in plants and animals or a moral imperative to protect them.

The language of H.R. 2351 masks reality. For example, in proclaiming that "we are rapidly proceeding" down a path that will deny future generations a rich biological endowment, it makes assertions not supported by evidence from the landscape. Americans are not sending domestic biological diversity into a tailspin (see chapter 4). When they highlight the agriculture and economic benefits of plants and animals, authors of the bill neglect to mention that these benefits do not come from natural areas; rather, they originate from places where people have manipulated the landscape to improve human well-being. Moreover, the gains come overwhelmingly from exotic species that were never part of the natural environment of the United States. The bill advances the biocentric cause by failing to acknowledge our moral obligation to humans (one that most Americans would agree is certainly higher than our stewardship responsibilities to nonhuman species or ecosystems) and by increasing the ability of Suckling and like-minded new paradigmists to intensify social stress until the social change they desire is achieved.

ESRA would give new opportunities to environmentalists to dictate land use by using the courts. The authors write that "all members of the public have a right to be involved in the decisions made to protect biological diversity" and give substance to their belief by amending the ESA to make any "violation of this Act, any regulation or permit issued under this Act, any statement provided by the Secretary under section 7(b)(3) [which relates to consultations with other federal agencies regarding the impacts of proposals on threatened or endangered species], or any agreement concluded under authority of this Act" actionable in a federal court.[77] Not only would Miller provide new paradigmists ever greater access to the courts, but he gives them new things to sue over. The legislation imposes several new requirements on the secretary and landowners. Many of the fresh mandates are vague, and thus their fulfillment is wide open to legal challenge if not performed to the satisfaction of "any member of the public." For example, there is a new demand that the secretary and landowners protect species from direct and indirect cumulative impacts of "past, present, and reasonably foreseeable future action, regardless of what person undertakes such other actions."[78] And impacts are to include "diminishment of the species' habitat, both qualitatively and quantitatively," "disruption of normal behavior patterns," and "impairment of the species' ability to withstand random fluctuations in environmental conditions."[79] Not only are the ideas that the sponsors embody in this language indefinite, but the information that officials might use to operationalize them is largely nonexistent and its gathering beyond our resources. Just what are "random fluctuations in environmental conditions" anyway? In order to protect species from them, are affected parties now required to defend species from hurricanes, unusually cold or dry or hot or wet periods, or disease infestations? Should Congress adopt such indefinite and imprecise language, it would significantly increase the potential for mischief in the courthouse.

Representative Miller guarantees additional societal tension by facilitating new paradigmist involvement in the management of substantially more public and private land than is now the case. The bill expands the definition of species to include "the last remaining distinct population segments in the United States of any plant or invertebrate species," thereby extending its geographic reach.[80] Unlike current law, under which many species do not yet have specific habitat mapped out for protection, Miller would require the secretary to designate a new kind of protected area, "interim habitat," at the time any species is listed.[81] The act forbids the secretary from considering the human impacts of interim habitat designation when preparing the maps; "the Secretary shall designate interim habitat of a species based only on biological factors, giving special consideration to habitat that is currently occupied by the species."[82] All activities on lands appearing on these maps would be instantly subject to scrutiny by environmentalists intent on protecting Mother Nature.

Finally, H.R. 2315 assumes that the motives and ethics of new paradigmists are superior to those who advocate on behalf of humans. The bill requires the secretary to consult with the National Academy of Sciences to select "independent scientists" for the purpose of determining objective criteria on species recovery. But it specifically prohibits the secretary from involving otherwise qualified scientists if they have ties to affected economic interests.[83] In other words, a university ecologist who consults for the Sierra Club could serve on a panel, but an ecologist in the same department who consults for a timber company would be disqualified. The sponsors believe that individuals and organizations seeking to protect Mother Nature are morally superior to other people and groups.

The 105th Congress held no hearings on the bills I discuss in this section, partly because of growing concern with the new paradigmist vision of ecosystem management and its significance for the nation. In the next section I review some of these concerns as expressed in congressional hearings in the 1990s.

Congressional Concerns with Ecosystem Management

Congressional skepticism of ecosystem management grew during the 1990s as it moved from concept to concrete land use management proposals. In July and September 1993, the House of Representatives held hearings on H.R. 1845, the National Biological Survey Act of 1993.[84] As introduced by Representative Gerry Studds of Massachusetts, the bill would have created the National Biological Survey as an agency within the Department of the Interior, in part to provide "information to be used in protecting and managing ecosystems, including their plant, fish, and wildlife components."[85] In his statement endorsing the ecosystem management, Representative Jim Saxton of New Jersey says that "it is time that resource management take a more long-term, ecosystem approach, and that our science look more comprehensively at the status and trends of natural systems and what causes declines in their health and abundance."[86] The act passed the House, but the Senate did not take up this bill. Some members expressed concerns with the bill at these hearings, but their focus was on private property, not on the applicability of the ecosystem concept to public policy. They feared that data gathered by the survey in the name of ecosystem management would be used in conjunction with the ESA to repress "people's rights" in favor of "the rights of woodpeckers and swamp creatures."[87] Jay Dickey of Arkansas was one of the few members to demonstrate any understanding of the ecosystem concept when he asked Secretary Babbitt, the administration witness, "What is an ecosystem, how will it be defined, and how will you differentiate one from another?"[88]

In March 1993, recognizing that congressional understanding of ecosystem management was shallow at best, influential supporters in the House of Representatives (George Miller, Gerry Studds, Charlie Rose, and Norm Dicks) requested that the General Accounting Office (GAO) examine ecosystem management and its status among federal land management agencies.[89] The GAO report reveals many of the public policy shortcomings of the ecosystem concept, and three committees held a joint hearing on it in September 1994.[90] At the hearing, Miller and Studds, while remaining supportive of ecosystem management, nonetheless acknowledge difficulties with its implementation. Miller says that, if done right, "ecosystem management has great potential for managing resources. . . . If done wrong, we could see added layers of bureaucracy," and Studds observes that "as the GAO report indicates, there are many serious questions to be dealt with prior to fully implementing any type of ecosystem management."[91]

At this hearing representatives go beyond concern for property rights and begin to seriously question the value of the ecosystem notion itself to natural resource and land use decision making. Congressman Charles Taylor of North Carolina says that "the concept is founded upon a misguided idea based more on early 19th century romanticism than 20th century scientific knowledge."[92] Ecosystem management, he thinks, is "nebulous" and "subjective," and "no one who is serious about resource management has a clue what it means."[93] Dickey wonders, "What does Ecosystem Management mean?" and when James Hansen of Utah asks whether anyone in the hearing room has a definition of ecosystems, Miller responds, "Well, you could probably get a hundred different definitions."[94] Critics exhibit a growing understanding of the arbitrary nature of ecosystems and their boundaries. Richard Pombo of California wonders, "Where does one ecosystem begin and another one end and what criteria do you establish to determine the range of an ecosystem?"[95] The new paradigm lost ground at the hearing and lost even more headway during a May 1997 joint Senate-House hearing on the Interior Columbia Basin Ecosystem Management Plan (ICMEMP; see chapter 7).

According to its supporters, the ICBEMP will bring environmental and social enlightenment to large portions of Oregon, Washington, Idaho, and Montana and will increase efficiencies in management decision making and resource use throughout the region. Thus, it is remarkable that not a single member of the congressional delegations from these states stepped forward at the hearing to support this proposal to implement ecosystem management for the region. At the hearing members criticized the effort on several fronts. Senator Slade Gorton faults its underlying philosophy that "the state of nature is the ideal situation and that everything human beings do to it is bad."[96] Senator Craig Thomas of Wyoming asks, "What is an ecosystem?" and gets the now familiar response, in this case from regional forester Bob Williams, that "it can be described in different ways depending on what the factors are that you use."[97] "The concept of ecosystem management itself is ill-defined," says Representative George Nethercutt of Oregon.[98] Members note that the plan requires new levels of management, offers multiple new opportunities for litigation, and will inevitably result in "meta gridlock" because the proposal is "impossible to implement."[99] For many new paradigmists this is precisely the point: to put something in place that is impossible to implement and then use it as a tool to force people to leave the region. Senator Dirk Kempthorne of Idaho says that "we have to ask ourselves, do we want viable industries supporting our rural communities or do we want to chase our population to urban areas in search of jobs?"[100] The answer that the senator would receive from Foreman, Noss, Suckling, and many other new paradigm champions is "chase the people off the land and into urban islands;" that, after all, is exactly what the Wildlands Project and so many other environmentalist efforts to forbid or limit human use of the land are all about.

New paradigmists are attempting to remake society to benefit Mother Nature. In the next chapter, I examine more closely what their agenda means for people.

Notes

1. *Wild Earth* (special issue, 1992). *Wild Earth* is the magazine of the Wildlands Project, which is the popular name not only for the initiative itself but also for the organization behind it. It is incorporated in the state of Arizona as a nonprofit educational, scientific, and charitable organization with offices in Tucson. The plan was also presented at the June 1993 meeting of the Society of Conservation Biology in Tempe, Arizona. The outline of the project is available at its Web page at http://www.wild-land.org/html/about.html.

2. Unless otherwise noted, the quotes in this paragraph are from "The Wildlands Project Mission Statement," *Wild Earth* (special issue, 1992), 3–4.

3. Reed Noss, "The Wildlands Project: Land Conservation Strategy," *Wild Earth* (special issue, 1992), 13.

4. "The Wildlands Project Mission Statement," *Wild Earth* (special issue, 1992), 4.

5. Dave Foreman, "Around the Campfire," *Wild Earth* (special issue, 1992), inside front cover.

6. Charles C. Mann and Mark L. Plummer, "The High Cost of Biodiversity," *Science* 260 (June 25, 1993): 1868.

7. Noss, "The Wildlands Project," 15.

8. "The Wildlands Project Mission Statement," 4.

9. "The Wildlands Project Mission Statement," 4.

10. Noss, "The Wildlands Project."

11. Noss, "The Wildlands Project," 11.

12. Noss, "The Wildlands Project," 11.

13. Noss, "The Wildlands Project," 13.

14. Noss, "The Wildlands Project," 14.

15. Mann and Plummer, "The High Cost of Biodiversity," 1868.

16. Mann and Plummer, "The High Cost of Biodiversity," 1868.

17. Southern Rockies Ecosystem Project, "Accomplishments of the Southern Rockies Ecosystem Project 1992–1996," downloaded from the project's Web page at http://csf.colorado.edu/srep/history.html, January 26, 1998.

18. Southern Rockies Ecosystem Project, "Accomplishments of the Southern Rockies Ecosystem Project 1992–1996."

19. Southern Appalachian Forest Coalition, "Index" and "Member Groups," downloaded from the coalition's Web page at http://www.safc.org/index.html and http://www.safc.org/members.html,January 20, 1998.

20. Southern Appalachian Forest Coalition, "Conservation Planning," downloaded from the coalition's Web page at http://www.safc.org/conplan.html, January 20, 1998.

21. Southern Appalachian Forest Coalition, "Conservation Planning." The coalition cites Noss and Cooperrider, *Saving Nature's Legacy,* 89–93, as the source of these objectives.

22. Southern Appalachian Forest Coalition, "Biological Reserves," downloaded from the coalition's Web page at http://www.safc.org/biores.html,January 20, 1998. See especially "Conservation Areas in the Southern Blue Ridge," map, at http://www.safc.org/images/fndmap.JPG, downloaded January 20, 1998.

23. Southern Appalachian Forest Coalition, "Calendar News," downloaded from the coalition's Web page at http://www.safc.org/calendar.html, on January 20, 1998.

24. U.S. Forest Service, "Questions and Answers—National Forest Transportation System," downloaded from the Forest Service's Web page at http://www.fs. fed.us/news/roads/19980121_qa.html, January 26, 1998. See also USDA Forest Service, "Forest Service Protects Roadless Areas and Announces Development of New Transportation Policies," press release, January 22, 1998, downloaded from http://www.fs.fed.us/news/roads/19980122_roads.html, January 26, 1998, and Department of Agriculture, Forest Service, "Advance Notice of Proposed Rulemaking—Administration of the Forest Development Transportation System," downloaded from http://www.fs.fed.us/news/roads/ab67.html, January 26, 1998.

25. As quoted by Elizabeth Pennisi, "Conservation's Ecocentrics," *Science News* 144 (September 11, 1993): 168.

26. Noss and Cooperrider, *Saving Nature's Legacy,* 121.

27. Rodger Schlickeisen, "The Argument for a Constitutional Amendment to Protect Living Nature," in *Biodiversity and the Law,* ed. William J. Snape III (Washington, D.C.: Island Press, 1996), 221–42.

28. Schlickeisen, "The Argument for a Constitutional Amendment to Protect Living Nature," 221, 230.

29. Schlickeisen, "The Argument for a Constitutional Amendment to Protect Living Nature," 223.

30. Schlickeisen, "The Argument for a Constitutional Amendment to Protect Living Nature," 234.

31. Schlickeisen, "The Argument for a Constitutional Amendment to Protect Living Nature," 234.

32. Dave Foreman, *Confessions of an Eco-Warrior* (New York: Harmony Books, 1991), 187–90.

33. David Brower, *Let the Mountains Talk, Let the Rivers Run* (New York: HarperCollinsWest, 1995), 56.

34. Noss and Cooperrider, *Saving Nature's Legacy,* 335. For a description of their Native Ecosystems Act, Noss and Cooperrider refer to Reed F. Noss, "From Endangered Species to Biodiversity," in *Balancing on the Brink of Extinction,* ed. Kathryn A. Kohm (Washington, D.C.: Island Press, 1991), 227–46. Support for an endangered ecosystem act is also offered by Anne H. Ehrlich and Paul R. Ehrlich, "Needed: An Endangered Humanity Act," in Kohm, ed., *Balancing on the Brink of Extinction,* 298–302, and Malcom Hunter Jr., "Coping with Ignorance: The Coarse-Filter Strategy for Maintaining Biodiversity," in Kohm, ed., *Balancing on the Brink of Extinction,* 266–81.

35. Noss and Cooperrider, *Saving Nature's Legacy,* 89.

36. Noss and Cooperrider, *Saving Nature's Legacy,* 168.

37. Noss and Cooperrider, *Saving Nature's Legacy,* 107.

38. Gary D. Meyers, "Old Growth Forests, the Owl, and Yew: Environmental Ethics versus Traditional Dispute Resolution under the Endangered Species Act and Other Public Lands and Resources Laws," *Boston College Environmental Affairs Law Review* 18, no. 4 (1991): 623–68.

39. Meyers, "Old Growth Forests, the Owl, and Yew," 660.

40. Meyers, "Old Growth Forests, the Owl, and Yew," 661.

41. Meyers, "Old Growth Forests, the Owl, and Yew," 662.

42. Robert B. Keiter, "Conservation Biology and the Law: Assessing the Challenge Ahead," *Chicago-Kent Law Review* 69, no. 4 (1994): 932–33.

43. Robert B. Keiter, "Beyond the Boundary Line," *University of Colorado Law Review* 65, no. 2 (1994): 293–333, and "Conservation Biology and the Law: Assessing the Challenge Ahead," *Chicago-Kent Law Review* 69, no. 4 (1994): 911–33.

44. Keiter, "Beyond the Boundary Line," 328.

45. Keiter discusses the content of his overall ecosystem protection legislation in "Beyond the Boundary Line."

46. Keiter, "Beyond the Boundary Line," 329.

47. Keiter, "Beyond the Boundary Line," 329.

48. Scott W. Hardt, "Federal Land Management in the Twenty-First Century: From Wise Use to Stewardship," *Harvard Environmental Law Review* 18, no. 2 (1994): 345–403; Michael C. Blumm, "Public Choice Theory and the Public Lands: Why 'Multiple Use' Failed," *Harvard Environmental Law Review* 18, no. 2 (1994): 405–32.

49. Hardt, "Federal Land Management in the Twenty-First Century," 392.

50. Blumm, "Public Choice Theory and the Public Lands," 429.

51. Blumm, "Public Choice Theory and the Public Lands," 432.

52. U.S. House, The Northern Rockies Ecosystem Protection Act of 1997, 105th Cong., 1st sess., 1997, H.R. 1425. This is not the first manifestation of the bill. The Northern Rockies Ecosystem Management Act of 1995 was introduced in the House, 104th Cong., 1st sess., 1995, as H.R. 852.

53. Northern Rockies Ecosystem Protection Act of 1997, sec. 2 (1).

54. Northern Rockies Ecosystem Protection Act of 1997, sec. 2 (4).

55. Northern Rockies Ecosystem Protection Act of 1997, sec. 2 (6).

56. Northern Rockies Ecosystem Protection Act of 1997, sec. 3 (b)(1–5).

57. Northern Rockies Ecosystem Protection Act of 1997, sec. 301 (b); 302 (b).

58. Northern Rockies Ecosystem Protection Act of 1997, sec. 502 (b).

59. Northern Rockies Ecosystem Protection Act of 1997, sec. 503 (a)(1).

60. U.S. House, Forest Biodiversity Act of 1997, 105th Cong., 1st sess., 1997, H.R. 1861.

61. Forest Biodiversity Act of 1997, sec. 2 (b)(10–11).

62. Forest Biodiversity Act of 1997, sec. 2 (b)(12).

63. Forest Biodiversity Act of 1997, sec. 4 (a)(2).

64. Forest Biodiversity Act of 1997, sec. 4 (b).

65. Forest Biodiversity Act of 1997, sec. 4 (b).

66. Forest Biodiversity Act of 1997, sec. 4 (b).

67. Forest Biodiversity Act of 1997, sec. 4 (b).

68. Forest Biodiversity Act of 1997, sec. 4 (b).

69. For a review of the science behind the act, see National Research Council, *Science and the Endangered Species Act* (Washington, D.C.: National Academy Press, 1995).

70. "Staff" downloaded from www.sw-center.org, February 4, 1998.

71. Tom Kenworthy, "In the Desert Southwest, a Vigorous Species Act Endangers a Way of Life," *Washington Post,* February 1, 1998, A3.

72. Environmental groups such as Defenders of Wildlife, National Audubon Society, Sierra Club, and Sierra Club Legal Defense Fund support an Endangered Natural Heritage Act to expand the reach of the ESA and the ability of private actors to use it to prevent human use of the land. See "The Endangered Natural Heritage Act," available at the National Audubon Society's Web site at http://www.audubon.org/campaign/esa/enha.html, downloaded February 5, 1998.

73. U.S. House, Endangered Species Recovery Act of 1997, 105th Cong., 1st sess., 1997, H.R. 2351.

74. See, for example, Sierra Club, "Finally, a Bi-Partisan Bill to Save Wild Places for Wild Life ESRA Takes First Positive Steps to Improve Endangered Species Act," press release, September 3, 1997, downloaded from the Sierra Club's Web site at http://www.sierraclub.com/news/releases/0015.html, February 4, 1998, and National Wildlife Federation "Take a Proactive Stance on Endangered Species Protection by Cosponsoring H.R. 2351," "Take Action" statement, downloaded from the National Wildlife Federation's Web site at http://www.nwf.org/nwf/endangered/leg/lz12hr23.html, February 4, 1998.

75. Endangered Species Recovery Act of 1997, sec. 2 (1).

76. Endangered Species Recovery Act of 1997, sec. 2 (2–3).

77. Endangered Species Recovery Act of 1997, secs. 2 (6), 109 (1).

78. Endangered Species Recovery Act of 1997, sec. 101 (3).

79. Endangered Species Recovery Act of 1997, sec. 101 (3).

80. Endangered Species Recovery Act of 1997, sec. 101 (3).

81. Endangered Species Recovery Act of 1997, sec. 102 (1).

82. Endangered Species Recovery Act of 1997, sec. 102 (1).

83. Endangered Species Recovery Act of 1997, sec. 105 (1).

84. U.S. Congress, House Subcommittee on Environment and Natural Resources of the Committee on Merchant Marine and Fisheries and the Subcommittee on National Parks, Forests, and Public Lands of the Committee on Natural Resources, *Joint Hearing on H.R. 1845, The National Biological Survey Act of 1993,* 103rd Cong., 1st sess., July 15, 1993, Merchant Marine Committee Print 103–56 (hereafter cited as *July 1993 Hearing*); U.S. Congress, House Subcommittee on Technology, Environment and Aviation, and the Subcommittee on Investigations and Oversight of the Committee on Science, Space, and Technology, *Hearing on H. R. 1845, The National Biological Survey Act of 1993,* 103rd Cong., 1st sess., September 14, 1993, Committee Print 56 (hereafter cited as *September 1993 Hearing*).

85. U.S. House, The National Biological Survey Act of 1993, 103rd Cong., 1st sess., 1993, H.R. 1845.

86. H. James Saxton, *July 1993 Hearing,* 4.

87. Dana Rohrabacher, *September 1993 Hearing,* 35

88. Jay Dickey, *July 1993 Hearing,* 15.

89. U.S. General Accounting Office, *Ecosystem Management: Additional Actions Needed to Adequately Test a Promising Approach,* GAO/RCED-94–111 (Washington, D.C.: U.S. General Accounting Office, August 1994). At the time of the request, Miller chaired the Natural Resources Committee, Studds chaired the Merchant Marine and Fisheries Committee, and Rose chaired the Subcommittee on Specialty Crops and Nat-

ural Resources of the Agriculture Committee. Miller's Natural Resources Committee held a series of field hearings on ecosystem management and, to praise the approach, issued a Majority Staff Report, "Ecosystem Management: Sustaining the Nation's Natural Resource Trust," on April 30, 1994.

90. U.S. Congress, House Subcommittee on Oversight and Investigation of the Committee on Natural Resources and the Subcommittee on Specialty Crops and Natural Resources of the Committee on Agriculture and the Subcommittee on Environment and Natural Resources of the Committee on Merchant Marine and Fisheries, *Joint Oversight on Ecosystem Management,* 103rd Cong., 2nd sess., September 20, 1995, Natural Resources Committee Print 103–118 (hereafter cited as *September 1995 Hearing*).

91. Miller, *September 1995 Hearing,* 5; Studds, *September 1995 Hearing,* 11.

92. Taylor, *September 1995 Hearing,* 18.

93. Taylor, *September 1995 Hearing,* 18, 61–65. The subjectivity of the concept is a theme that also runs through the questions Congressman Harold Volkmer of Missouri put to hearing witnesses.

94. Hansen and Miller, *September 1995 Hearing,* 55–56; Dickey, *September 1995 Hearing,* 15.

95. Pombo, *September 1995 Hearing,* 8.

96. U.S. Congress, Senate/House Subcommittee on Forests and Public Land Management of the Senate Committee on Energy and Natural Resources and the Subcommittee on Forests and Forest Health of the House Committee on Resources, *Joint Hearing on the Interior Columbia Basin Ecosystem Management Plan,* 105th Cong., 1st sess., May 15, 1997, 33 (hereafter cited as *May 1997 Hearing*).

97. *May 1997 Hearing,* 38.

98. *May 1997 Hearing,* 8.

99. Bob Smith, *May 1997 Hearing,* 11.

100. *May 1997 Hearing,* 3.

Chapter 9

Human Consequences of the New Paradigm

> He has erected a multitude of new offices, and sent hither swarms of officers to harass our people.
>
> —Declaration of Independence

Full implementation of the new paradigm will reach the foundations of the nation, and that is exactly what many of its backers intend. Recall political scientist Lynton Caldwell's conclusion that "a strategy common to the environmental movement everywhere is to revise the ground rules under which society operates."[1] For Vice President Al Gore that means we need a "wrenching transformation of society."[2] Caldwell views "human rights, economic interests, and national sovereignty" as barriers to the correct relationship between people and the earth.[3] In his book *The Shaping of Environmentalism in America,* environmentalist Victor Scheffer writes that "an economy that rests on private ownership of land . . . on 'healthy growth,' and on profits . . . is one that strains the laws of both nature and ethical society."[4] Philosophy professor Laura Westra considers the "sacred cow" of democracy as an impediment to reaching an appropriate human relationship with Mother Nature.[5]

Stripped of obfuscation, the new paradigmist effort to make protection and restoration of ecosystem integrity, health, or sustainability the centerpiece of federal natural resource and environmental policy translates into collectivized national land use planning and management.[6] The United States needs, according to Caldwell and philosophy professor Kristen Shrader-Frechette, a national land use system "with allocations of responsibility appropriate to all levels of government and to nongovernmental institutions" because we "cannot rely on individual self interest, local determination, or corporate judgment for optimal land-use decisions."[7] This new paradigm world means more centralized federal control, regulation, and intrusion into the lives of the citizenry. It means more authoritarian and less democratic decision making. It means a large new federal bureaucracy and the aggregation of power over land use and natural resource decision making into the hands of regulators and special interests unaccountable to the public. It means a significant erosion of property rights, individual freedoms, and liberty. Advocates of ecosystem management rarely identify such consequences, and many might strongly reject my claim that they would occur. Yet, as political scientist Michael Greve observes,

> At bottom environmentalism is an ideology . . . [that] views common law rights—such as private property and freedom of contract—as a menace to an imperiled planet. It therefore aims to eviscerate common-law rights and to replace them with a legal regime that would organize transactions among individual citizens for a single public purpose, environmental protection. Environmentalism thus pushes toward a centralized, unlimited political scheme. To the extent that this scheme allows for "rights," they are defined by public purposes. . . . But no property or any other right could limit or provide a defense against the regulatory system.[8]

One area where many new paradigmists are quite forthright in their demands for fundamental change is their attack on private property rights. Joseph Sax, the noted new paradigmist law professor and former counselor to Secretary of the Interior Bruce Babbitt, writes that "an ecological worldview presents a fundamental challenge to established property rights."[9] I begin this chapter by looking at ecosystem management and property rights.

Impacts on the Rights of Property Owners

Students of American history accept the proposition that concerns for property have a central role in the development of the United States.[10] We inherit the idea that the ability to acquire and hold property is a bedrock for individual freedom and liberty from the English. Blackstone, in his *Commentaries on the Laws of England,* writes that "the third absolute right, inherent in every Englishman, is that of property."[11] History professor Lawrence Leder finds that "there was an obvious interplay in the minds of eighteenth century Americans between the concepts of liberty and property."[12] He sums up the view by quoting an essayist writing in 1735: "As for liberty, it cannot be bought at too great a rate; life itself is well employed when 'tis hazarded for liberty. . . . As for property, it is so interwoven with liberty that whenever we perceived the latter weakened, the former cannot fail to be impaired."[13] The founders of the country believed that private property—the fruits of one's labor, diligence, and intellect—was so vital to freedom that they listed it along with life and liberty as fundamental aspects of human existence that the law is to protect.[14] Yale professor of political science Robert Dahl concludes that the framers of the Constitution constructed a "framework of government that would . . . rest on popular consent and yet ensure as best they knew how the preservation of certain basic rights to life, liberty and property that they held to be morally inalienable."[15] Thus, the Fifth Amendment to the Constitution (ratified December 15, 1791) directs that "No person shall . . . be deprived of life, liberty, or property, without due process of law; nor shall private property be taken for public use, without just compensation." Later commentators restate the importance of property. Nobel Prize-winning economist Friedrich von Hayek writes in his classic 1944 book *The Road to Serfdom* that "the system of private property, is the most important guaranty of freedom, not only for those who own property, but scarcely less so for those who do not."[16] More recently, Roger Meiners, University of Texas at Arlington professor of law and economics, writes that

> if traditional property rights are lost in favor of legislated control of property, a major cornerstone of liberty has been lost. We come closer to being like most peoples in the world—granted certain favors (called "rights") at the pleasure of the legislature, but having few rights that may not be invaded by the legislature and the agencies it creates to execute its wishes.[17]

This is not to say that property owners can do anything they please inside their property boundary, for they cannot now and never could. Our legal system fully acknowledges that the public and neighboring landowners have legitimate interests in activities on private land, and so it offers multiple avenues (e.g., laws of nuisance and use of police powers) for both public and private actors to influence those activities. The extensive body of environmental law at federal, state, and local levels, plus innumerable local land use and planning regulations across the county, offers ample proof that property owners do not have free rein within the boundaries of their own land. American society continuously seeks to find the proper balance between the rights of property owners and the interests of neighbors and the public.

New paradigmists view the tension between private rights and public interest as one of morality and survival, not as a question of balance. They see constitutional and other legal protections afforded to property owners as the major obstacles to the imposition of their agenda on society. Consequently, they attack property rights as anachronisms of a bygone day that Americans must once and for all subordinate to the higher good as proclaimed by champions of the new paradigm.

Syracuse University law and public policy professor James Karp offers a typical new paradigmist view of property rights. Believing that we face environmental calamity largely at the hands of private property owners, he claims that "the privileges of private property won in recent centuries have in significant manner created the problem. . . . The human community's interest in survival must be the privileged right. The natural systems upon which we as a community rely for survival must be protected, and if private property rights must be sacrificed to achieve that protection, so be it."[18] But as I show in chapter 4, the status of the environment in the United States is generally good and getting better. Americans do not face an environmental crisis. Conditions on the landscape of the United States in no way justify new paradigmists' efforts to drastically undermine constitutionally protected rights of property owners or to significantly redefine those rights to include a responsibility to protect ecosystems.

Notwithstanding the reality of the landscape, some new paradigmists want property rights taken away from individuals and vested in an activist state. Caldwell and Shrader-Frechette suggest abandoning traditional ideas of property ownership. Instead, they favor a new regime in which the state would allow people to own rights only to specific socially and ecologically acceptable uses of the land, but no one could own the land itself. They see the "ultimate repository" of property rights as being society at large, and "their administration would be through government."[19] According to Karp, Americans must "reorder the rights of the individual and the rights of the community regarding land."[20] He believes that reprioritization of property rights is required to free ourselves from the "tyranny of millions of people making small, insular decisions, for narrow, selfish purposes" that threatens our existence.[21] But who are these "millions of people"? The "tyrannizing masses," as Karp calls them, are citizens of yesterday, today, and tomorrow. They are you and I, our neighbors, our parents, and our children. What are "insular decisions"? They are decisions made without the approval of an as-yet-to-be-created centralized national planning and land use authority (most land use decisions already require approval of local planning bodies and in some cases state and federal agencies as well). What are "narrow, selfish purposes"? They are decisions that people make to improve the quality of their lives, such as a farmer clearing land to plant a crop in order to earn a living or a couple building a retirement home.

Altering property rights is also appropriate now because, in Karp's judgment, "the times have changed. This country is two hundred years beyond autocratic rule when individuals need broadly constructed private property rights to protect them from an overbearing government."[22] From personal experience I find this claim other-worldly. As a member of the White House Interagency Wetlands Task Force in the Bush administration, I participated in all-day public meetings around the country where hundreds of ordinary citizens as well as state and local elected officials often drove long distances and waited many hours to spend three minutes before a microphone to plead—often in anger and frustration and a few times in tears—that we make changes in wetland regulations "to protect them from an overbearing government," to use Karp's words.[23] The so-called Sage Brush Rebellion of the 1980s, the emergence of the property rights movement in the mid-1990s, and the level of the public complaints against the excesses of the Endangered Species Act (ESA) and wetlands regulators refute Karp's belief that government is not now or cannot be "overbearing."[24]

Karp and others propose gutting property rights through redefinition. They would make landowners first and foremost custodians of "ecosystem needs" and would make private use subordinate to public "demands for the maintenance of natural resources, *even where the private owner's property is left valueless* [emphasis in the original]."[25] His reordering of property rights requires government to place what he calls a new stewardship responsibility on property owners by demanding that they "use and maintain the land in a manner that will not interfere with any significant natural resource value."[26] New paradigmist law professor Eric Freyfogle proclaims that "property stills serves as a bulwark against an overreaching state," yet he would destroy the bulwark of property rights by including within them an ownership responsibility to "maintain a natural harmony" with the earth.[27] Property rights, in this view, must reach beyond concern with impacts on people to include effects on ecosystems and their component parts or the entire natural community. Even though researchers disclaim the notion of "harmony" and ecologist Robert Peters explains that the phrase "everything is connected to everything else" is an empty concept, not a statement of fact, Freyfogle nonetheless invokes ecologists and ecology as requiring a change in property rights.[28] He writes that "ecologists have explained . . . the world around us is intricate and intertwined. Each part of nature is attached to each other part," so in his new view of property rights, society and the law must see a person who owns land "as part of a natural as well as social community, with all of the obligations that accompany that status."[29] Appealing to the now-sacred precept of sustainability, Freyfogle tells us that the lawmakers' task is to require landowners to engage in only sustainable practices.

The Takings Clause of the Fifth Amendment to the Constitution stands in the way of new paradigmist undermining of property rights. As Washington University law professor Richard Lazarus puts it, "To a great extent the takings clause has been left to serve as the final constitutional barrier to the attack on private property in natural resources [law] prompted by expanding notions of nuisance doctrine and police power legislation."[30] The Takings Clause does not prevent the government from taking private property for public purposes, but it does require the government to pay for what it takes. The notion of paying for what you expropriate is a troubling idea for some people. "For those of us who are deeply interested in preserving open spaces," says Senator John Chafee of Rhode Island, "this idea of compensation could be ruinous."[31] New paradigmists and others so object to the implications of the government adhering to the intent of the Takings Clause that President Bill Clinton threatened to veto legislation that simply streamlined the process for allowing a landowner to get a takings claim into the courthouse.[32] Citing a Congressional Budget Office cost estimate of between $10 billion and $45 billion, the White House finds the suggestion that the federal government protect high-value wetlands by buying them to "be infeasible from a budgetary standpoint."[33] It is no wonder that new paradigmist attorney David Hunter, writing in the *Harvard Environmental Law Review,* finds that the greatest legal challenge to imposing their views on society "is the Takings Clause of the Fifth Amendment which has been labeled 'the major obstacle to effective environmental land-use regulation.' "[34] Like Rodger Schlickeisen (see chapter 8), Caldwell and Shrader-Frechette point to the need for a constitutional amendment to create what they term "a principled public policy on 'takings' " that would pull "environmental and transgenerational considerations" into Takings adjudications.[35] Of course, new paradigmists would appoint themselves to represent the environment and future generations in such proceedings.

Takings jurisprudence has a long history as the courts seek to find a balance between individual rights and the public interest.[36] The courts have always deemed the physical seizing of property by government to be a compensable taking under the Fifth Amendment. If the government, for example, takes private property so that it can build a road, then it must pay for the land. Matters become more cloudy when the issue is regulatory taking, that is, when government enacts laws and imposes regulations that restrict land use. The question becomes, At what point does the legitimate use of police power by government cross the line and become a taking? Supreme Court Justice Oliver Wendell Holmes, in writing for the court in *Pennsylvania Coal Co. v. Mahon* (1922), observes that "the general rule at least is, that while private property may be regulated to a certain extent, if regulation goes too far it will be recognized as a taking."[37] The vexing problem for the court is deciding where that line might be. In 1978, more than fifty years after Justice Oliver Wendell Holmes established the general rule, Justice William Brennan, in delivering the opinion of the court in *Penn Central Transp. Co. v. New York City,* writes,

> The question of what constitutes a "taking" . . . has proven to be a problem of considerable difficulty. While this Court has recognized that the "Fifth Amendment's guarantee [is] designed to bar Government from forcing some people alone to bear public burdens which, in all fairness and justice, should be borne by the public as a whole," this Court, quite simply, has been unable to develop any "set formula" for determining when "justice and fairness" require that economic injuries caused by public action be compensated by the Government, rather than remain disproportionately concentrated on a few persons.[38]

Despite such hand-wringing about "justice and fairness," over time the court has permitted a wide range of "public burdens" to be "disproportionately concentrated on a few persons." Justice John Paul Stevens, in his dissent in *Lucas v. South Carolina Coastal Council,* writes that "it is well established that a 50 percent diminution in value does not by itself constitute a taking."[39] But the court continues to look at the takings question. The most important recent takings case directly affecting the new paradigm is *Lucas,* decided in 1992.

David Lucas bought two beachfront lots in South Carolina in 1986 with the intention of building a house on each, one for himself and one for sale. In 1988 the state of South Carolina enacted legislation creating zones wherein no new beachfront construction could take place. The act covered Lucas's property. Forbidden to build, he sued the state for compensation. He won in the trial court, but the state supreme court reversed, and Lucas appealed to the U.S. Supreme Court. The Court held that since Lucas was denied 100 percent of the economic value of his land, a taking occurred, unless preexisting laws in South Carolina would otherwise prevent his building. The Court remanded the case to the South Carolina Supreme Court to determine whether such laws existed; they did not, and Lucas ultimately settled for $850,000 in compensation plus $725,000 in interest and costs. Now the owner of two beachfront lots, the state of South Carolina, so willing to deprive Lucas of the value of the property to pursue a public benefit, was not willing to spend public money in the same cause and sold the lots to a developer.[40]

Legal scholars of all turns of mind analyzed *Lucas*. In his examination of *Lucas,* Sax argues that Justice Antonin Scalia, writing the opinion for the court, wants to clearly convey the message that "states may not regulate land use solely by requiring landowners to maintain their property in its natural state as part of a functioning ecosystem. In this sense, while the *Lucas* majority recognizes the emerging view of land as part of an ecosystem, rather than purely as private property, the Court seeks to limit the legal foundation for such a conception."[41] In Sax's judgment, the *Lucas* Court broadly confronts laws intended to preserve ecological services provided by land in a natural state. He concludes that "if the South Carolina regulation had been sustained, the decision would have constitutionalized a broad panoply of laws requiring landowners to leave their property in its natural condition. The opinion recognizes that, in the name of environmental protection, an entirely new sort of regulation could be imposed."[42] The Court acts "to prevent such a result" and "effectively reverses" a 1972 decision of the Wisconsin Supreme Court that rejected a takings claim because it found that the state of Wisconsin had a proper police power interest in preserving "nature, the environment, and natural resources as they were created and to which the people have a present right."[43]

Sax faults the Court for not adopting the new paradigm. Its "outdated view of property . . . is not satisfactory in an age of ecological awareness" that requires people to "accommodate ecosystem demands" and recognize that "any disruption of natural conditions can be viewed as harmful."[44] Sax's legal reasoning about the implications of *Lucas* might well be correct; however, his claim that science now requires Americans to make drastic revision to takings jurisprudence is not. In the first instance, Sax incorrectly presupposes that ecosystems are living objects on the landscape possessing needs. Mark Sagoff, in his analysis of ecology and takings, points out that ecosystems are just "a pointless hodgepodge of constantly changing associations of organisms and environments."[45] As I demonstrate in earlier chapters, ecosystems are but arbitrary human constructs. They are neither alive nor real entities on the landscape, so how can they make demands that must (or can) be recognized in the law? His second point, that deviation from natural conditions is harmful, is equally unsupportable by appeals to science and instead is a fixture of the myths of balance and harmony in nature (see chapter 6).[46]

Greve concurs with Sax's analysis that the *Lucas* Court realized that their affirming the South Carolina Supreme Court would give constitutional endorsement to the new paradigm and would fundamentally change property rights.[47] Unlike Sax, however, Greve applauds the Court's decision because, in part, "instead of being taken in by environmental rhetoric . . . Lucas treats it as a subterfuge for extortion."[48] He sees the decision as evidence that the Court is intentionally rejecting an environmentalist perspective it embraced in recent decades.[49] Greve poses the question that new paradigmists do not want to answer: "If ecological values are so important and so widely shared, why would the public not be willing to pay" to protect them as required by the Takings Clause?[50]

But suppose that new paradigmists could find a way to have the federal government protect ecosystems without triggering the Takings Clause. And suppose that they could find a way to force the federal government to protect ecosystems even if Congress and a non-new-paradigmist president declared that doing so was not in the national interest. Now suppose that they could find a way to combine these two conditions. From a new paradigmist's perspective, this would create a nearly perfect legal and policy situation in which they could force their agenda on society without regard to the Constitution, formalities of democratic decision making, or impacts on people. Because the public trust doctrine offers an opportunity to accomplish these things, it attracts strong support from purveyors of ecosystem management.

The New Paradigm and the Public Trust Doctrine

University of Washington law professor William Rodgers Jr. describes the public trust doctrine as "resoundingly vague, obscure in origin and uncertain of purpose" but possessing "both radical potential and indifferent prospects."[51] Essentially, the doctrine declares that government possesses an inalienable trust responsibility to protect certain natural resources for the public. This duty cannot be avoided or delegated to others, and because it is an inherent function of government, it pre-dates and supersedes any claim of rights that a property owner might make. David Hunter finds that "the public trust doctrine permits states to avoid traditional takings inquiries when they are merely fulfilling their obligations as trustees of the public's interest in private lands. The doctrine is important because it empowers states to place onerous regulations on certain lands—most frequently tidelands—while successfully avoiding the takings problem."[52] In other words, wherever it applies geographically, the government must fulfill its trust responsibility, and actions done in the name of meeting public trust obligations do not require compensation under the Takings Clause.[53] As Hunter suggests, American jurisprudence spatially links the public trust doctrine to water, but less firmly than in times past. Rodgers observes that the doctrine "remains confined restlessly to . . . the bed of the sea, and other navigable waters, which can be understood to mean fresh waters of any consequence."[54]

In order to effectively use the public trust doctrine to trump the Takings Clause, new paradigmist legal scholars work to free it from its watery shackles. Apart from arguments over legal theories (or lack thereof), they invariably fall back on what I contend throughout this book is the high-sounding but nebulous, scientifically weak, earth-worshiping rhetoric of biocentrism. Gary Meyers, for example, submits that natural objects, such as "rivers, forests, and wildlife," have rights that are not presently represented in law or land use management.[55] The resulting inequality between people and Mother Nature is best rectified by society appointing guardians to represent nature in a public trust regime that extends to the ecosystems on which wildlife depends. Other than urban centers, it is hard for one to imagine what land and water in the United States would escape oversight by nature's watchdogs if the nation adopts Meyers's idea.[56] University of Maine law professor Alison Rieser argues for a public property right necessary to maintain the integrity of naturally functioning ecosystems.[57] In Reiser's view, the public interest in protecting ecosystem integrity is best defended through use of the public trust doctrine. Hunter believes that the only thing that prevents the public trust doctrine from applying to "all ecologically important lands" is "ancient rhetoric."[58] New ecological understanding demands that society assert "its right to the ecological integrity of the land," and since there really is no land that Hunter and his fellow supporters of ecosystem management will ever judge to be ecologically unimportant, Hunter would turn the public trust doctrine loose across the entire landscape. Sax, the originator and most prominent advocate of the public trust doctrine in natural resource and environmental law and policy, also would release the doctrine from any association with aquatic ecosystems. He says that no one has a property right allowing them to "destroy or to impair the productivity of our [natural] endowment" and that all acquired rights must be "subordinate to the public trust obligation" to protect and restore biological diversity (including ecosystems) and to ensure sustainable production.[59] Sax and the like-minded fabricate an argument from ecological buzzwords to justify government taking private property for public use without paying for it.

Getting something for nothing is a temptation that many people find hard to resist. Backers of ecosystem management endeavor to add limitation on limitation to the prohibition of taking private property without compensation found in the Fifth Amendment in order to do just that. Justice Holmes anticipates their efforts. Writing in 1922, he observes "that the natural tendency of human nature is to extend the qualification [using police powers to evade takings] more and more until at last private property disappears."[60]

National Land Use Planning and Centralizing Power

Only through national land use planning and management can the federal government ensure that the public obtain newly fabricated rights to ecosystem integrity, biological diversity, sustainability, and all the rest that new paradigmists claim is necessary for survival of the planet, establishment of a morally solvent society, and the creation of a sustainable domestic economy. But Americans have rejected national land use planning and management throughout the history of the United States, so the new paradigm indeed means a new order.[61]

Traditional societal disapproval of national land use control poses a serious problem for advocates of ecosystem management. Backers can try to persuade the people to change their minds or force the new paradigm on them via new statutes or favorable court rulings. Many scientists and others believe that a properly educated citizenry is the answer. Biologists Hal Mooney and Clifford Gabriel ask, "How do we inform and educate people to the nature and extent of the crisis?" while fellow scientist Tom Lovejoy says that "the real challenge is how we as biologists can create that sense of urgency" with the public that will cause it to support policies to save the planet.[62] Ecologist Stephen Kellert likewise advocates education to overcome the American public's present "degree of ignorance and indifference" to the siege of nature caused by "greed, arrogance, and apathy."[63] Ed Grumbine, director of the Sierra Institute at the University of California–Santa Cruz Extension, worries that even with education, ecosystem management might not "survive the inevitable checks and balances of American politics."[64] Wondering whether our "political system rooted in human-centered values and the private property rights of individuals can be transformed to foster sustainability in less than a generation," he sees one way to improve the prospects for the new paradigm in the political arena as "avoid[ing] the democratic trap of giving equal weight to all interest groups."[65] The voice of those who would "destroy biological diversity [including ecosystems] for short-term economic gain" would be given less weight in the political process than advocates for Mother Earth.[66] Yet education is a two-way street. Suppose that well-informed Americans repudiate the environmental crisis (and other) arguments advanced by new paradigmists. Or perhaps they might agree that our society should improve its handling of environmental issues but that adoption of the new paradigm is not in our overall best interests. Then what?

Grumbine fears that "decision makers may resort to an eco-fascism that attempts to force people into ecologically responsible behavior."[67] This is "doomed to failure," he says, because a "moral dictatorship" cannot successfully impose ecologically correct behavior on society. But other new paradigmists brush aside concerns for democratic processes, public debate, and the importance of individual choice. According to philosopher George Sessions, "The immediacy of the environmental crisis does not afford us the luxury of waiting for the rest of mankind to make the radical paradigm shift to a . . . Leopoldian ethical posture. . . . Nature must be protected from the . . . ecologically ignorant by a vast interlocking system of national and international law."[68] Hunter also feels that there is too much at stake to allow the public to make its own decisions. He writes that we in the United States "cannot afford to wait for social value changes to be reflected more boldly in the political process. Ecologists see the scientific imperative of their view as legitimating the imposition of new laws on a society that perhaps has not yet been steeped in the environmental sciences."[69] Whether the public concurs or not, Karp would use government to "create a social ethic of ecocentrism" and put it in statutes because the "civil religion of law must be used as arbiter of correct behavior."[70] Absent government guidance, people might return to sinful ways, subjecting us and Mother Nature to the adverse consequences of decisions made by the "tyrannizing masses." "Some individuals," writes Karp, "left unguided . . . are apt to favor short-term, selfish interests."[71] Professor Rieser feels that "decisionmaking authority must be vested in an entity with a frame of reference broader, both spatially and temporally, than may be common among private actors."[72] In her judgment, proper management of ecosystems requires a single decision maker. Laura Westra wonders whether a democratic society can shift to an "ecosystem approach" and muses that placing governance in the hands of a "Platonic philosopher-queen" might be a better way to accomplish that goal.[73] In one way or another, land use decisions must be removed from individual landowners and managers and given to government and powerful self-selected special interests operating through government.

The new paradigm requires an omnipresent new federal bureaucracy (which I hereby christen the National Land Use and Planning Agency, or NLUPA) with enormous power and influence over American's lives and lifestyles. As ecosystems blanket the nation, so will the rules and regulations that NLUPA establishes to protect and restore ecosystems. The agency will reach every acre of the country. Its rules and regulations will directly impact every landowner, every land manager, and every land user. The NLUPA will evaluate the permissibility of all human activity on the landscape primarily on the basis of its effect on a deified or romanticized Mother Nature. The new agency will, among other activities, put out regulations creating land use planning and management guidelines, procedures, and standards; develop a national land use permitting system; and devise means to hold landowners, managers, and users accountable to the federal government (and new paradigmists via the courts) for their actions on the landscape.

Since the landscape constantly changes, the NLUPA must continuously monitor human actions on the land and water to ensure that all public and private landowners and managers meet new national ecosystem health, integrity, and sustainability standards. The NLUPA's scrutiny and concurrence will by no means be limited to lands far removed from population centers or confined to the activities of farmers, ranchers, foresters, and miners. Local governments wishing to build schools, clear lots for soccer fields, or construct public water supply systems will need to obtain federal government approval. Home owners fertilizing their lawns, building decks or swimming pools, or changing their landscaping can easily come under the purview of the NLUPA. The constant flux on the land also means that the work of the NLUPA will never be completed. Officials could never declare that all ecosystems possessed integrity, health, and sustainability. Consequently, a public policy of federal management and protection of ecosystems requires that the NLUPA become a permanent fixture in peoples' lives and permanently dictate land use in every state.

The NLUPA's actions will be arbitrary from the outset. That is because they are rooted in ideas that, as I illustrate in earlier chapters, are inherently vague and poorly defined. The NLUPA, for example, must identify and locate on a map the single group of specific ecosystems across the nation that its rules and regulations are to protect or restore. Chapter 2 makes it clear there is no way to produce such a map other than by a bureaucratic decree. Since the agency must consider the ecosystem boundaries as fixed in time for regulatory purposes, they will be scientifically indefensible, for scholars agree that ecosystems constantly change in space and time. Likewise the NLUPA must give legally enforceable and operational definitions to new paradigm notions such as ecosystem health, integrity, restoration, diversity, and sustainability. They will do so by fiat since such terms continue to elude generally accepted understanding within the scientific, land management, and public policy communities. Indeed, some prominent observers conclude that these notions should be abandoned entirely. Recall ecologist Robert MacArthur's observation that "perhaps the word 'diversity' like many words in the early vocabulary of ecologists . . . should be eliminated from our vocabularies as doing more harm than good."[74]

The notion that a Washington bureaucracy can rationally plan or direct land use decision making to achieve what Caldwell and Shrader-Frechette refer to as "optimal land use decisions" rests on a false assumption and posits in government a wisdom and an ability that neither our government nor any others have ever demonstrated.[75] One cannot determine an optimal land use decision, for no such thing exists. Optimal for whom? Over what period of time? And in what regional setting? Participants in land use decisions often have quite different views about the best use of given tract and use divergent and unreconcilable measures to judge outcomes. There is no dispassionate methodology by which anyone can integrate the land use decision preferences of, for example, advocates for low-income family housing, county supervisors desiring to build new schools, an owner wishing to expand a factory site, a nature preservation society wanting to establish a preserve, and a couple wanting to build their dream house, in order to objectively judge a particular decision or set of decisions to be "optimal."

What evidence is available that a Washington bureaucracy is capable of either making or guiding intelligent land use decisions for the nation? The total amount of knowledge and information that people use in making such decisions is beyond calculation. The United States, after all, is the third most populous nation on earth. Its economy is the world's largest and most diverse. The nation encompasses an expansive range of physical environments as befits its status as the fourth-largest country on the planet. People make land use decisions that reflect constantly changing economic, social, and environmental circumstances. Their judgments factor in new developments in science, technology, and other ingredients that contribute to the functioning of society. Decisions reflect individual choices and ideas about how to better the lives of those making them. How is it possible for a central authority to accumulate all the knowledge and data needed to make sound land use decisions, much less comprehend what it collects? How can a bureaucracy sort out the needs of tens of millions of individuals and tens of thousands of communities and make benevolent land use decisions on that basis? Because of the enormity and complexity of the task, they could not do so even if they wanted to. Ultimately, the planners' decisions would be driven by the political signals they receive from the interests that keep them in power and by their own desires to expand their programs and budgets rather than by the views of the public from which they would be isolated and to which they would be unaccountable.

History teaches us that the new bureaucracy will become increasingly Byzantine and dysfunctional as it fortifies itself behind thousands of pages of rules, regulations, memorandums of understanding, and other administrative appurtenances.[76] The fortification will enable new paradigmists to impose their values on a public that they variously judge as too greedy, arrogant, apathetic, or ill-informed to be trusted with the responsibility of decision making. "Environmental policy making is too important to be left to the policy makers," says Reed Noss, who prefers a policy situation in which unaccountable scientists would determine so-called critical biological thresholds to determine what land use and management decisions are biologically acceptable.[77] I suggest, however, that the more fundamental matters of liberty and freedom are too important to be left to new paradigmists. The federal effort to protect wetland ecosystems using Section 404 of the Clean Water Act is a case in point.

Protection of Wetland Ecosystems

For centuries, most Americans viewed swamps, bogs, and marshes as wastelands, obstacles to important human endeavors, and frequently sources of killing and debilitating diseases. Public policies supported the draining and filling of these lands in order to build cities, homes, and roads and to create productive farmland and improve public health. As a result, researchers estimate that today the conterminous states contain about 100 million acres of wetlands compared to approximately 220 million acres in colonial times.[78] While some scientists were documenting the decline in wetland acreage, others were finding that these lands can function as important habitat for many species and also identified a role for wetlands in the flux and flow of water, biochemical cycling, and biological productivity.[79] The attitude of Americans toward wetlands changed over time, and so did public policies. Now multiple federal, state, and local laws are in place to preserve a legacy of wetlands.[80] Wetlands losses are much reduced from times past, and at least one wetlands analyst even estimates the nation is presently undergoing a net gain in wetlands.[81]

Several federal laws address wetland protection in some manner, but observers generally regard Section 404 of the Clean Water Act as the most controversial.[82] Section 404 uses the police power of the federal government to elevate the protection of Mother Nature (as wetland ecosystems) over the consideration of human well-being. The current 404 program comes to us from the 1972 Federal Water Pollution Control Act Amendments (FWPCA) and the 1977 Clean Water Act (CWA).[83] Section 404 of the FWPCA requires the U.S. Army Corp of Engineers (COE) to issue permits for the deposition of dredge or fill materials into "navigable waters," which the act defines only as "the waters of the United States, including its territorial seas," and Congress gives the Environmental Protection Agency veto authority over the COE's permitting decisions. Without regard to human impacts, the COE will not issue a permit unless the proposed action cannot be relocated (regardless of cost) outside a wetland.[84]

The FWPCA does not contain the word *wetlands* but does encourage agencies to interpret the phrase *navigable waters* as broadly as possible. Federal agencies (with the concurrence of the Supreme Court) read these laws and their legislative histories as including wetlands as navigable waterways.[85] By operating in a geographic never-never land, the program drifted further from the reality of the landscape and into a convoluted labyrinth of rules and regulations. For example, just what is a wetland? Writing in the journal *Environmental Science and Technology,* Julian Josephson notes that "wetlands mean one thing to a farmer, another to an ecology professor, and yet something else to a government official."[86]

Scientists and government agencies added the term *wetlands* to their lexicon in the 1950s. The term began as a collective word, meant to cover watery landscapes such as swamps, bogs, and marshes as reflected in the 1956 Fish and Wildlife Service (FWS) definition of wetlands as "lowlands covered with shallow and sometimes temporary or intermittent waters."[87] The agency thought of wetlands as land covered for some part of the year by standing water. This first understanding of the term—once shared by scientists, bureaucrats, and the public alike—did not last long in the realm of either science or government but continues to dominate public understanding of the word, as evidenced by Webster's definition of wetlands as "swamps or marshes," which in turn are defined as "a piece of wet spongy land that is permanently or periodically covered by water" and "a tract of low, wet, soft land that is temporarily or permanently covered with water."[88] Scientists quickly descended on the expression and during the last forty years have proposed more than fifty definitions of wetlands.[89] In 1979 researchers at the FWS wrote that "there is no single, correct, indisputable ecologically sound definition for wetlands."[90] One and a half decades later, the National Research Council concurred, saying that "scientists have not agreed on a single commonly used definition of wetlands."[91]

This is not a failure of science or scientists but rather the predictable result of the reality that wetlands, like all other ecosystems, are mental constructs. The concept lost clarity as more and more researchers—soil scientists, hydrologists, ecologists, biologists, geologists, and others—with diverse training and objectives entered discussions about how to define a wetland. What began as a term to cover distinctive and easily recognizable inundated landscapes became lost in an arcane world of soil and vegetation classification schemes and hydrologic regimes. Periodic inundation is no longer a prerequisite for researchers to declare someplace a wetland. Scientists and federal regulators can now label as wetlands the Everglades, normally dry riverbeds, and so-called vernal pools (which are routinely parched, small depressions that are well removed from any water body). It is no wonder that wetland researchers William Mitsch and James Gosselink conclude that "wetland definitions and terms are many and are often confusing or even contradictory."[92]

The word *wetlands*—like the word *field* or *forest*—describes the look of the landscape. It is a term of art, not science, and does not lend itself to an ecologically correct scientific definition, as the scientific community freely acknowledges. Scientists' views regarding what constitutes a wetland are certainly important, but there is no compelling reason for anyone to suppose that scientists' perceptions of wetlands are presumptively more valuable in the realm of public policy than those of any other group, such as hunters, fisherman, farmers, foresters, or even the public at large. Scientists' opinions of wetlands might be more technical than those of other people, but that does not necessarily make them more insightful or useful in policymaking. In 1977, after considering twenty-two definitions of wetlands for use with the 404 program, the COE finally settled on "areas that are inundated or saturated by surface or ground water at a frequency and duration sufficient to support, and that under normal circumstances do support, a prevalence of vegetation typically adapted for life in saturated soil conditions. Wetlands generally include swamps, bogs, marshes and similar areas."[93] This definition is now used by all federal agencies for 404 program purposes. By including the reference to swamps, bogs, and marshes, the COE tries to maintain the aura of traditional understandings of wetlands. The new definition represents a significant departure from earlier meanings, however, because it abandons the prerequisite that land must be periodically inundated by water to qualify as a wetland.[94] The government thus replaced a regulatory definition of wetlands that is consistent with public understanding with one supported by scientists but that conflicts with citizen perceptions.

Not surprisingly, when government agencies developed procedures for identifying newly defined wetlands, the result was a continuing "firestorm of controversy."[95] Members of the scientific community declare that only they are capable of identifying wetlands for federal land use management and regulatory purposes under current policies. The National Research Council concludes that

> wetland delineation requires scientific education at the college level, combined with specialized training in delineation methods and practices. Some knowledge of or familiarity with several fields is needed for wetland delineation. Delineators should have some knowledge of plant taxonomy, botany, soil science, surface water hydrology, general ecology, wetland (or aquatic) ecology, sampling methodol-

ogy, and plant morphology. Knowledge of ground water hydrology, geology, plant physiology, and perhaps other disciplines is also desirable.[96]

Perhaps they are right, but, if so, what does it say about the complexity of wetland policies when intelligent and otherwise well-educated (but nonscientist) citizens cannot determine whether an area is a wetland? How much more incomprehensible will public policies be under the new paradigm with its dependence on notions far less substantial and significantly more controversial than those associated with wetland delineation? One can assume that a self-anointed elite will claim that only they can give those notions meaning for policy purposes so that the citizenry must stand aside and simply accept their pronouncements. The ability to define terms represents a gathering of power into the hands of those doing the defining. Moreover, in such a policy and regulatory environment, words mysteriously take on unusual meanings.

The FWCPA makes no reference to wetlands, referring only to "navigable waters," which it defines as "the waters of the United States."[97] Anyone asserting that swamps, bogs, or marshes are navigable waters is committing geography without a permit.[98] In doing so, they capriciously give arbitrary new meanings to words. In response to the FWPCA, the COE proposed a regulatory interpretation of the law allowing the 404 program to reach wetlands.[99] The COE initially attempted to maintain a minuscule tie between the idea of navigable waters and federal regulations. It limited the wetlands included in the 404 program to those touched by the ebb and flow of the tide or wetlands adjacent to waters that were, are, or could be used in interstate commerce. Environmentalists found unacceptable even this effort to keep a tiny thread of rationality between the words in the law and actuality of the landscape. In 1975 the Natural Resources Defense Council successfully sued the COE, arguing that the agency must adopt a more aggressive geographic interpretation of the act. The federal district court determined that the legislative history of the FWPCA demonstrated that Congress did not really mean what is normally understood by the phrase *navigable waters* when Congress used it in the law. The court ordered the COE to issue new rules that would apply to additional portions of the landscape.[100] In its 1985 decision in *United States v. Riverside Bayview Homes,* the Supreme Court similarly finds that Congress did not really mean "navigable waters" in the normal linguistic connotation of those words.[101] They write that the COE can regulate wetlands (at least those adjacent to other waters of the United States) under the FWCPA and that the COE can also strike any requirement for surface inundation from a regulatory definition of wetlands.[102]

It is not the role of the Supreme Court to save Congress from itself or the citizenry from a bureaucracy willing to turn its back on the public's understanding of words describing the landscape. In reviewing congressional action, the

Riverside Court did not pass judgment on the merits of congressional reasoning or the depth of its knowledge; rather, the justices simply considered what members of Congress said and wrote as part of a process leading to passage of the FWPCA. Thus, even if the record is replete with ecological and policy muddleheadedness, the record is the record, and that is what constitutes the basis for judicial action. In reaching its decision, the Court specifically highlights the objective of the act to "restore and maintain the chemical, physical, and biological integrity of the Nation's waters" and House Report language pointing out that "the word integrity . . . refers to a condition in which the natural structure and function of ecosystems [are] maintained."[103] The Court also makes it plain that it will not second guess agency decision making on substantive grounds. The COE is certainly not required by the FWPCA or the CWA to adopt a wetland definition driven by views of specialists at the expense of lay understandings, but they may do so.

Out of all this comes a highly controversial program. Federal district court judge Roger Vinson describes it as a "regulatory hydra" emerging from "Alice in Wonderland" interpretations of the English language that enables bureaucrats to make a "quantum leap" and regulate dry land as waters of the United States.[104] In order to protect the integrity and natural functioning and structure of a wetland ecosystem, the federal government successfully prosecuted (and a court sentenced to prison) people for placing, in the words of Judge Vinson, "clean fill dirt on a plot of subdivided dry land" because it had been declared a wetland by the federal government for purposes of Section 404.[105] Despite these shortcomings, Congress has been unable to reform Section 404. The new paradigmists are a formidable political force.

Many members of Congress recognize that Section 404 cries out for major changes, but after more than twenty years of trying, reformists have been unable to make significant changes in the law. Congress debated the scope of the 404 program extensively in the two years prior to its passage of the CWA in 1977 but ultimately did not make substantive alterations affecting the geographic reach of Section 404 or clarify the meaning of key ideas. It did not, for example, define wetlands for policy or regulatory purposes.[106] Congress' more recent efforts to revise Section 404 have also failed. Reacting to an outcry from large and small private landowners and elected public officials from around the country, the House of Representatives considered H.R. 961—the Clean Water Act Amendments of 1995—in 1995.[107] The bill would have overhauled the 404 program by, among other things, requiring periodic inundation before land could be designated a wetland and providing compensation for landowners when their property was taken for the public purpose of preserving wetlands.[108]

Many wetland scientists, the Clinton administration, and others faulted H.R. 961, often singling out for special criticism its requirement that land must sometimes have standing water at the surface in order to be deemed a wetland for public policy purposes.[109] On May 16, 1995, the House passed H.R. 961 by a vote of 240 to 185.[110] The vote reflects public rejection of the government's adoption of the dominant scientific definition of wetlands to guide policy and the public's insistence that the government compensate people when it takes their land to accomplish a civic purpose. The Senate, however, took no action on H.R. 961. In the summer and fall of 1995, the Senate did hold subcommittee hearings on S. 851, the Wetlands Regulatory Reform Act of 1995, introduced by Senator Bennett Johnston of Louisiana along with twenty cosponsors. The bill was less far reaching than H.R. 961 but did propose significant changes to the Section 404 program, including a periodic inundation requirement for wetlands. As with H.R. 961, witnesses representing the administration, environmental organizations, and some business interests strongly opposed the changes, while many others vigorously supported the revisions.[111] Senator John Chafee, chairman of the full committee with jurisdiction over wetlands, opposes major reforms to Section 404, so S. 851 died in the subcommittee. Chafee finds especially troubling the revisions that may limit the reach of federal regulators, make legislative definitions of wetlands more consistent with public perceptions of the term, or provide that the public pay for what it takes from individuals.[112]

With the existing 404 program, new paradigmists can exert considerable influence over any land proclaimed to be a wetland. They are not about to relinquish that power. Indeed, they are pushing to expand it by insisting that Congress give them the right to intervene in the authorization of every activity requiring a 404 permit no matter how spatially small or environmentally insignificant.[113] In the world of the new paradigm, the Sierra Club, Greenpeace, the Wilderness Society, Earth First!, and other nature-worshiping groups would sit in judgment on public and private land use decisions throughout the country.

The New Paradigm and Existing Regulations

Government implementation of the new paradigm requires an enormous array of new rules, regulations, and intrusions into the lives of the American public. Nonetheless, some people might hold out hope that there would be trade-offs elsewhere resulting in a net reduction of regulatory burdens and costs or a streamlining of bureaucratic processes to the benefit of human endeavors. This thinking supposes that if society adopts ecosystem management, then some of the regulatory hydras that one finds elsewhere in federal law may be slain or at least wounded. Such hopes are unfounded. As attorney Rebecca Thompson concludes in her examination of ecosystem management, "The policy of ecosystem management will be yet another overlay on top of existing environmental and public land law and processes."[114] New paradigmists openly advocate government protection of ecosystems in addition to, not in place of, laws such as the Endangered Species Act or Section 404. "Ecosystem conservation," write new paradigm ecologists Noss, La Roe, and Scott, "is not a replacement for existing conservation measures such as the Endangered Species Act."[115]

The interplay between ecosystem management and existing environmental statutes became evident during a 1997 hearing on the Draft Environmental Impact Statements for the Interior Columbia Basin Ecosystem Management Plan (see chapter 7). Senator Larry Craig questioned administration witnesses closely on the impact of ecosystem management in relation to enforcement of other environmental statutes. He noted that federal regulatory agencies—the Environmental Protection Agency, the National Marine Fisheries Service, and the FWS—demanded that the draft environmental impact statements be recalled from the printer because none of the alternatives meet the requirements of the ESA in their view.[116] Craig points out that over the next six months "dozens of new standards, conditions, and analysis [sic]" were added to drafts.[117] He goes on to ask the witnesses whether they believe that the revised management alternative selected by government decision makers for implementation "will meet the requirements of the Endangered Species Act, the Clean Water Act, and the Clean Air Act."[118] After receiving an affirmative response, he then asks whether the land managers could implement the selected alternative without further consultation with regulators. The regulators say no, that they must still be consulted.[119]

The cumulative impact of shifting to ecosystem management in the Interior Columbia Basin is the creation of "new layers of administrative activity . . . [that] will result in the creation of meta-gridlock," according to Oregon Senator Gordon Smith.[120] Shifting to ecosystem management means that development will be forgone, exactly the intent of dedicated new paradigmists. The public and decision makers alike should understand that this scenario would be repeated throughout the nation should ecosystem management become the law of the land. We cannot forget that the objective of new paradigmists is to defend Mother Earth from violation by people. Their objective is not to reduce regulatory burdens, improve human well-being, or advance freedom and democracy. New paradigmists measure success by such things as the number of projects stopped, the number of acres placed into more restrictive management categories, and the number of people indoctrinated into the faith. For them, every step toward removing existing signs of human activity from the landscape is a step in the right direction, no barrier to preventing intrusions into lightly impacted areas can be high enough, and no cost to individual people or to society at large is too much to pay in the defense of Mother Earth.

Conclusion

Empowering the federal government to manage and protect ecosystems across the country is bad public policy. Advocates of ecosystem management use sectarian arguments built on false claims of pending environmental doom. They cloak themselves in a garment woven from ill-defined scientific buzzwords that are valueless for the purpose of guiding government decision making. They employ religious arguments that confuse the Creator with creation. New paradigmists would lessen human liberty and improvements in human well-being through assaults on constitutionally protected property rights. They would establish a vast new centralized bureaucracy to manage land use throughout the nation, using a maze of rules and regulations that would greatly surpass in complexity and cost any set of environmental regulations previously seen in the United States.

Throughout this book I argue that the new paradigm is fatally flawed as a guide for public policy, but if Americans are to become even better stewards of the environment, how might we go about it? I address this question in the next chapter.

Notes

1. Lynton Caldwell, *Between Two Worlds: Science, the Environmental Movement, and Policy Choice* (Cambridge: Cambridge University Press, 1990), 97.

2. Al Gore, *Earth in the Balance: Ecology and the Human Spirit* (New York: Penguin, 1993), 274.

3. Caldwell, *Between Two Worlds,* 173.

4. Victor Scheffer, *The Shaping of Environmentalism in America* (Seattle: University of Washington Press, 1990), 168–69.

5. Laura Westra, *An Environmental Proposal for Ethics: The Principle of Integrity* (Lanham, Md.: Rowman & Littlefield, 1994), 193. For a review of thinking on the relationship between environment and democracy, see Daniel Press, *Democratic Dilemmas in the Age of Ecology* (Durham, N.C.: Duke University Press, 1994).

6. For a review of federal land use controls under existing law, see Robert H. Nelson, "Federal Zoning: The New Era in Environmental Policy," in *Land Rights: The 1990s Property Rights Rebellion,* ed. Bruce Yandle (Lanham, Md.: Rowman & Littlefield, 1995), 295–317.

7. Lynton K. Caldwell and Kristin Shrader-Frechette, *Policy for Land: Law and Ethics* (Lanham, Md.: Rowman & Littlefield, 1993), 254.

8. Michael S. Greve, *Demise of Environmentalism in American Law* (Washington, D.C.: AEI Press, 1996), 1–2.

9. Joseph L. Sax, "Property Rights and the Economy of Nature: Understanding Lucas v. South Carolina Coastal Council," *Stanford Law Review* 45 (1993): 1439.

10. Donald Lutz, *A Preface to American Political Theory* (Lawrence: University of

Kansas Press, 1992), and Richard Hofstadter, "The Founding Fathers: An Age of Realism," in *The Moral Foundations of the American Republic,* ed. Robert H. Horowitz (Charlottesville: University of Virginia Press, 1979), 73–85.

11. Blackstone, as quoted in Robert R. Wright and Morton Gitelman, *Case and Materials on Land Use,* 4th ed. (St. Paul: West Publishing, 1991), 373.

12. Lawrence H. Leder, *Liberty and Authority: Early American Political Ideology 1689–1763* (Chicago: Quadrangle Books, 1968), 125.

13. Leder, *Liberty and Authority,* 125. Quote cited in the *New York Weekly Journal,* June 16, 1735.

14. James Madison, *The Federalist Papers: Number Ten* (New York: Mentor, 1961), 77–84.

15. Robert A. Dahl, "On Removing Certain Impediments to Democracy in the United States," in Horowitz, ed., *The Moral Foundations of the American Republic,* 238.

16. Friedrich von Hayek, *The Road to Serfdom,* 50th anniversary ed. (Chicago: University of Chicago Press, 1994), 115.

17. Roger E. Meiners, "Elements of Property Rights: The Common Law Alternative," in Yandle, ed., *Land Rights,* 272.

18. James Karp, "A Private Property Duty of Stewardship: Changing Our Land Ethic," *Environmental Law* 23, no. 3 (1993): 737.

19. Caldwell and Shrader-Frechette, *Policy for Land,* 104.

20. Karp, "A Private Property Duty of Stewardship," 754.

21. Karp, "A Private Property Duty of Stewardship," 736.

22. Karp, "A Private Property Duty of Stewardship," 753.

23. Hearings were held in Anchorage, Alaska; Peoria, Illinois; New Orleans, Louisiana; North Dakota; Providence, Rhode Island; and Olympia, Washington.

24. See Yandle, ed., *Land Rights,* and William L. Graf, *Wilderness Preservation and the Sagebrush Rebellions* (Lanham, Md.: Rowman & Littlefield, 1990).

25. Sax, "Property Rights and the Economy of Nature," 1453; see also Alison Reiser, "Ecological Preservation as a Public Property Right: An Emerging Doctrine in Search of a Theory," *The Harvard Environmental Law Review* 15, no. 2 (1991): 393–433, and David Hunter, "An Ecological Perspective on Property: A Call for Judicial Protection of the Public's Interest in Environmentally Critical Resources," *Harvard Environmental Law Review* 12, no. 2 (1988): 311–83.

26. Karp, "A Private Property Duty of Stewardship," 748.

27. Eric T. Freyfogle, *Justice and the Earth* (Urbana: University of Illinois Press, 1993), 53; see also Eric Freyfogle, "The Owning and Taking of Sensitive Lands," in *Land Use & Environmental Law Review 1996,* ed. Stuart L. Deutsch and A. Dan Tarlock (Deerfield, Ill.: Clark, Boardman, Callaghan, 1996), 123–84, and "Ownership and Ecology," *Case Western Reserve Law Review* 43 (1993): 1269–97.

28. Robert H. Peters, *A Critique for Ecology* (Cambridge: Cambridge University Press, 1991), 98.

29. Freyfogle, *Justice and the Earth,* 52.

30. Richard J. Lazarus, "Shifting Paradigms of Tort and Property in the Transformation of Natural Resources Law," in *Natural Resources Policy and Law,* ed. Lawrence J. MacDonnell and Sarah F. Bates (Washington, D.C.: Island Press, 1993), 208.

31. John Chafee, quoted in Keith Schneider, "Environmental Laws Face a Stiff Test from Landowners," *New York Times,* January 20, 1992, A15.

32. Helen Dewar, "Property Rights Bill Blocked," *Washington Post,* June 14, 1998, A4. State and local governments also generally oppose changes in takings law that might expose them to takings claims.

33. White House Office of Environmental Policy, "Protecting America's Wetlands: A Fair, Flexible, and Effective Approach," White House Office of Environmental Policy, August 24, 1993.

34. Hunter, "An Ecological Perspective on Property," 316.

35. Caldwell and Shrader-Frechette, *Policy for Land,* 98.

36. The issue of takings and the courts is the subject of numerous books and scholarly papers, among them Mark Sagoff, "Muddle or Muddle Through? Takings Jurisprudence Meets the Endangered Species Act," *William and Mary Law Review* 38, no. 3 (March 1997): 825–993; Carol M. Rose, "Property as the Keystone Right?" in Deutsch and Tarlock, eds., *Land Use & Environmental Law Review 1997,* 275–316; R. Prescott Jaunich, "The Environment, the Free Market, and Property Rights: Post-Lucas Privatization of the Public Trust, " *Public Land Law Review* 15 (1994): 167–97; and Richard Epstein, *Takings: Private Property and the Power of Eminent Domain* (Cambridge, Mass.: Harvard University Press, 1985).

37. As quoted in Wright and Gitelman, *Case and Materials on Land Use,* 391.

38. As quoted in Wright and Gitelman, *Case and Materials on Land Use,* 397; citations embedded in Brennan's writing are not included.

39. 112 Sup. Ct. 2886 (1992).

40. Greve, *Demise of Environmentalism in American Law,* 36.

41. Sax, "Property Rights and the Economy of Nature," 1438.

42. Sax, "Property Rights and the Economy of Nature," 1440. *Just v. Marinette County,* 201 N.W. 2d 761, 768 (Wis. 1972).

43. *Just v. Marinette County,* 201 N.W. 2d 761, 768 (Wis. 1972); quote from Wright and Gitelman, *Case and Materials on Land Use,* 522.

44. Sax, "Property Rights and the Economy of Nature," 1441, 1442, 1445, 1446.

45. Sagoff, "Muddle or Muddle Through?" 901.

46. Daniel B. Botkin, *Discordant Harmonies* (New York: Oxford University Press, 1990).

47. Greve, *Demise of Environmentalism in American Law,* 23–41.

48. Greve, *Demise of Environmentalism in American Law,* 129.

49. Writing in 1989 prior to *Lucas,* Pramaggiore concludes the "recent Supreme Court decisions demonstrate a shift in position from one of extreme deference to state and local regulatory entities to one of increased protection for the private property owner from destructive government regulatory schemes." See Anne Pramaggiore, "The Supreme Court's Trilogy of Takings: Keystone, Glendale, and Nolan," *DePaul Law Review* 38 (1989): 483.

50. Greve, *Demise of Environmentalism in American Law,* 129.

51. William H. Rodgers Jr., *Environmental Law,* vol. 1 (St. Paul: West Publishing, 1986), 155. The modern effort to use public trust doctrine in environmental and natural resource law and policy began with Joseph L. Sax, "The Public Trust Doctrine in Natural

Resources Law: Effective Judicial Intervention," *Michigan Law Review* 68 (1970): 471–526.

52. Hunter, "An Ecological Perspective on Property," 367.

53. James L. Huffman, "A Fish Out of Water: The Public Trust Doctrine in a Constitutional Democracy," *Environmental Law* 19, no. 3 (spring 1989): 527–72; James L. Huffman, "Avoiding the Takings Clause through the Myth of Public Rights: The Public Trust and Reserved Rights Doctrines at Work," *Journal of Land Use and Environmental Law* 3, no. 2 (fall 1987): 171–211.

54. Rodgers, *Environmental Law,* 158.

55. Gary D. Meyers, "Variation on a Theme: Expanding the Public Trust Doctrine to Include Protection of Wildlife," *Environmental Law* 19, no. 3 (spring 1989): 723–35.

56. Urban areas contain wildlife, but I do not read Meyers to include downtown pigeons and Norway rates among the species of interest.

57. Alison Rieser, "Ecological Preservation as a Public Property Right: An Emerging Doctrine in Search of a Theory," *The Harvard Environmental Law Review* 15, no. 2 (1991): 393–418.

58. Hunter, "An Ecological Perspective on Property," 377.

59. Joseph L. Sax, "The Search for Environmental Rights," *Journal of Land Use and Environmental Law* 6 (1990): 105.

60. As quoted in Wright and Gitelman, *Case and Materials on Land Use,* 391.

61. See generally Rutherford H. Platt, *Land Use and Society: Geography, Law, and Public Policy* (Washington, D.C.: Island Press, 1996); Caldwell and Shrader-Frechette, *Policy for Land,* 147–58.

62. Hal Mooney and Clifford J. Gabriel, "Preface," *BioScience* (Special Supplement on Science & Biodiversity Policy, June 1995); Thomas E. Lovejoy, "Will Expectedly the Top Blow Off," *BioScience* (Special Supplement on Science & Biodiversity Policy, June 1995): S6.

63. Stephen R. Kellert, *The Value of Life* (Washington, D.C.: Island Press, 1996), 210.

64. Ed Grumbine, "Introduction," in *Environmental Policy and Biodiversity,* ed. R. Edward Grumbine (Washington, D.C.: Island Press, 1994), 13.

65. Grumbine, "Introduction," in *Environmental Policy and Biodiversity,* 13; Ed Grumbine, "Protecting Biological Diversity through the Greater Ecosystem Concept," *Natural Areas Journal* 10, no. 3 (1990): 117.

66. Grumbine, "Protecting Biological Diversity," 117. A fascinating glimpse into new paradigmist compartmentalization of players in the political process at the grassroots level into "villains, victims, and heroes" is provided by Pat Veitch and Alita Wilson, "A Plan, a Plan, Gotta Have a Plan," in the Sierra Club's magazine *The Planet* (March 1998), downloaded from the Sierra Club's Web site at http://www.sierraclub.org/plantet/199803/plan.html, February 26, 1998.

67. Grumbine, "Protecting Biological Diversity," 118.

68. George Sessions, "Anthropocentrism and the Environmental Crisis," *Humboldt Journal of Society and Religion* (fall-winter 1974), as quoted in Hunter, "An Ecological Perspective on Property," 317.

69. Hunter, "An Ecological Perspective on Property," 316.

70. Karp, "A Private Property Duty of Stewardship," 754–55.

71. Karp, "A Private Property Duty of Stewardship," 755.

72. Rieser, "Ecological Preservation as a Public Property Right," 421.

73. Westra, *An Environmental Proposal for Ethics,* 196.

74. Robert H. MacArthur, *Geographical Ecology* (Princeton, N.J.: Princeton University Press, 1972), 197.

75. Caldwell and Shrader-Frechette, *Policy for Land,* 254.

76. Robert Heilbroner, "Socialism," in *The Fortune Encyclopedia of Economics,* ed. David R. Henderson (New York: Warner Books, 1993), 161–64.

77. Reed Noss, "Sustainable Forestry or Sustainable Forests?" in *Defining Sustainable Forestry,* ed. Gregory H. Aplet, Nels Johnson, Jeffrey T. Olson, and V. Alaric Sample (Washington, D.C.: Island Press, 1993), 39.

78. T. E. Dahl, *Wetland Losses in the United States 1780s–1980s* (Washington, D.C.: U.S. Department of the Interior, U.S. Fish and Wildlife Service, 1990).

79. National Research Council, *Wetlands: Characteristics and Boundaries* (Washington, D.C.: National Academy Press, 1995); Curtis J. Richardson, "Ecological Functions and Human Values in Wetlands: A Framework for Assessing Forestry Impacts," *Wetlands* 14, no. 1 (March 1994): 1–9.

80. William B. Meyer, "When Dismal Swamps Became Priceless Wetlands," *American Heritage* 45, no. 3 (May/June 1994): 108–16; Jon A. Kusler, *State Wetland Regulation: Status of Programs and Emerging Trends* (Berne, N.Y.: Association of State Wetland Managers, 1993); National Wetlands Policy Forum, *Protecting America's Wetlands: An Action Agenda* (Washington, D.C.: The Conservation Foundation, 1988).

81. Thomas E. Dahl and Craig E. Johnson, *Status and Trends of Wetlands in the Conterminous United States, Mid-1970's to Mid-1980's* (Washington, D.C.: Department of the Interior, Fish and Wildlife Service, 1991), estimate annual wetland losses at 290,000 acres. Heimlich and Melanson place annual losses of nonfederal wetlands at between 70,000 and 90,000 acres in the period 1982–92; see Ralph Heimlich and Jeanne Melanson, "Wetlands Lost, Wetlands Gained," *National Wetlands Newsletter* 17, no. 3 (May–June 1995): 1–23, 25. Tolman argues that expected acreage gains via wetlands restoration already exceed expected losses, so the United States has essentially already reached a condition of no net loss of wetlands; see Jonathon Tolman, "Achieving No Net Loss," *National Wetlands Newsletter* 17, no. 3 (May–June 1995): 5–7, and Jonathon Tolman, "Gaining More Ground: Analysis of Wetland Trends in the United States," Competitive Enterprise Institute, Washington, D.C., October 1994. Virtually all wetlands researchers agree that the rate of wetland loss has declined significantly over the last decade.

82. For a summary of major federal wetlands legislation, see U.S. General Accounting Office, *Wetlands Overview,* GAO/RCED-92–79FS (Washington, D.C.: U.S. General Accounting Office, November 1991), 19–29; for a review of the results of federal wetland programs, see U.S. Department of the Interior, *The Impact of Federal Programs on Wetlands. Vol. II: A Report to Congress by the Secretary of the Interior* (Washington, D.C.: Department of the Interior, March 1994).

83. Numerous articles on wetlands law and regulation are available; see Sam Kalen, "Commerce to Conservation: The Call for a National Water Policy and the Evolution of

Federal Jurisdiction over Wetlands," *North Dakota Law Review* 69, no. 4 (1993): 873–913, and Michael C. Blumm and Bernard Zaleha, "Federal Wetlands Protection under the Clean Water Act: Regulatory Ambivalence, Intergovernmental Tension, and a Call for Reform," *University of Colorado Law Review* 60 (1989): 695–772.

84. 33 CFR §§ 325 and 40 CFR §§ 1508.20.

85. *United States v. Riverside Bayview Homes,* 474 U.S. 121.

86. Julian Josephson, "Status of Wetlands," *Environmental Science and Technology* 26, no. 3 (March 1992): 422.

87. As quoted in William Mitsch and James Gosselink, *Wetlands* (New York: Van Nostrand, 1993), 25.

88. Victoria Neufeldt, ed., *Webster's New World College Dictionary* (New York: Macmillan, 1996).

89. Donal Hook, "Wetlands: History, Current Status, and Future," *Environmental Toxicology and Chemistry* 12 (1993): 2157–66; Jon Kusler, "Wetlands Delineation: An Issue of Science or Politics?" *Environment* 34, no. 2 (March 1992): 7–11.

90. Lewis Cowardin, Virginia Carter, Francis Golet, and Edward LaRoe, *Classification of Wetlands and Deepwater Habits of the United States* (Washington, D.C.: U.S. Department of the Interior, U.S. Fish and Wildlife Service, 1979), 3.

91. National Research Council, *Wetlands: Characteristics and Boundaries* (Washington, D.C.: National Academy Press, 1995), 36.

92. Mitsch and Gosselink, *Wetlands,* 21.

93. 33 C.F.R. § 328.3(b); Kusler, "Wetlands Delineation."

94. Kalen, "Commerce to Conservation," 897. In 1974 the COE published a definition of wetlands as "those land and water areas subject to regular inundation by tidal, riverine, or lacustrine flowage." See *Federal Register* 39, no. 12121 (1974). The interim final rule defining wetlands at 33 CFR § 209(d)(2)(h) (1976) included a requirement for periodic inundation. The COE repudiated this requirement at *Federal Register* 42, no. 37128 (1977).

95. Ken Bierly, representing the Association of State Wetland Managers, testimony on Implementation of Section 404 of the Clean Water Act, before Congress, Senate Subcommittee on Environmental Protection of the Committee on Environment and Public Works, 102nd Cong., 1st sess., June 20, July 10, and November 22, 1991, Committee Print S. Hrg. 102–450, 224. Multiple views of the wetlands controversy can be found in U.S. General Accounting Office, *Wetlands Protection: The Scope of the Section 404 Program,* GAO-RCED 93–26 (Washington, D.C.: U.S. General Accounting Office, April 1993); Douglas A. Thompson and Thomas G. Yocum, "Uncertain Ground," *Technology Review* 96, no. 6 (August/September 1993): 22–29; Joseph Alper, "War over the Wetlands: Ecologists v. the White House," *Science* 257 (August 21, 1992): 1042–43; Michael J. Olivier, "The Wetlands Dilemma," *Economic Development Review* 10 (summer 1992): 56–57; Jean Seligmann, "What on Earth Is a Wetland?" *Newsweek,* August 26, 1991, 48–49; Jerome Cramer and J. Madeleine Nash, "War over the Wetlands," *Time,* August 26, 1991, 53; Betsy Carpenter, "In a Murky Quagmire," *U.S. News and World Report,* June 3, 1991, 45–46; Rick Henderson, "The Swamp Thing," *Reason,* April 1991, 30–35; Richard Miniter, "Muddy Waters," *Policy Review,* no. 56 (spring 1991): 70–77; and Curtis C. Bohlen, "Controversy over Federal Definition of Wetlands," *BioScience* 41, no. 3 (March 1991): 139.

96. National Research Council, *Wetlands,* 186.

97. 33 U.S.C. § 1362 (7).

98. For a discussion of the disconnect between geography and the law, see Rutherford H. Platt and James M. Kendra, "The Sears Island Sage: Law in Search of Geography," *Economic Geography* (extra issue, 1998): 46–61.

99. According to *Webster's New World College Dictionary* (1996), for a water body to be navigable it must be "wide or deep enough, or free enough from obstructions, for the passage of ships"; ships are "any vessel of considerable size navigating deep water . . . larger than a boat"; and a boat is a "small, open vessel or watercraft propelled by oars, paddles, sails or engine."

100. *Natural Resources Defense Council v. Callaway,* 392 F. Supp. 685 (D D.C., 1975).

101. 474 U.S. 121.

102. The issue of the regulation of isolated wetlands, those not obviously adjacent or connected to navigable waters or their tributaries, was not addressed by the Riverside Court. On December 23, 1997, the U.S. Court of Appeals for the Fourth Circuit struck down 33 C.F.R. § 328.3(a)(3) (1993), which allows the COE to reach such wetlands and remarks that even if Congress specifically sought to reach such wetlands by statute, their doing so "would present serious constitutional difficulties." See *United States v. James Wilson* (1997 U.S. App. LEXIS 35971).

103. 474 U.S. 132; 33 U.S.C. § 1251; House Report 92–911 (1992), 76, on the Federal Water Pollution Control Act.

104. *United States v. Ocie Mills and Carrie C. Mills,* 817 F. Supp. 1546, 1993 WL 112105 (N.D. Fla.).

105. *United States v. Ocie Mills and Carrie C. Mills.*

106. In the CWA, Congress did provide relief for farmers and ranchers by exempting normal farming practices, such as maintaining stock ponds or clearing drainage ditches from the need to obtain 404 permits, and it did furnish some paperwork streamlining by authorizing general permits. The COE could use general permits to specify circumstances and conditions under which a person could receive a permit. If a proposed project fits within those parameters, the sponsor is not required to obtain an individual permit for that project. The COE grants most of its project approvals via general permits.

107. The excesses and human harm caused by the existing Section 404 program are amply documented in congressional hearing records; see U.S. Congress, Senate Subcommittee on Clean Air, Wetlands, Private Property, and Nuclear Safety of the Committee on Environment and Public Works, *Hearing on S. 851, The Wetlands Regulatory Reform Act of 1995,* 104th Cong., 1st sess., July 19, August 2, and November 1, 1995, Environment and Public Works Committee Print S. Hrg. 104–643.

108. H.R. 961, The Clean Water Act Amendments of 1995, § 803, completely revising Section 404 by enacting an earlier COE definition of wetlands as lands that, among other things, "are inundated by surface water" and requiring that the water be "found to be present at the surface of such lands for 21 consecutive days in the growing seasons in a majority of the years for which records are available."

109. Richard Stone, "Wetlands Reform Bill Is All Wet, Say Scientists," *Science* 268 (May 19, 1995): 970.

110. For a concise discussion of H.R. 961 and the House debate, see Bob Benenson, "Rewrite of Clean Water Law on Verge of House Passage," *Congressional Quarterly* (May 1995): 1321–22; and Bob Benenson, "Water Bill Wins House Passage, May Not Survive in Senate," *Congressional Quarterly,* May 20, 1995, 1413.

111. See generally U.S. Congress, Senate Subcommittee on Clean Air, Wetlands, Private Property, and Nuclear Safety of the Committee on Environment and Public Works, *Hearing on S. 851, The Wetlands Regulatory Reform Act of 1995,* 104th Cong., 1st sess., July 19, August 2, and November 1, 1995.

112. John H. Chafee, opening statement in Congress, Senate Subcommittee on Clean Air, Wetlands, Private Property, and Nuclear Safety of the Committee on Environment and Public Works, *Hearing on S. 851, The Wetlands Regulatory Reform Act of 1995,* 7–9.

113. National Wildlife Federation, "Wetlands: Where NWF Stands in the 105th Congress," downloaded from the National Wildlife Federation's Web page at http://www.nwf.org/wetlands/facts/wet105.html, March 4, 1998. The National Wildlife Federation wants authority for public review of all projects authorized under general permits.

114. Rebecca W. Thompson, "Ecosystem Management: Great Idea, but What Is It, Will It Work, and Who Will Pay?" *Natural Resources & Environment* 9, no. 3 (winter 1995): 72.

115. Reed F. Noss, Edward T. LaRoe III, and J. Michael Scott, *Endangered Ecosystems of the United States: A Preliminary Assessment of Loss and Degradation,* Biological Report No. 28 (Washington, D.C.: U.S. Department of the Interior, National Biological Service, February 1995), 18.

116. Congress, Senate/House Subcommittee on Forests and Public Land Management of the Senate Committee on Energy and Natural Resources and the Subcommittee on Forests and Forest Health of the House Committee on Resources, *Joint Hearing on the Interior Columbia Basin Ecosystem Management Plan,* 105th Cong., 1st sess., May 15, 1997 (hereafter cited as *Joint Hearing*), 29.

117. *Joint Hearing,* 29.

118. *Joint Hearing,* 45.

119. *Joint Hearing,* 46–47.

120. *Joint Hearing,* 10.

Chapter 10

Improving Environmental Stewardship

> Humans cannot live on the earth without altering and using natural resources. Our responsibility is to be good stewards of the environment.
>
> —Dixie Lee Ray, with Lou Guzzo, *Trashing the Planet,* 171

The overall condition of the environment in the United States at the end of the twentieth century gives compelling testimony that Americans are generally good environmental stewards. Nonetheless, we can do better. The key to enhancing stewardship is a shift to policies that emphasize the use of markets, incentives, and local decision making within a framework that emphasizes the fundamental importance of private property rights. By enhancing stewardship, I mean further minimization of waste in the use of natural resources and additional lessening of unnecessary impacts on other species consistent with our primary responsibility to improve human well-being. It is well for us to remember that human use of the land is not synonymous with destruction and degradation. The tens of millions of visitors to Yosemite Valley have not destroyed it any more than the production of oil from the outer continental shelf has created an environmental ruin in our coastal waters. The idea is to strengthen stewardship while we turn aside threats to liberty inherent in the new paradigm. Americans can build a better mousetrap.

In this chapter, I examine the importance of balance in developing politically sustainable environmental policies. I show how property rights and markets are fundamental to improved stewardship and illustrate how they can be applied in achieving a national environmental goal of maintaining a legacy of wetland landscapes. The chapter then turns to the management of federal lands and concludes by noting that future generations will judge us by the amount of liberty we leave them rather than by the size of our wilderness areas.

The Need for Balance

Students of public policy understand that in a democracy change is normally accomplished incrementally.[1] Radical alteration in policy direction can be justified only in extraordinary circumstances. Advocates of revolutionary policy shifts must be able to convincingly demonstrate that a preeminent national problem has arisen that society cannot effectively resolve using existing laws or policies even if they are adjusted to meet the new threat. Natural resource policy is no different. In chapter 1, I indicate how the current body of environmental and natural resource policy is an amalgam of laws reaching back over 100 years. It places concerns with improving human well-being at the center of government actions while recognizing a responsibility to other living things. New paradigmists want to jettison our traditional policies in favor of revolutionary new directives that would subordinate enhancement of the human condition to the protection of Mother Nature. Yet because the United States faces no significant environmental crisis and the geographic targets of their revolution—ecosystems—are figments of the human mind, they fail to make a case for their radical policy changes. Defending illusions from make-believe crises is not an appropriate foundation for government actions that undermine constitutional guarantees of liberty.

Our experience does, however, justify continued incremental improvements in policies and practices that lead to better stewardship. A number of analysts, members of the public, and elected officials increasingly point to the significant human harms caused by federal environmental laws, such as the Endangered Species Act (ESA) and Section 404 of the Clean Water Act (CWA).[2] They erode human freedoms; impose high financial costs on communities, businesses, and individuals alike; and cause injury by depriving people of livelihoods, opportunities, and even their homes. Other observers note that both the ESA and the Section 404 program achieve minimal environmental benefits and that incentive-based programs are far more successful in protecting wetlands than is the 404 program.[3] Consequently, many people believe that the time has come to make some important changes in existing statutes and to clarify the rights of property owners. Nonetheless, their proposals fall far short of the dramatic restructuring of the relations between government and the citizenry envisioned by new paradigmists, as I illustrate in chapter 9.

Formulation of a successful long-term public policy in a democratic society requires that decision makers balance competing policy goals. Leaders must consider costs, benefits, and trade-offs associated with the various policy alternatives available for reaching desired policy outcomes. During the rush to leap on the environmental bandwagon in the 1970s, Congress turned a blind eye to these components of good policymaking. In the Clean Air Act, the CWA, and the ESA,

Congress prohibits consideration of costs in achieving legislative objectives.[4] In the ESA, Congress goes so far as to declare that even the most ecologically insignificant of species is beyond price, according to the Supreme Court. Utah State University political science professor Randy Simmons observes that "the Supreme Court declared in its Tellico Dam decision that the act defines 'the value of endangered species as incalculable,' that endangered species must 'be afforded the highest of priorities,' and that 'whatever the cost' species losses must be stopped."[5] While few people quarrel with the goal of improving environmental quality, Congress' earlier dismissal of human costs in this endeavor is largely responsible for widespread calls for change in natural resource statutes. Critics of these statutes do not favor dirty air, foul water, or the indiscriminate eradication of other life forms; rather, they want balanced policies—a level playing field on which protection of Mother Nature is not the presumptive best "use" of the land.

Advancement of national environmental policymaking requires congressional abandonment of new paradigmist dogmas. For example, contrary to their assertions, every component of the natural environment is not of equal ecological value, and each kind of living thing is not beyond price, requiring its protection at any cost. The human immunodeficiency (HIV) virus is not priceless (and neither are grizzly bears despite their magnificence). Endangered desert pupfish do not have the same biological significance as coyotes. Ten academic biologists and ecologists writing in *BioScience* observe that "many ecologists believe that all species were not created equal."[6] Rutgers University biology professor David Ehrenfeld points out that

> the species whose members are fewest in number, the rarest, the most narrowly distributed—in short, the ones most likely to become extinct—are obviously the ones least likely to be missed by the biosphere. Many of these species were never common or ecologically influential; by no stretch of the imagination can we make them vital cogs in the ecological machine. If the California condor disappears forever from the California hills, it will be a tragedy: but don't expect the chaparral to die, [or] the redwoods to wither . . . they won't.[7]

Congressional requirements that humans make herculean financial and personal sacrifices to protect ecologically inconsequential species is bad public policy. At least some resources spent in saving such species could often yield far greater environmental and human benefits if invested in other ways. Current policy, however, forbids the citizenry from considering such investments. By embracing the precepts of a narrow constituency at the expense of the broader public interest, Congress frustrates efforts to make much needed reforms to existing policies and keeps the door open for the adoption of nature-worshiping new paradigmist policies.

If Americans are to improve our stewardship, then we must introduce a balancing and prioritizing mechanism into our environmental policies, reduce regulatory complexity, and provide incentives for landowners and managers to take actions that benefit the public. The question is, How can we best do it?

Toward More Effective and Equitable Natural Resource Policies

Constitutionally guaranteed property rights protect home owners from the government insisting that they house the homeless in spare bedrooms and feed the hungry in their kitchens in order to achieve a public good. There is no reason for government to short-circuit the Constitution when the target of interest is an endangered species or a wetland instead of the homeless and ill-fed. There is no reason for it to short-circuit the Constitution when the property involved is a tract of land instead of a single family residence. The ideal of fairness—that is, doing what is just and impartial (to say nothing of the Constitution)—demands that the government treat the owners of habitats of particular plants or animals and the owners of wetlands with no less respect than owners of suburban homes.

Government protection of private property rights is a prerequisite to American's efforts to improve environmental stewardship.[8] Wheaton College economics professor Seth Norton finds that "environmental quality and economic growth rates are greater in regimes where property rights are well defined. Therefore, property rights and growth should be viewed . . . as favorable to environmental quality and conservation of resources."[9] Property rights protect the efforts of all—from those driven purely by ethical considerations to those motivated solely by profit seeking to all those who find a foot in each camp—who would invest in conservation and preservation. The same property rights protecting someone's right to build a home on their own land also protect the rights of an environmental organization to buy land for a nature reserve without fear that government will force them to build a shopping mall instead in order to meet a public policy goal of providing jobs.

Secure, enforceable, and transferable property rights make it possible for the Nature Conservancy (TNC) to manage 1,600 preserves in the United States covering 10 million acres.[10] The Nature Conservancy owns some of this acreage outright. On much of the remainder, TNC owns conservation easements purchased from landowners who made a market-based decision to sell the easement rather than to use the land for other purposes. In the Adirondack Mountains, for example, large numbers of bats use abandoned mines as wintering sites that TNC wants to protect. One major site is part of a 1,900-acre parcel that its owner recently put up for sale. Kathy Regan, head of the Adirondacks chapter of TNC, says that "the Conservancy definitely did not wish to own an abanded mine, and there was no need to buy the entire property."[11] Instead, when International Paper bought the land, TNC purchased a conservation easement from it that prohibits underground work and establishes a "quiet zone" around the mine opening. According to Regan, "the arrangement worked out well for everybody."

The work of other conservation organizations is also made possible by the existence of property rights. Over the years, Ducks Unlimited (DU) purchased 375,000 acres of wetlands and holds fifty conservation easements on 84,000 acres. The organization is in the process of raising $600 million to protect and restore 1.7 million acres of wetlands.[12] Donors to the effort—Habitat 2000: Campaign for a Continent—are assured their monies can result in the perpetuation of wetlands because of the security offered by property rights.

Without secure ownership and assurance that they can benefit from investments in conservation, there is little incentive for many property owners to do so. If you cannot sell timber, why invest in scientific forestry? If you cannot plant and grow crops, why spend money on practices to reduce soil erosion? Conversely, if you can sell timber, you invest in practices that keep your forests productive, and if you can sell crops, you take care of the soil because it is essential to your agricultural enterprise. Property rights improve stewardship in other ways. Property rights underpin the common law of nuisance that allows public officials and individual landowners to bring action against polluters.[13] Property rights and the market enhance conservation.

Winston Churchill observed that democracy is the worst form of government except for all the other kinds people have tried. Likewise, a system in which people make environmental and natural resource decisions on the basis of market signals is also imperfect. But it is far superior to the new paradigmists' central environmental planning route. Implementation of the new paradigm rests on the same conceit that doomed central economic planning—namely, the assumption that a self-annointed government elite can become all-knowing, all-wise, and altruistic—so decision making must rest with them instead of the citizenry.

Many scholars and policy analysts argue that market-based policies—rooted in clearly defined, defendable, and transferable property rights—are the most effective way to achieve improvements in both human well-being and environmental quality.[14] University of Minnesota law professor Daniel Farber writes that the solution to better environmental protection lies in "more use of markets and federalism" coupled with "regulatory reform."[15] Biologists Ray Hilborn (University of Washington), Carl Walters (University of British Columbia), and Don Ludwig (University of British Columbia) conclude that "the most successful institution for promoting sustainable exploitation of fish, wildlife, and forests has been private ownership. The use of private property rights in traditional tropical fisheries, private forests around the world, private game parks in Africa and Europe, and private freshwater fishing preserves illustrates that long-term sustainability of renewable resources is possible."[16] Market mechanisms bring a fairness and discipline in achieving environmental goals that are absent in federal authoritarian command-and-control methods. Emphasis on markets better ensures that the people who obtain the benefits of policy choices also pay the costs of providing them and thereby forces prioritization. Markets can provide incentives for landowners to manage lands so as to yield public benefits. They unleash creativity to help reach environmental goals at the least overall cost. Markets are the best means to resolve Justice William Brennan's concerns about the "justice and fairness" of government forcing a comparatively small number of people to carry "public burdens."[17] The size of those burdens is considerable.

Analysts are unable to estimate the cumulative societal costs of the ESA or Section 404 with any degree of accuracy because costs are of many kinds and data are routinely unavailable.[18] A sampling of these costs suggests that they are very high and include an extraordinary amount of human anguish and loss. The majority staff of the Resources Committee of the House of Representatives estimates that federal expenditures just to implement the ESA were about $4 billion between 1991 and 1997.[19] Their estimate does not include lost federal revenues, nonfederal costs, or unquantifiable human costs, such as the distress associated with loss of savings and jobs. Simmons writes that "current plans to save the Sacramento delta smelt are likely to cost billions."[20] Nelson Wolff, the mayor of San Antonio, Texas, notes in congressional testimony that the effort to protect the fountain darter will cost state taxpayers $4 billion to develop a new water supply. He says that perhaps 100,000 of the approximately 700,000 jobs in the San Antonio area will be lost because of endangered species listings.[21] Such losses easily translate into more than a billion dollars annually in lost wages alone. Observers see these losses repeated across the country. Congressman Wes Cooley of Oregon points out that "I have 64,000 people out of work [in his district] because of the Endangered Species Act and the way it has been interpreted."[22] In Travis County, north of San Antonio, appraisers found that land values declined from $335 million to $57 million largely as the result of endangered species listings.[23] Margaret Rector, a septuagenarian from Austin, Texas, saw the value of fifteen acres of land she bought near the city in 1973 drop from $900,000 to $30,000.[24] The Fish and Wildlife Service scared away potential buyers by determining that her land contained habitat suitable for the endangered golden-cheeked warbler and thereby wiped out the investment she made to secure her retirement.

According to the House Resources Committee majority staff, the Fish and Wildlife Service approved some 200 habitat conservation plans (HCPs) to protect living space for threatened and endangered species by the spring of 1998. They report estimates of the nonfederal cost of the HCP for San Diego at $650 million; that to protect the Stephens kangaroo rat in Riverside County, California, at $45 million; and that to protect the Balcones Canyonlands Conservation Plan in Travis County at $160 million.[25]

Section 404 imposes similar costs on individuals and communities. In congressional testimony the executive director of the Carteret County (North Carolina) Economic Development Council cites a study finding that strict enforcement of Section 404 could prevent the development of some $50 billion in urban land around the Hampton Roads area of Virginia.[26] More than opportunity costs are at stake. Current authoritarian regulatory policies routinely waste enormous amounts of peoples' time, patience, and energy as well as their money. Gaylord Enns, pastor of the Pleasant Valley Assembly of God Church in Chico, California, tells Congress,

> In 1978, the congregation I serve purchased a 19 acre site for relocation and expansion of our church facilities. Now, over sixteen years later and after spending $300,000 in mitigation attempts, environmental consulting fees, attorney fees, taxes, assessments, site work and construction plans, the land still lays unused . . . our congregation was unable to proceed with the project because of the cumulative effects, both financially and emotionally, of nine years of trying to navigate through environmental regulations and to withstand the harassment of environmental extremists.[27]

The bulk of ESA and Section 404 burdens fall on a small segment of the population: those who own lands reached by these programs, those who live near or on such lands, and those whose livelihoods depend directly or indirectly on the ability of people to use these lands for economic purposes.

Current programs likewise disproportionately distribute their benefits among the citizenry. Gains fall mostly to new paradigmists and to a few groups, such as commercial fishermen. New paradigmists effectively use statutes to advance their cause of nature worship while the fishermen employ them for economic gain. These beneficiaries pick up little of the tab. When the Natural Resources Defense Council (NRDC) won its suit forcing the Corps of Engineers to expand the geographic scope of wetland regulations, landowners and users across the country—not the NRDC—paid the price of the court's decision. Describing the Sierra Club's motivation in pursuing endangered species listings, San Antonio Mayor Wolff says that "it is very clear what their agenda is, clear as clear can be—stop this community from growing. . . . Stick an economic dagger right in the heart of San Antonio."[28] Sierra Club members and other new paradigmists gain at the expense of the people of San Antonio.[29]

Commercial fishermen join with new paradigmists in free riding. They enjoy a free ride at the expense of wetland owners because the fishermen cannot be prevented from catching wetland-dependent fish. At the same time, the fishermen do not pay taxes on those same wetlands or pick up other expense associated with wetland ownership or maintenance. Testifying before Congress in general support of the present command-and-control 404 program, Glen Spain, executive director of the Pacific Coast Federation of Fishermen's Association, claims that commercial fishing "generates about $111 billion in personal income" and that "71 percent of that income is based on species that are wetland dependent during part of their life cycle."[30] Since the fishermen do not have an ownership interest in these landscapes, they get something for nothing when regulations force wetland owners to maintain habitat for commercial fish species. Spain, however, shows little sympathy for the plight of those burdened for the benefit of his group. "I get tired of private land owners thumping on takings," he says.[31] One wonders what his view of property rights would be if fishing boats were appropriated (without compensation) by government for the benefit of other business interests. For example, suppose that the government required captains to give tourists free rides in order to attract customers to coastal motels and restaurants. The blatant unfairness associated with the ESA and Section 404 makes the current policies unsustainable in a democratic society.

Benefits of Market-Based Policies

University of Maryland public affairs professor Robert Nelson characterizes current centralized and authoritarian federal land use policies as "economically and administratively irrational, and . . . probably not politically viable."[32] They are wasteful of human resources and, in the case of the ESA, actually thwart achievement of espoused environmental goals. Policies based on markets and property rights overcome these notable deficiencies. They would put in place a politically sustainable means to improve stewardship by more equitably distributing costs and benefits as well as by emphasizing adherence to constitutional principles. Doing so eliminates much of the human anguish and widespread public resentment associated with present policies. The market/property rights approach introduces rationality into natural resource policies by leveling the playing field, providing for choice, and requiring prioritization. It reduces the waste of resources that occurs when regulators and free riders are able to spend someone else's time, money, and energy in pursuit of their own agendas. After all, when you are using your own resources, human nature pushes you in the direction of receiving the greatest value for your investment. One environmentalist involved in acquiring land for conservation purposes notes that "every thicket in the country is priceless to someone but you have to make trades, you have to recognize that some things are more important than others."[33]

Spain claims that wetland losses cost the industry $27 billion annually in fisheries production and that 450,000 jobs have been lost.[34] Yet this wound is largely self-inflicted. If fishermen had employed the market to acquire protections offered by property rights, they could have avoided many of these losses. They had but to follow the example of DU and buy wetlands or conservation easements thereon or secure other enforceable agreements to preserve wetlands that they value. If DU can protect 1.7 million acres of wetlands with $600 million, imagine how much wetland acreage commercial fishermen could protect by each year investing a fraction of their $111 billion annual income in conservation. By depending on the market and property rights instead of authoritarian regulations, the fishermen gain, wetland owners (for whom the presence of wetlands is transformed from a curse into a blessing) gain, and so do other living things that thrive in wetlands. Stewardship improves.

Under the current command-and-control approach, landowners are penalized, often severely, if regulators find wetlands, threatened or endangered species, or habitat favorable to such species on their property. Not only do owners have no incentive to maintain or create the kind of landscape conditions that government policies seek to perpetuate, but they have every reason to do just the opposite, to the detriment of the wildlife that policymakers purport to care about. If the federal government determines that an endangered species exists on your land or that it contains habitat suitable to such species, government agents gain control of land use decisions. If they list a species as threatened, their degree of control is less although still significant. What landowner wants government agents to dictate land use, especially when the agents have no motivation to consider the impacts of their actions on the owner?

Dave Cameron, a retired biology and genetics professor and a Montana rancher, tells Congress of his experience trying to reintroduce a native fish into a stream on his land:

> Hoping to continue in the family tradition of ecological reconstructive activity, I implored upon my colleague, Dr. Calvin Kaya, a fish biologist with special interest in the Montana grayling, to examine the tributaries of our trout stream for sites suitable for the reintroduction of grayling—a native species which for some reason has disappeared from much of its former habitat, including our ranch. A suitable site was found . . . we were about to proceed when word arrived that the Federal Wildlife Service was seriously considering listing the Montana grayling as an endangered species. At this point people knowledgeable about the heavy-handed approach of the Feds counseled me to forget the experiment. We might lose the right to graze our pastures. . . . I sadly bowed out. It seemed a good deed would probably be punished. . . . Reasonable property owners are frightened and angry at . . . the government for managing with brick bats. Why does the hosting of a rare

> or troubled creature have to be a threat to their livelihood rather than a source of pride and pleasure? It doesn't.[35]

Cameron is correct, but getting there requires Congress to rethink its approach to protecting endangered species. It needs to take advantage of basic human nature and get the incentives right if the goal is to protect species as opposed to stopping human use of the land.

Nearly everyone will protect species if they can benefit from doing so. Some others, like Cameron, will care for species if doing so does not cause overt harm to themselves. But very few people will protect other living things at the expense of their own subsistence. Congress can exploit this situation by changing to policies that allow landowners to profit from having rare species on their land.[36] The government can buy conservation easements or otherwise pay landowners to manage land use so as to maintain or create habitat for species of government interest. It can provide matching funds to conservation organizations to allow them to buy land or easements. The government can offer tax benefits for landowners who maintain or create desired habitat. It can permit the sale of threatened or endangered species or other uses that allow an owner to profit by the maintenance of a population of designated species on their land. Should a landowner not wish to use these options, then the government can acquire title to land that it wishes to protect via cash purchase, land exchange, or some other means negotiated with the landowner. If the government and the landowner cannot agree on price, then the government can condemn the land and have its fair market value determined by the courts as provided for by existing law.[37]

As I note in chapter 9, some people complain that land acquisition (or acquiring easements) could cost a lot money, and so it could. Randal O'Toole, a natural resource economist who heads the Thoreau Institute, counters by writing that

> shifting the burden of protecting habitat to a few landowners doesn't reduce the cost, it just makes a few people pay for something that benefits everyone. If society as a whole benefits from endangered species protection, then society as a whole, not just a few unlucky land owners, should pay the costs. This is true not just because it is fair, but because it will work better [to protect species] than a system that threatens to take people's land from them without compensation.[38]

A policy whereby the government bears the cost of its actions forces it to prioritize. Congress must make choices in how it spends tax dollars in other areas of public policy, from defense to Medicare. There is no reason that budgetary discipline should not likewise apply to its actions relating to natural resources and land management.

Positive incentives and nonregulatory approaches work. In order to maintain a legacy of wetland landscapes, Congress authorizes the nonregulatory methods, such as the Wetlands Reserve Program, the North American Waterfowl Management Plan, and the Partners for Wildlife Program.[39] These programs use easements, cost sharing, cost matching, and partnering with landowners and third parties to preserve and restore wetlands on private property. They are more cost effective and more likely to achieve desired environmental goals than are regulations intended to prohibit legitimate uses of land. Jonathan Tolman, a policy analyst at the Competitive Enterprise Institute, points out that in 1995 nonregulatory federal programs resulted in the restoration of over 200,000 acres of wetlands while 46,000 acres were restored through the command-and-control-based Section 404 program.[40] Moreover, according to Tolman, the Section 404 program costs the taxpayer five times more per wetlands acre restored than did nonregulatory methods. As useful as these incentive programs are, however, they still do not take full advantage of the power of the market to generate desired environmental benefits. In the next section I explore how that can be done. I take it as a given that widespread agreement exists that the public benefits by the presence of wetlands in the overall landscape mosaic of the United States. I also take it as a given that the public purpose of Section 404 is to help ensure that the landscape contains wetlands rather than be an instrument to prevent people from using the land.[41] Finally, by wetlands I mean swamps, bogs, marshes, and similar wet landscapes consistent with the original scientific and present-day lay understanding of the term (see chapter 9).

Using the Market to Maintain a Reservoir of Wetlands

In the 404 program, Congress assumes that maintaining an existing wetland is the highest and best use of the land. In brief, the program operates this way: Anyone desiring to put material into a wetland must obtain a permit to do so from the U.S. Army Corps of Engineers (COE).[42] The underlying assumption of the preeminent value of wetlands is manifest in that part of the permitting process known as sequencing: first, avoid impacting a wetland; second, minimize any unavoidable impacts; and third, mitigate unavoidable consequences.[43] In deciding whether to issue a permit, the COE first asks whether the proposed project can be done outside the wetland. The COE does not consider cost in making this determination. If relocation to an upland site is possible, the COE will not issue the permit. Next, if the project cannot be built elsewhere, then a permit will be issued only after the COE and the Environmental Protection Agency (which can veto COE permits) agree that all appropriate design steps have been taken to minimize adverse wetland impacts. Finally, the COE can then condition the permit with a requirement that the unavoidable impacts be mitigated by restoring, enhancing, or (occasionally) preserving wetlands elsewhere. Government regulators, not the landowner, become the ultimate land use decision maker at the expense of liberty.

All sides want to change the Section 404 program. Tolman argues that it should be abandoned as too costly, ineffective, and unnecessary.[44] He concludes that the United States has reached a condition of no net loss of wetlands and has done so with little help from Section 404. Bernie Goode, an environmental consultant who formerly headed the COE's national 404 program, is one of many calling for major streamlining and simplification. He tells Congress that "my business thrives on the complexity, chaos, and overreach of the Section 404 wetland program."[45] New paradigmists want to expand it. Margie Vicknair of the Sierra Club says that "we place the highest priority on the protection of existing natural wetlands . . . therefore, any activity which destroys or degrades a previously functioning wetland should not be tolerated."[46] Vicknair's position represents the classic new paradigmist view that undisturbed landscapes are sacred space (see chapter 5). Congress will change (but not end) the 404 program. Various bills introduced in Congress during the 1990s would overcome one or more of the basic flaws in the existing program. One looks in vain, however, for congressional proposals that would aggressively harness the power of the market to achieve the public good of ensuring that the nation contains wetlands while reducing the present government assault on liberty found in the 404 program.

If the goal of Section 404 is to ensure the public benefits from a legacy of wetlands, then it should apply to all wetland-destroying activities, not just the placing of dredge or fill material into a wetland, as is presently the case. If Americans wish to prevent sacrificing the fundamental principle of liberty in achieving their wetland policy goal, then the government must return land use decisions to landowners and assume most of the costs for providing public benefits.

The landscape presents Americans with a significant opportunity to expand wetland acreage. The present Section 404 program, however, inhibits our taking advantage of it. The White House observes that American farmers own some 53 million acres of former wetlands.[47] In addition, scientists can sometimes create wetlands where none previously existed.[48] Together, such lands provide an enormous opening for Americans to increase the amount of wetlands in the contiguous states. Yet the current structure of the 404 program provides little incentive for owners to undertake the expense of restoration and subsequent maintenance of a wetland or to accept the opportunity cost of doing so. Indeed, it offers a significant disincentive to restoration because the presence of a wetland invites intrusive government regulators. Enter a market-based approach to Section 404.

Suppose that Congress changes Section 404 so that the act of restoring a degraded wetland creates a marketable good. Suppose that a landowner receives a wetland protection certificate (WPC) for each acre restored if that owner enters into an enforceable agreement to keep it as wetland for some period of time. Suppose that the WPCs could be bought and sold. But where is the market for the WPCs? Congress can simultaneously establish a market for WPCs, replace the authoritarianism of Section 404, and better reach the policy goal of expanding our wetland landscapes. They accomplish these tasks by 1) eliminating sequencing, 2) requiring permits for a wider array of wetland-destroying activities, and 3) stipulating that regulators must issue permits whenever a permittee presents them with their share of the requisite number of WPCs. In a reformed 404 program, development in most wetlands cannot be denied if the developer agrees to offset wetland losses.

How will the COE, as the permitting authority, and the developer know how many WPCs are needed to secure project approval? The answer depends on the class of wetland involved. Congress must provide that wetlands be classified, preferably at the level of states or groups of states, and provide a classification framework. For example, stipulate the use of three or four categories, such as vital wetlands in which no development can take place and must therefore be purchased at fair market value; important wetlands wherein development is permitted if, say, two WPCs are presented for each acre disturbed; ordinary wetlands requiring WPCs at a 1:1 ratio; and wetlands of little value requiring WPCs at a .5:1 ratio.[49] State and local officials would determine the criteria for classification within their region. They could agree to allow WPCs from outside their region to offset losses.

Who pays for the WPCs? Since decision makers protect wetlands for public benefits, it is only reasonable that the public shoulder most of the burden of providing WPCs. Some might argue that the public pay the total cost, while many free riders want landowners to continue to bear all the expense of preserving wetlands.[50] Bills on takings offer some guidance as to where a reasonable apportionment of responsibility might lay. Utah Senator Orrin Hatch's Omnibus Property Rights Act of 1997 stipulates that a Fifth Amendment taking occurs if federal regulators deprive an owner of more than 33 percent of the value of that owner's land.[51] Senator Richard Shelby of Alabama, in the Private Property Owners Bill of Rights, puts the takings figure at 20 percent and also calls for compensation for any loss over $10,000.[52] Perhaps landowners should provide somewhere between 20 and 33 percent of the required WPCs or pay a fixed amount into a fund that buys WPCs. In any event, policymakers must cap landowner expenses to avoid placing inordinate burdens on the private sector by inflating the number of WPCs required to obtain a permit.[53]

The reform of Section 404 advances liberty while achieving desired environmental goals. Land use decisions reside with landowners and users, as regulators cannot deny a permit if owners provide their share of offsets. The acquisition requirement for nondevelopable wetlands means that government can no longer forbid the use of wetlands without paying for its regulatory taking of land. The condition that the public pay for most of the WPCs shifts the burden of supplying a public benefit from the private sector to the general public, where it belongs. The market-based Section 404 program better accomplishes the goal of preserving a legacy of wetland landscapes by encompassing a full array of wetland-destroying activities and then requiring that more acres of wetlands are restored than are lost.

Dependence on decentralized market forces rather than on a centralized command-and-control regulatory structure yields additional benefits. Decision makers will produce better-informed judgments and will do so more quickly than federal bureaucrats (no matter how well intentioned the latter might be). Decisions made closer to the resource are far more likely to reflect local conditions and incorporate local specialized expertise and understanding than are those guided by direction from Washington. Rational land use decisions require the timely gathering and processing of a wide range of information, as the private sector does every day. In contrast, federal land-managing agencies routinely take years to produce a land management plan for lands that they have controlled for decades. Finally, a market for WPCs will stimulate the development of better wetland restoration methods and techniques because it will generate a demand for these services. Overall, the market approach does a better job of guaranteeing that Americans use both human and natural resources wisely.

Policymakers are already adopting the power of the market to achieve environmental policy goals in some circumstances. In April 1995 the state of California approved the use of conservation banks as a means of creating credits for developers to use to mitigate environmental impacts of their projects.[54] The policy offers landowners a market-based incentive to manage land for conservation purposes since doing so creates a saleable conservation credit. Within fourteen months after the program began, landowners put together thirty-six banks in twelve counties encompassing land worth $40 million.[55] The Clinton administration makes a gesture of support to markets by endorsing mitigation banking as part of the 404 program. People can restore degraded wetlands to obtain credits that can be used to fulfill mitigation requirements arising out of the sequencing provisions of the present program.[56] Government agencies (e.g., state highway departments) or large developers operate most banks. They do so, however, chiefly to meet mitigation needs associated with their own projects rather than to produce a WPC for sale to another party. On the surface, the administration's position seems clearly market oriented, but if one looks more closely it is evident that its new paradigmist philosophy stifles entrepreneurial wetland restoration. By staunchly defending sequencing—which reinforces the new paradigmist agenda by preventing or limiting development—the administration allows federal regulators to tightly control and limit the market for WPCs as well as fill it with uncertainty. At present the owner of credit in a mitigation bank can spend it only with the permission of a federal regulator. This is one reason that by the middle of 1998 the private sector created only about sixty entrepreneurial banks, each probably averaging less than 3,000 acres.[57]

The advantages and disadvantages of a market-based Section 404 do not distribute themselves equally. Living things that prefer wetlands benefit at the expense of living things that prefer uplands because wetland acreage expands and upland acreage declines. Among people, in one sense everyone (including advocates of the new paradigm) gains by a reduction in the threats to freedom and liberty inherent in 404 reform. The public at large gains in that the market approach offers a politically sustainable means to reach the goal of a national legacy of wetland landscapes. Property owners and users of land gain because they are no longer forced to carry an unfair share of the costs for producing a public good. In the human population, the only losers are new paradigmists, who could no longer use Section 404 to control land use on some 75 million acres of privately owned wetlands in defense of Mother Nature. But what about land use on federal lands? I look at that question in the next section.

Federal Lands

The federal lands contain many things that the public values. One can find vast quantities of energy and mineral resources essential to our quality of life as well as places of scenic beauty and solitude. Forests provide timber for human homes and homes for wildlife. On their federal lands, Americans produce oil, hunt and fish, hike, cut trees, dam streams, protect endangered species, graze cattle, and engage in a host other activities. Some uses are geographically compatible, and some are not. On most private land, owners balance applicable costs and benefits in making land use decisions. Owners reap the rewards of good decisions and pay the costs for bad ones. On federal lands, however, the discipline and prioritizing imposed by market mechanisms is lacking.

Decision makers at all levels sort out what land uses will be allowed by a political process. Congress assigns some lands to specific land management categories, such as national parks, wilderness, and wildlife refuges, in which human use is greatly restricted and preservation is the dominant management directive. It calls for managers to supervise other lands according to principles of multiple use and sustained yield (see chapter 7). Observers once considered the political horse trading behind such apportioning of land as the best means to balance competing interests and public values.[58] Scholars now question this approach because it has degenerated into acrimony and stalemate with the political ascendancy of uncompromising new paradigmists.[59] They constantly press to exclude or diminish the mark of humans from public land while rejecting efforts to keep the public domain available to generate societal wealth through the use and development of its resources. The Sierra Club, for example, advocates "an end to commercial logging on all federally owned public lands in the United States."[60] New paradigmists are particularly fanatical in their opposition to new energy development on federal lands, especially in places such as the Arctic National Wildlife Refuge or the outer continental shelf (OCS).[61]

In 1982, Congress began imposing moratoria on the development of new exploration and development of oil and natural gas deposits beneath the federally controlled OCS.[62] It did so after environmentalist propaganda whipped up public fears of ecological disasters if the nation developed more OCS energy resources. Then, as now, claims that production of oil and natural gas constitute a grave threat to coastal environments are inconsistent with the evidence from the oceans. In 1985 the National Research Council noted that "over the last 40 years, offshore oil and gas technology has evolved that allowed the design and construction of facilities and equipment capable of extracting oil and gas from the continental shelf in a safe and pollution free manner."[63] Five years later the Congressional Research Service concluded that "it is widely recognized that the spill risk for domestically produced OCS oil is far less than that for tankered oil. . . . The volume of oil spilled into U.S. waters will likely increase as tankered imported oil is substituted for OCS production."[64] Past decision makers cast these efforts in terms of environmental protection. When President George Bush jumped on the moratoria bandwagon in June 1990, he said, "My desire is to achieve a balance between the need to provide energy for the American people and the need to protect unique and sensitive coastal and marine environments."[65] He notes that "I continue to believe that there are significant offshore areas where we can and must go forward with resource development." President Bill Clinton fully supports moratoria but takes them into new philosophical territory by justifying them on the basis of a need to protect sacred space. While past moratoria were done in the name of preventing pollution, Clinton does it to defend Mother Earth. In June 1998 he told the attendees at the National Oceans Conference that we must "preserve our living oceans as a sacred legacy for all time to come" and lauds those "standing firm against off-shore oil drilling . . . in California and around the nation."[66]

A prime example of present-day difficulty with the political approach to federal land use management is Clinton's creation of the Grand Staircase-Escalante National Monument.[67] On September 18, 1996, Clinton invoked the

Antiquities Act of 1906 to establish the 1.7-million-acre monument from public land in Utah.[68] He shored up his credentials with environmentalists by preventing development of the largest deposit of low-sulfur coal in the United States with an estimated value of $1 trillion.[69] New paradigmists, from Vice President Al Gore to the heads of the National Audubon Society, Friends of the Earth, Defenders of Wildlife, the Sierra Club, and the Wilderness Society, heaped praise on the president.[70]

Clinton's method in establishing the monument demonstrates new paradigmists' self-righteousness and contempt for other views that so often cripple effective use of political processes to achieve balanced results in the public interest. His policy affects a very large area of public land, including about 175,000 acres of state lands.[71] Nonetheless, he did not put out a plan or concept for debate, discussion, or review by the public or their elected representatives. Many members of the new 105th Congress were angry over the administration's secrecy and its snub of themselves as elected lawmakers, the Congress as the institution charged with directing the management of federal lands, and the public. Members introduced several bills to amend the Antiquities Act to prevent what many felt was an egregious misuse of the authority it grants to the president.

The House Resources Committee set an oversight hearing for April 29, 1997. On March 18 they requested copies of relevant administration documents, and the Department of the Interior and the White House produced some of these in a timely manner. The White House sought, however, to withhold others and went so far as to defy a congressional subpoena until committee staff began to prepare a contempt-of-Congress resolution.[72] The administration delivered the withheld documents on October 22, 1997. Had members of Congress seen them prior to the April hearing, their indignation would have been heightened. Consider a March 26, 1996, e-mail prepared by Kathleen McGinty, chair of the Council on Environmental Quality (CEQ). In her capacity as head of the CEQ, McGinty served as the chief White House staffer on the issue. She says "i'm [*sic*] increasingly of the view that we should just drop these utah [*sic*] ideas . . . especially because these lands are not really endangered."[73]

Utah officials first learned of the plan in a September 9 story leaked to the *Washington Post,* nine days before the president's announcement. Utah Senator Hatch explains to congressional colleagues that

> what was particularly and still is galling to me . . . is the fact the both Secretary Babbitt and CEQ Director McGinty assured myself and [Utah] Senator Bennett in a meeting just a week prior to the President's announcement that the leaks concerning a designation of a monument in Utah were not true, and that no such action was contemplated. If it were true, we were told, the Utah Delegation would be fully apprised and consulted.

> But as we all know this promise was not kept. The biggest Presidential set aside in almost 20 years was done . . . without any notification, let alone consultation or negotiation with our Governor or State Officials . . . let me emphasize this point. There was no consultation, no hearings, no town meetings, no TV or radio discussion shows, no nothing.[74]

Utah Governor Bob Leavitt shares this view, telling Congress that the White House and the Department of the Interior (DOI) operated "under a cloak of secrecy."[75] Leavitt provides Congress with a detailed chronology of events between September 9 and 18. Throughout the period, the White House and the DOI maintained that no decisions had been made, yet he notes that on September 13

> my office became aware through the news media that an important environmental announcement was planned by the President at the Grand Canyon the following week. Preparations were being made by environmental organizations to transport groups from Utah. When we inquired directly of the Administration about the time, place and subject of the event they were not willing to even confirm the event would occur. Mr. Babbitt [and] his office . . . continued to indicate they had no information, insisting the matter was being handled by the White House. When we called the White House we were referred to the Interior Department.[76]

Leavitt met with White House Chief of Staff Leon Panetta in Washington during the afternoon of September 17. The president called Leavitt at 2:00 A.M. on September 18. After a thirty-minute conversation, Leavitt prepared a memo, which he faxed to Clinton (then in Illinois) at 4:00 A.M. The president made his announcement ten hours later. This, and an "emergency meeting" (that included the Utah congressional delegation) called by DOI Secretary Bruce Babbitt late in the afternoon of Friday the thirteenth for the next day, is the extent of involvement by the elected officials of Utah in the decision. Leavitt observes that at the time of the announcement

> no member of Congress, local official or the Governor were ever consulted, nor was the public. As the Governor, I had not seen a map, read the proclamation or for that matter even been invited [to the announcement ceremony]. This is not about courtesy, it is about process and public trust. A major land decision, the biggest in the last two decades, was being made. Obviously, this is not the way public land decisions should, or were ever intended to be made.[77]

Few can doubt the truth of Leavitt's conclusion.

On October 7, 1997, the House of Representatives passed H.R. 1127, the National Monument Fairness Act. It requires the president to give the governor of the affected state thirty days to review and comment on any proposed monument larger than 50,000 acres. It further stipulates that any such monument established under the Antiquities Act will cease to exist unless Congress concurs in its designation within two years.

Gridlock now characterizes federal land management. That condition serves new paradigmists quite well but not the interests of the remainder of the public.[78] The way out is not absolutely clear, but some basic ideas suggest themselves. Foremost is the need to enhance public understanding of the new paradigm. If the public wishes to embrace it, it is free to do so. But people should fully understand the intent of new paradigmists, the substance (or lack thereof) of their argument, and the human impact of their policies before they stroll too far down a path paved with mirages of healthy, sustainable, and government-protected ecosystems on or off federal lands.

The traditional notion of Congress apportioning use of federal lands still makes sense if people can free it from the grip of new paradigmist zealotry. Our present system of national parks, wildlife refuges, and wilderness areas serves us well. Conversely, economic use of federal lands to produce wealth in the furtherance of human well-being also serves us well, as does the ability of Americans to recreate in the outdoors. What does not serve us well is the incessant spatial expansion of the no-development, no-roads, no-people, no-use land management mind-set. Such absolutism spread over so much federal land not only deprives the nation of choice and opportunities to enhance the quality of human life but also adversely impacts large numbers of people. The General Accounting Office tells us that in 1964 approximately 9 percent of federal lands were managed primarily for preservation and conservation purposes, a figure that rose to about 43 percent in 1994 and will continue to grow under new paradigm managers.[79] As I point out in chapter 7, there is no legal authority for any federal agency to make protection of ecosystems its prime objective, and in recent years Congress specifically rejected efforts to enact that idea into law. Yet on March 2, 1998, in unveiling a new policy agenda for the U.S. Forest Service, Chief Mike Dombeck tells Forest Service employees that "our first priority is to maintain and restore the health of our ecosystems and watersheds."[80] Congress needs to reassert its authority over federal lands and clarify its priorities for federal land managers. Does Congress want multiple use and sustained yield or not? Does Congress really want the protection of endangered or threatened species to trump all other potential uses of public land all the time? Is perpetuation of a wetland always the best public use of public land? Should wilderness areas continue to be totally off limits to the artifacts of modern humans, or should some development be permitted? Do we need more wilderness areas? Do we need all the ones we have now? The new paradigmist effort to convert public lands into a vast temple dedicated to the worship of Mother Nature provides Americans with a critical focal point for a national debate on the management and disposition of public lands.[81]

Assuming that Congress retains control of public lands and retains traditional divisions between single-purpose and multiple-use land, it can restore more balance to federal land use by adopting market-based policies and incentives. Policies that require (and reward) land managers to generate a percentage of the operating cost of their unit from fees, leases, royalties, and so on could more closely align management decisions with public desires and congressional intent.[82] Congress should allow managers to keep some or all of the revenues that they generate to support operation of the lands they manage. Managers of multiple-use lands should have to generate a greater percentage of their operating costs than would a national park superintendent because, by definition, managers of multiple-use lands have more ways to produce income. The congressionally established purpose of a national park limits superintendents' potential income sources to fees charged visitors intent on sightseeing or outdoor recreation and, in some cases, to fees paid by concessioners. They cannot obtain revenue by selling timber or grazing rights, issuing oil and gas leases, permitting mining, or similar resource-dependent economic activities. Forest supervisors or Bureau of Land Management land managers—as managers of multiple-use lands—can and should do those things as well as provide for a variety of outdoor recreational pursuits while fulfilling their stewardship responsibilities.

The Essence of a Better Mousetrap

Future generations will judge us by the amount of liberty we leave them and by the quality of human life they inherit. Our future greatness turns on protecting human freedoms envisioned by the founding fathers, not on setting aside more wilderness. In framing environmental and natural resource policies, we must keep first things first. New paradigmists want to reorder society. Their policy goal—government-mandated maintenance and restoration of ecosystem health, integrity, and sustainability in service to their Earth Goddess—assaults liberty and freedom. They seek to defy history by resurrecting the failed notion that authoritarian central government planning leads to a better society. Future generations will judge us harshly indeed if we surrender their freedom to an eco-priesthood.

Americans can best improve stewardship by drawing on our societal strengths: democratic institutions, individual liberty, decentralized market-based decision making, and guaranteed rights for owners of private property. This is the most effective way for us to allocate resources and achieve policy goals agreed to by the people. Markets and incentives unleash entrepreneurism and creativity, which produce improved technologies and methods. These, in turn, allow us to use resources more efficiently, reduce pollution and other impacts on the environment, and generate the wealth we use to remediate past environmental damage. People win, as do other living things.

Notes

1. Michael T. Hayes, *Incrementalism and Public Policy* (New York: Longman, 1992).

2. In 1995 the Committee on Resources of the House of Representatives formed the Task Force on the Endangered Species Act and the Task Force on Wetlands. They held a series of hearings around the county from March to May. See U.S. Congress, House of Representatives, Committee on Resources, *Oversight Hearings on the Impact of the Endangered Species Act on the Nation—Part I, II, III* before the Task Force on Endangered Species Act, 104th Cong., 1st sess., May 10 (serial 104–10), May 18 (serial 104–14), and May 25 (serial 104–18), 1995; *Hearing on the Impact of the Endangered Species Act on the Area of Bakersfield, California* before the Task Force on Endangered Species, 104th Cong., 1st sess., April 17, 1995, serial 104–13; *Hearing on Local Impacts of the Endangered Species Act* before the Task Force on Endangered Species, 104th Cong., 1st sess., April 24, 1995, serial 104–15; *Hearing on The Impact of the Endangered Species Act on Northern California* before the Task Force on Endangered Species, 104th Cong., 1st sess., April 28, 1995, serial 104–16; *Oversight Hearing on the Impact of the Endangered Species Act on the Area Around Riverside, California* before the Task Force on Endangered Species, 104th Cong., 1st sess., April 26, 1995, serial 104–11; *Oversight Hearing on The Impacts of Endangered Species Act and Wetlands on the States of Louisiana and Texas* before the Task Force on the Endangered Species Act and the Task Force on Wetlands, 104th Cong., 1st sess., March 13,1995, serial 104–6; *Oversight Hearing on The Impacts of Endangered Species Act and Wetlands on the State of North Carolina* before the Task Force on the Endangered Species Act and the Task Force on Wetlands, 104th Cong., 1st sess., April 1, 1995, serial 104–7; and *Hearing on The Impact and Cost of Wetlands Regulations* before the Task Force on Wetlands, 104th Cong., 1st sess., April 19, 1995, serial 104–9.

3. A. Alan Moghissi, "Editorial—Management of the Endangered Species Act by the U.S. Government: Urgent Need for Corrective Action," *Environment International* 23, no. 3 (November 1997): 269–72; Robert E. Gordon, James K. Lacy, and James R. Streeter, "Conservation under the Endangered Species Act," *Environment International* 23, no. 3 (November 1997): 359–419; Jonathan Tolman, *Swamped: How America Achieved "No Net Loss"* (Washington, D.C.: Competitive Enterprise Institute, April 1997); Jonathan Adler, *Environmentalism at the Crossroads* (Washington, D.C.: Capital Research Center, 1995).

4. Robert H. Nelson, "How Much Is Enough? An Overview of the Benefits and Costs of Environmental Protection," in *Taking the Environment Seriously,* ed. Roger E. Meiners and Bruce Yandle (Lanham, Md.: Rowman & Littlefield, 1993), 1–23.

5. Randy T. Simmons, "Fixing the Endangered Species Act," in *Breaking the Environmental Policy Gridlock,* ed. Terry L. Anderson (Stanford, Calif.: Hoover Institution Press, 1997), 82.

6. Mary E. Power, David Tilman, James A. Estes, Bruce A. Menge, William J. Bond, L. Scott Milles, Gretchen Daily, Juan Carlos Castilla, Jane Lubchenco, and Robert T. Paine, "Challenges in the Quest for Keystones," *BioScience* 46, no. 8 (September 1996): 609.

7. David Ehrenfeld, "Why Put a Value on Biodiversity," in *Biodiversity,* ed. E. O. Wilson (Washington, D.C.: National Academy Press, 1988), 212. For a different view, see Oliver A. Houck, "Reflections on the Endangered Species Act," in *Land Use & Environmental Law Review 1996,* ed. Stuart L. Deutsch and A. Dan Tarlock (Deerfield, Ill.: Clark, Boardman, Callaghan, 1996), 513–26.

8. Thirty years ago, ecologist Garrett Harden demonstrated that the incentives for conservation are weak when property rights are held in common or when resources are unowned; Garrett Harden, "Tragedy of the Commons," *Science* 162 (December 13, 1968): 1243–48.

9. Seth Norton, "Property Rights, the Environment, and Economic Well-Being," in *Who Owns the Environment,* ed. Peter J. Hill and Roger E. Meiners (Lanham, Md.: Rowman & Littlefield, 1998), 51.

10. Data downloaded from the Nature Conservancy's Web site at http://www.tnc.org/infield/map.html, May 12, 1998.

11. Quote and data downloaded from the Nature Conservancy's Web site at http://www.tnc.org/infield/preserve/adirondack/adirondack.htm, May 12, 1998.

12. Personal communication with Stephen E. Adair, manager of habitat development for Ducks Unlimited, May 14, 1998 (copy on file with the author). Information on Ducks Unlimited Campaign 2000 downloaded from its Web site at http://www.ducks.org/habitat/index.html, May 15, 1998.

13. See generally Bruce Yandle, *Common Sense and Common Law for the Environment* (Lanham, Md.: Rowman & Littlefield, 1997); Roger E. Mieners, "Elements of Property Rights: The Common Law Alternative," in Yandle, ed., *Land Rights,* 269–94.

14. Economists and others working at universities and think tanks such as the Cato Institute, Competitive Enterprise Institute, Heartland Institute, Heritage Foundation, and Political Economy Research Center have championed market-based approaches to federal policies for several years. See, for example, Anderson, *Breaking the Environmental Policy Gridlock;* Meiners and Yandle, *Taking the Environment Seriously;* Fred L. Smith and Kent Jeffreys, "A Free Market Environmental Vision," in *Market Liberalism,* ed. David Boaz and Edward H. Crane (Washington, D.C.: Cato Institute, 1993), 389–402; and Terry Anderson and Donald R. Leal, *Free Market Environmentalism* (San Francisco: Pacific Research Institute, 1991). For a concise explanation of free-market environmentalism, see Richard Stroup, "Environmentalism, Free-Market," in *The Fortune Encyclopedia of Economics,* ed. David R. Henderson (New York: Warner Books, 1993), 442–45.

15. Daniel A. Farber, "Environmental Protection as a Learning Experience," *Loyola of Los Angeles Law Review* 27, no. 3 (April 1994): 806.

16. R. Hilborn, C. J. Walters, and D. Ludwig, "Sustainable Exploitation of Renewable Resources," in *Annual Review of Ecology and Systematics* (Palo Alto, Calif.: Annual Reviews, 1995): 60.

17. As quoted in Robert R. Wright and Morton Gitelman, *Case and Materials on Land Use,* 4th ed. (St. Paul: West Publishing, 1991), 397.

18. U.S. Congress, House Committee on Resources, "The Endangered Species Act:

How Much Does It Cost the Taxpayer?" Majority Staff Report, April 2, 1998, available from the Office of Chief Counsel (copy on file with the author).

19. U.S. Congress, House Committee on Resources, "The Endangered Species Act," 24.

20. Simmons, "Fixing the Endangered Species Act," 84.

21. Nelson Wolff, testimony at *Oversight Hearings on Endangered Species and Wetlands before Task Force on Endangered Species and the Task Force on Wetlands,* U.S. Congress, House of Representatives, Committee on Resources, 104th Cong., 1st sess., March 13, 1995, serial 104–6, 85, 188 (hereafter cited as *ESA and Wetlands Oversight Hearing,* serial 104–6).

22. Wes Cooley, testimony at *ESA and Wetlands Oversight Hearing,* serial 104–6, 68.

23. Maury Hood, testimony at *ESA and Wetlands Oversight Hearing,* serial 104–6, 107.

24. Margaret Rector, testimony at *ESA and Wetlands Oversight Hearing,* serial 104–6, 106.

25. U.S. Congress, House Committee on Resources, "The Endangered Species Act," 20.

26. Donald A. Kirkman, testimony at *Oversight Hearings on Endangered Species and Wetlands—Part II* before Task Force on Endangered Species and the Task Force on Wetlands, U.S. Congress, House of Representatives, Committee on Resources, 104th Cong., 1st sess., April 1, 1995, serial 104–7, 18. See generally U.S. Congress, Senate Subcommittee on Clean Air, Wetlands, Private Property, and Nuclear Safety of Committee on Environment and Public Works, *Wetlands Regulatory Reform Act of 1995: Hearing before the Subcommittee on Clean Air, Wetlands, Private Property, and Nuclear Safety of Committee on Environment and Public Works,* 104th Cong., 1st sess., July 19, August 2, and November 1, 1995. S. Hrg. 104–643 (hereafter cited as *Wetlands Regulatory Reform Act of 1995 Hearing*).

27. Gaylord Enns, testimony at *Hearing on The Impact and Cost of Wetlands Regulations* before the Task Force on Wetlands, U.S. Congress, House of Representatives, Committee on Resources, 104th Cong., 1st sess., April 19, 1995, serial 104–9, 108.

28. Nelson Wolff, testimony at *ESA and Wetlands Oversight Hearing,* serial 104–6, 86.

29. New paradigmists and others tell us that everyone benefits by preventing species extinction and preserving wetlands. Protecting species, in their view, makes each one of us a better person for doing so and is ethically and morally correct. They say that we might also gain utilitarian benefits down the road when some preserved species turns out to hold the key to the cure for cancer, human immunodeficiency virus (HIV), or another deadly disease. Even if you accept these benefits as real, the population as a whole bears few of the costs required for their creation.

30. Glen Spain, testimony at *Wetlands Regulatory Reform Act of 1995 Hearing,* 61.

31. Spain, testimony at *Wetlands Regulatory Reform Act of 1995,* 62.

32. Robert H. Nelson, "Federal Zoning: The New Era in Environmental Policy," in Yandle, ed., *Land Rights,* 311. See also Bruce Yandle, *Political Limits of Environmental Regulation: Tracking the Unicorn* (New York: Quorum Books, 1989).

33. Denis Hayes, quoted in Lou Cannon, "Environmentalists Try Power of the Purse," *Washington Post,* June 20, 1998, A10.

34. Spain, testimony at *Wetlands Regulatory Reform Act of 1995 Hearing,* 62.

35. David G. Cameron, testimony at *Oversight Hearings on The Impact of the Endangered Species Act on the Nation—III* before the Task Force on Endangered Species Act, U.S. Congress, House of Representatives, Committee on Resources, 104th Cong., 1st sess., May 25, 1995, serial 104–18, 165.

36. For a variety of ideas on how to use the market and private property to protect endangered species, see Simmons, "Fixing the Endangered Species Act"; Lee Ann Welch, "Property Rights Conflicts under the Endangered Species Act: Protection of the Red-Cockaded Woodpecker," in Yandle, ed., *Land Rights,* 151–97; Ike Sugg, testimony at *Oversight Hearings on The Impact of the Endangered Species Act on the Nation—Part II* before the Task Force on Endangered Species Act, U.S. Congress, House of Representatives, Committee on Resources, 104th Cong., 1st sess., May 18, 1995, serial 104–14, 141–58.

37. Uniform Relocation Assistance and Real Property Act of 1970 as amended, 42 U.S.C. §§ 4651, 4652 (1994).

38. Randal O'Toole, "Analyzing the Endangered Species Act," downloaded from the Thoreau Institute's Web site at http://www.teleport.com/rot/Analysis.html, June 4, 1998, 12.

39. Detailed descriptions of these programs are available from the Web sites of the federal agencies most involved. For the Wetland Reserve Program, see the Natural Resource Conservation Service's site at http://www.ftw.nrcs.usda.gov/pl1566/WRP.html; for the North American Waterfowl Management Plan and the Partners in Wildlife, see the Fish and Wildlife Service's Web site at http://www.fws.gov/r9nawwo/news.html and http:/www.fws.gov/r9endspp/esb/96/pfw.html, respectively.

40. Tolman, *Swamped.*

41. Government prevention of wetland destruction is consistent with the protection of property rights in cases in which such destruction would wrongfully harm other property owners. Just as government can prevent me from staging an unending series of rock concerts in my backyard because doing so would cause identifiable harm to identifiable parties in my neighborhood, government can legitimately prevent a landowner from destroying wetlands if it can demonstrate that doing so would flood land downstream.

42. For applicable COE regulations, see generally 33 C.F.R. 320–30. For EPA regulations, see generally 40 C.F.R. 230–33.

43. The idea of sequencing comes from regulations published by the President's Council on Environmental Quality at 40 C.F.R. 1508.20.

44. Tolman, *Swamped.*

45. Bernard N. Goode, *Wetlands Regulatory Reform Act of 1995 Hearing,* 151.

46. Margie Vicknair, testimony before the White House Interagency Wetlands Task Force, New Orleans, Louisiana, August 27, 1990 (transcript on file with the author).

47. White House Office of Environmental Policy, "Protecting America's Wetlands: A Fair, Flexible, and Effective Approach," White House Office of Environmental Policy, Washington, D.C., August 24, 1993.

48. Most wetland scientists consider creating wetlands to be more difficult than restoring a degraded wetland or enhancing an existing wetland.

49. These ratios are suggestive and are intended to illustrate the argument. I recognize that offsets based on acreage do not always capture the importance of a specific wetland

and that in some cases development should be permitted without offsets, as with a home owner making improvements or in Alaska, where more than 99 percent of pre-European settlement wetlands remain intact.

50. The idea that the public should pay all the costs of obtaining public benefits gained from ensuring a legacy of wetlands has considerable merit. It is, however, politically impossible to implement a policy based on this premise at this time.

51. U.S. Congress, Senate, *Omnibus Property Rights Act of 1997,* 105th Cong., 1st sess., S. 781, sec. 204 (a)(D).

52. U.S. Congress, Senate, *Private Property Owners Bill of Rights,* 105th Cong., 1st sess., S. 953, sec. 9 (a).

53. For two views on requiring landowners to buy offsets in the case of endangered species habitat, see Gregg Easterbrook, "Getting around the 'Takings' Problem?" *PERC Reports* 16, no. 2 (June 1998): 3–5, and Jane S. Shaw and Richard L. Stroup, "Response: Not the Free Market," *PERC Reports* 16, no. 2 (June 1998): 5–6.

54. Douglas P. Wheeler and James M. Strock, the respective heads of the California Resources Agency and the California Environmental Protection Agency, "Official Policy on Conservation Banks," April 7, 1995; California Resources Agency, "Market Created for Habitat Improvements: Conservation Banks Integrate Environmental, Economic Goals," press release, April 7, 1995, downloaded from the California Resources Agency's Web site at http://ceres.ca.gov, June 11, 1998.

55. California Resources Agency, "Conservation Banks Flourish in California: State Policy Promotes Selling of Conservation Credits," press release, June 13, 1996, downloaded from the California Resources Agency's Web site at http://ceres.ca.gov, June 11, 1998.

56. Robert W. Brumbaugh, "Wetland Mitigation Banking: Entering a New Era?" *The Wetlands Research Bulletin* 5, nos. 3–4 (October–December 1995), downloaded from http://www.wes.army.mil/el/wrtc/wrp/bulletins/v5n3/brum.html, June 10, 1998; White House Office of Environmental Policy, "Protecting America's Wetlands." For analyses of wetlands mitigation banking, see Institute for Water Resources, "National Wetlands Mitigation Banking Study: Commercial Wetland Credit Ventures—1995 National Survey," IWR Report No. 96-WMB-9, U.S. Army Corps of Engineers, Alexandria, Virginia, August 1996; Institute for Water Resources, "National Wetland Mitigation Banking Study: Commercial Wetland Mitigation Credits—Theory and Practice," IWR Report No. 95-WMB-7, U.S. Army Corps of Engineers, Alexandria, Virginia, November 1995; and Institute for Water Resources, "National Wetland Banking Study: Wetland Mitigation Banking," IWR Report No. 94-WMB-6, U.S. Army Corps of Engineers, Alexandria, Virginia, February 1994.

57. Conversation with Robert Brumbaugh, study manager for the Army Corps of Engineers' National Mitigation Banking Study, June 26, 1998.

58. For views on federal land management, see Brent Steel, ed., *Public Land Management in the West: Citizens, Interest Groups, Values* (Westport, Conn.: Praeger, 1997), and Robert H. Nelson, ed., *The Failure of Scientific Management: Rethinking the Public Lands* (Lanham, Md.: Rowman & Littlefield, 1995).

59. Terry L. Anderson, ed., *Multiple Conflicts over Multiple Uses* (Bozeman, Mont.: Political Economy Research Center, 1994).

60. "Sierra Club Members Choose New Directors, New Forest Policy," *The Planet* 3, no. 5 (June 1996). According to its cover, *The Planet* is "The Sierra Club Activist Resource." Downloaded from the Sierra Club's Web site at http://www.sierraclub.org/planet/199606/clubelec.html, June 18, 1998.

61. For example, see Bruce Babbitt, "Letter from the Secretary: The Arctic National Wildlife Refuge, As Close as Your Own Back Yard," and "Letter from the Secretary: Drilling in the Arctic Refuge, What's in It for You?" downloaded from the Department of the Interior's Web site at http://www.doi.gov/doibboc1.html and http://www.doi.gov/doibboc2.html, June 17, 1998. Environmentalists strongly support H.R. 900, the Morris K. Udall Memorial Wilderness Act (105th Cong., 1st sess.), which would give wilderness designation to the potentially energy-rich portions of the Arctic National Wildlife Refuge. See also testimony by Brooks Yeager, then president of the National Audubon Society, and Mike Matz, chairman of the environmentalist Alaska Coalition, on the *National Energy Security Act of 1991* before Congress, Senate Committee on Energy and Natural Resources, 102nd Cong., 1st sess., March 12, 1991. S. Hrg. 102–5, pt. 8.

62. Robert H. Nelson, "Greasing the Skids for a New Federal Oil and Gas Leasing System," in Anderson, ed., *Multiple Conflicts over Multiple Uses,* 77–100.

63. National Research Council, *Our Seabed Frontier* (Washington, D.C.: National Academy Press, 1989), 27. See also National Research Council, *Oil in the Sea: Inputs, Fates and Effects* (Washington, D.C.: National Academy Press, 1985).

64. Congressional Research Service, "Impact of Persian Gulf Crisis on Federal Offshore Leasing," *CRS Review,* November/December 1990, as quoted in Department of the Interior, Minerals Management Service, "Outer Continental Shelf Natural Gas and Oil Resource Management: Comprehensive Program 1992–1997, Proposed Final Summary and Decision," Washington, D.C., April 1992, 16.

65. George Bush, "Statement by the President," press release, Office of the Press Secretary, The White House, June 26, 1990 (copy on file with the author).

66. William J. Clinton, "Remarks by the President to the National Oceans Conference," June 12, 1998, Monterey, California, downloaded from http://library. whitehouse.gov, June 16, 1998, and "Memorandum for the Secretary of the Interior; Subject: Withdrawal of Certain Areas of the United States Outer Continental Shelf from Leasing Disposition," press release, Office of the Press Secretary, The White House, June 12, 1998, downloaded from http:/www.pub.whitehouse.gov, June 16, 1998.

67. William J. Clinton, "Establishment of the Grand Staircase–Escalante National Monument by the President of the United States of America, a Proclamation," press release, Office of the Press Secretary, The White House, September 18, 1996, and "Remarks by the President in Making Environmental Announcement," press release, Office of the Press Secretary, The White House, September 18, 1996.

68. 16 U.S.C. § 431–33 (1994).

69. Charles Fernandes, "Idaho Legislators Counter Clinton Act Would Curb Power," *Coeur d'Alene Press,* September 20, 1996; Orrin Hatch, statement at *Oversight Hearings on Establishing the Grand Staircase–Escalante National Monument* before Congress, House Subcommittee on National Parks and Public Lands of the Resources Committee, 105th Cong., 1st sess., April 29, 1996, Committee Print 105–20, 31 (hereafter cited as *Grand Staircase–Escalante Hearing*). The acrimony regarding the secrecy and

what critics see as the illegal use of raw political power are amply revealed in the hearing record. Secretary Babbitt and CEQ Chairman McGinty defend the administration by arguing that the public debate over public lands in Utah is an old one.

70. Stephen J. Siegel, "Even Democrat Orton Joins the Chorus against Clinton Move," *Ogden Standard-Examiner,* September 19, 1996; Jim Woolf, "Clinton Creates 1.7 Million Acre National Monument," *Salt Lake Tribune,* September 19, 1996; Marina O'Neill, "Leavitt Aims to Stall Monument Plan: Governor Offering Alternative Proposal in Meeting with Clinton," *Ogden Standard-Examiner,* September 17, 1996. No elected official from Utah—including Democrat Bill Orton, in whose congressional district Clinton established the monument—attended the announcement ceremony. Three-term congressman Orton lost his reelection bid in what some observers believe was an expression of voter disapproval of the monument designation.

71. Michael O. Leavitt, *Grand Staircase–Escalante Hearing,* 54.

72. U.S. Congress, House Committee on Resources, "Behind Closed Doors: The Abuse of Trust and Discretion in the Establishment of the Grand Staircase–Escalante National Monument—Staff Report," 105th Cong., 1st sess., November 7, 1997, serial 105-D, 1–2. Staff reports can be highly political documents. In this case the report contains complete copies of the documents on which it is based, and readers can draw their own conclusions.

73. Reprinted in U.S. Congress, House Committee on Resources, "Behind Closed Doors," 28.

74. Hatch, *Grand Staircase–Escalante Hearing,* 29.

75. Leavitt, *Grand Staircase–Escalante Hearing,* 68.

76. Leavitt, *Grand Staircase–Escalante Hearing,* 50.

77. Leavitt, *Grand Staircase–Escalante Hearing,* 53.

78. S. 1253, The Public Lands Management Improvement Act of 1997 (105th Cong., 1st sess.), would clarify the mission of the Forest Service by reasserting a multiple-use mandate, simplify its planning procedures, and limit the extent to which special interests could tie up decision making via endless appeals. On June 17, 1998, James Lyons, the Department of Agriculture's undersecretary for natural resources, told the Subcommittee on Forests and Public Land of the Senate Committee on Energy Natural Resources that "the Administration strongly opposes S. 1253." Testimony downloaded from the Forest Service's Web site at http://www.fs.fed.us/intro/testimony/19980617.html, July 9, 1998.

79. U.S. General Accounting Office, *Land Ownership: Information on the Acreage, Management, and Use of Federal and Other Lands,* GAO/RCED-96–40 (Washington, D.C.: U.S. General Accounting Office, March 1996), 4.

80. Mike Dombeck, "A Gradual Unfolding of a National Purpose: A Natural Resource Agenda for the 21st Century," speech delivered to U.S. Forest Service employees in Washington, D.C., March 2, 1998, downloaded from the Forest Service's Web site at http://www.fs.fed.us/news/agenda/sp30298.htm, July 10, 1998.

81. Terry L. Anderson and Donald R. Leal, "Rekindling the Privatization Fires: Political Lands Revisited," in *Breaking the Environmental Policy Gridlock,* 53–81.

82. Congress originally expected the national parks to be self-supporting with monies coming from charged concessioners and visitors. See John Ise, *Our National Park Policy: A Critical History* (Baltimore: The Johns Hopkins University Press, 1961), 619–38.

Conclusion

Knowledge and Responsibility

> In [knowledge's] light, we must think and act not only for the moment but for our time.
>
> —John F. Kennedy, "Speech at the University of California at Berkeley, 23 March 1962," in *Public Papers of the Presidents of the United States: John F. Kennedy,* 266

In his book, *No Turning Back, Dismantling the Fantasies of Environmental Thinking,* environmentalist and science writer Wallace Kaufman observes,

> Over the past twenty years the environmental meetings I attend as a journalist have seemed more and more like church meetings. Every year for five years I have sat with a group of widely published nature writers, most of them respected lecturers and professors. These writers compile the anthologies that schools and universities use in courses on literature and the environment. . . . In our meetings, they talk about "converting" America and not letting science interfere with a "spiritual understanding" of nature . . . their language is not the language of reason and does not invite democratic debate.[1]

Kaufman finds that "the environmental movement is quickly charting a new way for American thinking" rooted in "adversarial science," a belief that economic growth is "the enemy," and support for "centralized government management."[2] He includes among their ideals "pristine wilderness," "redistributing land and income," and "living as co-equals with nature."

Democratic societies and individual freedom advance with the acquisition and spread of knowledge among the citizenry. They retreat in the face of ignorance. We all have a responsibility to gain and use understanding for the betterment of ourselves and the nation. The classroom is a good place to start.

The June 21, 1998, edition of the *Washington Times* contains a full-page special feature highlighting two local student winners in a national environmental essay contest.[3] It reprints the winning works, which focus on cleaning up urban neighborhoods. The seventh-grade national winner writes that "our land and ocean get acid rain whenever it rains," and a fourth-grade regional victor claims that "city dumps are so overcrowded, we are actually running out of places to put our trash." Both statements are incorrect. I have no quarrel with the students; after all, they are but twelve and nine years old and only repeating what they gather from teachers and other sources of information. But why did the teachers, principals, and other reviewers that selected these essays not know that the statements were wrong? The students chant the dogma "We are polluting the land, the water and the air" and "Let's Now Save Mother Earth!" I applaud environmental awareness in the classroom, but more teachers need to gird youthful enthusiasm with a level of understanding that runs deeper than a collection of environmentalist bumper stickers. Of course, better environmental cognition must reach beyond our schools and into all segments of society.

The public, decision makers, and opinion leaders have a responsibility to improve their knowing on multiple fronts. They must clearly grasp the motives behind the new paradigm and its meaning for American society. They should recognize the hollowness of buzzwords such as *ecosystem health, ecosystem integrity,* and *sustainable ecosystems*. They ought to comprehend that ecosystems are products of the human mind, not the hand of God. Armed with knowledge, they can put some very pointed questions to new paradigmists seeking to use government power to protect some ecosystem that they claim is a threatened actual living object on the landscape. "Where," they might ask, "is this ecosystem?" "How did you determine where this ecosystem stops and its neighbor begins?" "When did it arrive there?" "Where did it come from, and where will it go when it leaves?" "You say it is sick and may die; how will we know when it is dead?" "Did it replace an earlier ecosystem that died in the same place?" "Where is its body?" "Why should we relinquish freedom so you can better defend your illusions or worship Mother Nature?" The answers that advocates give to such questions would be studies in the obscure.

Americans need not feel guilt about our stewardship of the environment, our democratic institutions, or our defense of individual liberty. "Cardinal Green" of the eco-church and his minions threaten freedom and improvement in human well-being at this moment and in our time, just as King George and his officials threatened them at our nation's founding. As we once rejected the king, I suggest that we rebuff the "cardinal."

Notes

1. Wallace Kaufman, *No Turning Back, Dismantling the Fantasies of Environmental Thinking* (New York: Basic Books, 1994), 7–8.
2. Kaufman, *No Turning Back,* 3–4.
3. *Washington Times,* June 21, 1998, A5.

Selected Bibliography

Adler, Jonathan. *Environmentalism at the Crossroads*. Washington, D.C.: Capital Research Center, 1995.

Agee, James K., and Darryll R. Johnson, eds. *Ecosystem Management for Parks and Wilderness*. Seattle: University of Washington Press, 1988.

Alper, Joseph. "War over the Wetlands: Ecologists v. the White House." *Science* 257 (August 21, 1992): 1042–43.

———. "Yellowstone Ecosystem: 'Win-Win' Solution." *Science* 255 (February 1992): 685–86.

Amundson, Ronald, and Hans Jenny. "On a State Factor Model of Ecosystems." *BioScience* 47, no. 8 (September 1997): 536–43.

Anderson, Jay E. "A Conceptual Framework for Evaluating and Quantifying Naturalness." *Conservation Biology* 5, no. 3 (1993): 347–52.

Anderson, Terry L., ed. *Breaking the Environmental Policy Gridlock*. Stanford, Calif.: Stanford University Press, 1997.

———, ed. *Multiple Conflicts over Multiple Uses*. Bozeman, Mont.: Political Economy Research Center, 1994.

Anderson, Terry, and Donald R. Leal. *Free Market Environmentalism*. San Francisco: Pacific Research Institute, 1991.

Angermeier, Paul L., and James R. Karr. "Biological Integrity versus Biological Diversity as Policy Directives." *BioScience* 44, no. 10 (November 1994): 690–97.

Aplet, Gregory, H. Nels Johnson, Jeffrey T. Olson, and V. Alaric Sample, eds. *Defining Sustainable Forestry*. Washington, D.C.: Island Press, 1994.

Ashton, Peter S. "Species Richness in Plant Communities." In *Conservation Biology,* edited by Peggy L. Fielder and Subodh K. Jain. New York: Chapman & Hall, 1992.

Atlanta Journal and Atlanta Constitution. "Georgia Cited among Top States Where Mother Nature Is at Risk, Wake-Up Call: A 'Report Card . . . on Ecological Health:' Ranks the State the Third Most Imperiled in the Nation," December 21, 1995, A20.

Babbitt, Bruce. "Between the Flood and the Rainbow: Our Covenant: To Protect the Whole of Creation." Article from the secretary, downloaded from the Department of the Interior's Web page at http://www.doi.gov, December 19, 1995.

———. "Between the Flood and the Rainbow: Stewards of Creation." *Christian Century* 113, no. 16 (May 8, 1996): 500–2.

———. "A Coordinated Campaign: Fight Fire with Fire." Remarks at Boise State University, February 11, 1997, downloaded from "The Secretary's Alcove" at the Department of the Interior's Web site at http://www.doi.gov.

———. Testimony before Congress on H.R. 1845, The National Biological Survey Act of 1993, House of Representatives, before the Subcommittee of Technology, Environment and Aviation, and the Subcommittee on Investigations and Oversight of the Committee on Science, Space, and Technology, 103rd Cong., 1st sess., September 14, 1993,Committee Print 56.

———. Testimony before the Congress on his nomination as secretary of the interior, Senate Committee on Energy and Natural Resources, 103rd Cong., 1st sess., January 19 and 21, 1993, S. Hrg. 103–3.

———. U.S. Department of the Interior, Secretarial Order No. 3173, September 29, 1993.

Bailey, Robert G. "Delineation of Ecosystem Regions." *Environmental Management* 7, no. 4 (1983): 365–73.

———. *Description of the Ecoregions of the United States*. Miscellaneous Publication No. 1391. Washington, D.C.: U.S. Department of Agriculture, U.S. Forest Service, 1980.

———. *Description of the Ecoregions of the United States*. 2nd ed. Miscellaneous Publication No. 1391. Washington, D.C.: USDA/Forest Service, 1995.

———. "Ecoregions of the United States." Map. Ogden, Utah: U.S. Department of Agriculture, U.S. Forest Service, 1976.

———. "Ecoregions of the United States." Map. Washington, D.C.: U.S. Forest Service, 1994.

———. *Ecosystem Geography*. New York: Springer, 1996.

———. *U.S. Ecoregions of the United States—RARE II Map B*. Washington, D.C.: U.S. Department of Agriculture, U.S. Forest Service, 1976.

Bailey, Robert G., Peter E. Avers, Thomas King, and Henry McNab, eds. *Ecoregions and Subregions of the United States*. Washington, D.C.: U.S. Department of Agriculture, U.S. Forest Service, 1994.

Bailey, Ronald, ed. *The True State of the Planet*. New York: The Free Press, 1995.

Bain, Ling, and Stephen J. Walsh. "Scale Dependencies of Vegetation and Topography in a Mountainous Environment of Montana." *Professional Geographer* 45, no. 1 (February 1993): 1–11.

Baker, William L., and Gillian M. Walford. "Multiple Stable States and Models of Riparian Vegetation Succession on the Animas River, Colorado." *Annals of the Association of American Geographers* 85, no. 2 (June 1995): 320–38.

Barton, John H. "Biodiversity at Rio." *BioScience* 42, no. 10 (November 1992): 773–76.

Bast, Joseph L., Peter J. Hill, and Richard C. Rue. *Eco-Sanity: A Common-Sense Guide to Environmentalism*. Lanham, Md.: Madison, 1994.

Bean, Michael J. "A Policy Perspective on Biodiversity Protection and Ecosystem Management." In *The Ecological Basis for Conservation,* edited by S. T. A. Pickett, R. S. Ostfeld, M. Shachak, and G. E. Likens. New York: Chapman & Hall, 1997.

Bear, Dinah. "The Promise of NEPA." In *Biological Diversity and the Law,* edited by William Snape III. Washington, D.C.: Island Press, 1996.

Beattie, Mollie. "An Ecosystem Approach to Fish and Wildlife Conservation." *Ecological Applications* 6, no. 3 (August 1996): 696–99.

Beckerman, Wilfred. *Green-Colored Glasses: Environmentalism Reconsidered.* Washington, D.C.: Cato Institute, 1996.

Bengston, David N. "Changing Forest Values and Ecosystem Management." *Society and Natural Resources* 7 (1994): 515–33.

Berry, Thomas. "The Viable Human." In *Deep Ecology for the 21st Century: Readings in the Philosophy and Practice of the New Environmentalism,* edited by George Sessions. Boston: Shambala, 1995.

Berry, Wendall. "Christianity and the Survival of Creation." In *Sacred Trusts: Essays on Stewardship and Responsibility,* edited by Michael Katakis. San Francisco: Mercury House, 1993.

Birch, Charles, William Eakin, and Jay B. McDaniel, eds. *Liberating Life: Contemporary Approaches to Ecological Theology.* Maryknoll, N.Y.: Orbis Books, 1990.

Bisson, Peter A., Thomas P. Quinn, Gordon H. Reeves, and Stanley V. Gregory. "Best Management Practices, Cumulative Effects, and Long-Term Trends in Fish Abundance in Pacific Northwest River Systems." In *Watershed Management,* edited by Robert J. Naiman. New York: Springer-Verlag, 1992.

Bloch, Ben, and Harold Lyons. *Apocalypse Not: Science, Economics, and Environmentalism.* Washington, D.C.: Cato Institute, 1993.

Blumm, Michael. "Public Choice Theory and the Public Lands: Why 'Multiple Use' Failed." *Harvard Environmental Law Review* 18, no. 2 (1994): 405–32.

Blumm, Michael C., and Bernard Zaleha. "Federal Wetlands Protection under the Clean Water Act: Regulatory Ambivalence, Intergovernmental Tension, and a Call for Reform." *University of Colorado Law Review* 60 (1989): 695–772.

Bohlen, Curtis C. "Controversy over Federal Definition of Wetlands." *BioScience* 41, no. 3 (March 1991): 139.

Bosselman, Fred P., and A. Dan Tarlock. "The Influence of Ecological Science on American Law: An Introduction." *Chicago-Kent Law Review* 69, no. 4 (1994): 847–74.

Botkin, Daniel. *Discordant Harmonies: A New Ecology for the Twenty-First Century.* New York: Oxford University Press, 1990.

———. *Our Natural History: The Lessons of Lewis and Clark.* New York: G. P. Putnam and Sons, 1995.

Boul, S. W., F. D. Hole, and R. J. McCracken. *Soil Genesis and Classification.* 3rd ed. Ames: Iowa State University Press, 1989.

Bowler, Peter J. *The Environmental Sciences.* New York: W. W. Norton, 1992.

Bowyer, Jim. "Fact vs. Perception." *Forest Products Journal* (November/December 1995): 31–36.

Bradley, Raymond S., and Philip D. Jones, eds. *Climate since A.D. 1500.* New York: Routledge, 1992.

Bramwell, Anna. *The Fading of the Greens.* New Haven, Conn.: Yale University Press, 1994.

Breuilly, Elizabeth, and Martin Palmer, eds. *Christianity and Ecology*. London: Cassell Publishers, 1992.

Brower, David. *Let the Mountains Talk, Let the Rivers Run*. New York: HarperCollinsWest, 1995.

Brown, Dwight A. "Early Nineteenth Century Grasslands of the Midcontinent Plains." *Annals of the Association of American Geographers* 83, no. 4 (December 1993): 589–612.

Brown, Michael. "Earth Worship of Black Magic?" *The Amicus Journal* 14, no. 4 (winter 1993): 32–34.

Brown, Ralph H. *The Historical Geography of the United States*. New York: Harcourt, Brace & World, 1948.

Brumbaugh, Robert W. "Wetland Mitigation Banking: Entering a New Era?" *The Wetlands Research Bulletin* 5, nos. 3–4 (October–December 1995).

Budiansky, Stephen. "Extinction or Miscalculation?" *Nature* 370 (July 14, 1994): 104.

———. *Nature's Keepers: The New Science of Nature Management*. New York: The Free Press, 1995.

Bush, George. "The President's Message on Environmental Quality." In *Environmental Quality: The Twenty-First Annual Report of the Council on Environmental Quality*. Washington, D.C.: U.S. Government Printing Office, 1991.

———. "Remarks to Members of Ducks Unlimited." In *Public Papers of the Presidents of the United States,* vol. 1. Washington, D.C.: U.S. Government Printing Office, 1989.

Butzer, Karl W. "The Americas before and after 1492: An Introduction to Current Geographical Research," *The Annals of the Association of American Geographers* 82, no. 3 (September 1992): 345–68.

———. "The Indian Legacy in the American Landscape." In *The Making of the American Landscape,* edited by Michael P. Conzen. New York: Routledge, 1994.

Caldwell, Lynton Keith. *Between Two Worlds: Science, the Environmental Movement, and Policy Choice*. Cambridge: Cambridge University Press, 1990.

Caldwell, Lynton K., and Kristin Shrader-Frechette. *Policy for Land: Law and Ethics*. Lanham, Md.: Rowman & Littlefield, 1993.

Callicott, J. Baird. "The Scientific Substance of the Land Ethic." In *Aldo Leopold: The Man and His Legacy,* edited by Thomas Tanner. Ankeny, Iowa: Soil Conservation Society of America, 1987.

Calow, Peter. "Ecosystem Health—a Critical Analysis of Concepts." In *Evaluating and Monitoring the Health of Large Scale Ecosystems,* edited by David J. Rapport, Connie L. Gaudet, and Peter Calow. New York: Springer, 1995.

Campbell, John. *Map Use and Analysis*. Dubuque, Iowa: Wm C. Brown, 1991.

Carlin, A., and the Environmental Law Institute. *Environmental Investment: The Cost of a Clean Environment—a Summary*. EPA-230-12-90-083. Washington, D.C.: U.S. Environmental Protection Agency, December 1990.

Carlin, Alan, Paul F. Scodari, and Don H. Garner. "Environmental Investments: The Costs of Cleaning Up." *Environment* 34, no. 2 (1992): 12–20, 38–44.

Carpenter, Betsy. "In a Murky Quagmire." *U.S. News and World Report,* June 3, 1991, 45–46.

Chadwick, D. H. "Mission for the 1990s." Defenders of Wildlife Special Report. Washington, D.C.: Defenders of Wildlife, 1990.

Chase, Alton. *In a Dark Wood*. Boston: Houghton Mifflin, 1995.

———. "Prophets for the Temple of Green." *Washington Times,* January 26, 1996, A16.

Chial, Douglas L. "The Ecological Crisis: A Survey of the WWC's Recent Responses." *Ecumenical Review* 48 (January 1996): 53–61.

Christensen, James, E. K. Jeffrey Danter, Debra L. Griest, Gary W. Mullins, and Emmalou Norland. *U.S. Fish and Wildlife Service Ecosystem Approach to Fish and Wildlife Conservation: An Assessment*. Columbus: Ecological Communications Lab, School of Natural Resources, Ohio State University, January 1998.

Christensen, Norman L., Ann M. Bartuska, James H. Brown, Stephen Carpenter, Carla D'Antonio, Robert Frances, Jerry F. Franklin, James A. MacMahon, Reed F. Noss, David J. Parsons, Charles H. Peterson, Monica G. Turner, and Robert G. Woodmansee. "Report of the Ecological Society of America Committee on the Scientific Basis for Ecosystem Management." *Ecological Applications* 6, no. 3 (1996): 665–91.

Clark, Stephen R. L. "Global Religion." In *Philosophy and the Natural Environment,* edited by Robin Attfield and Andrew Belsey. Cambridge: Cambridge University Press, 1994.

Clark, Tim, and Steven Minta. *Greater Yellowstone's Future*. Moose, Wyo.: Homestead Publishing, 1994.

Clark, Tim W., and Dusty Zaunbrecher. "The Greater Yellowstone Ecosystem: The Ecosystem Concept in Natural Resources Policy and Management." *Renewable Resources Journal* (summer 1987): 8–16.

Clinton, William J. "Message from the President of the United States Transmitting the Convention on Biological Diversity, with Annexes, Done at Rio de Janeiro June 5, 1992, and Signed by the United States in New York on June 4, 1993." Senate, 103rd Cong, 1st sess., treaty doc. 103–20.

———. "Remarks by the President in Earth Day Speech." Press release. The White House, Office of the Press Secretary, April 21, 1993.

———. "Remarks by the President in Making Environmental Announcement Outside the El Tovar Lodge in Grand Canyon National Park." Press release. The White House, Office of the Press Secretary, September 18, 1996.

Clawson, Marion. "Forests in the Long Sweep of American History." *Science* 204 (June 15, 1979): 1168–74.

Cockburn, Andrew. *An Introduction to Evolutionary Ecology*. Oxford: Blackwell Scientific Publications, 1991.

Congressional Research Service. *Yellowstone: Ecosystem, Resources, and Management* (86–1037 ENR), by M. Lynn Corn and Ross W. Gorte. Washington, D.C.: Library of Congress, December 12, 1986.

Conroy, Michael J., and Barry Noon. "Mapping Species Richness for Conservation of Biological Diversity: Conceptual and Methodological Issues. *Ecological Applications* 6, no. 3 (1996): 763–73.

Cortner, Hanna J., and Margaret A. Moote. "Setting the Political Agenda: Paradigmatic Shifts in Land and Water Policy." In *Environmental Policy and Biodiversity,* edited by R. Edward Grumbine. Washington, D.C.: Island Press, 1994.

Costanza, Robert. "Toward an Operational Definition of Ecosystem Health." In *Ecosystem Health: New Goals for Environmental Management,* edited by Robert Costanza, Bryan G. Norton, and Benjamin D. Haskell. Washington, D.C.: Island Press, 1992.

Costanza, Robert, Ralph d' Arge, Rudolf de Groot, Stephen Farber, Monica Grasso, Bruce Hannon, Karin Limburg, Shahid Naeem, Robert V. O'Neill, Jose Paruelo, Robert G. Raskin, Paul Sutton, and Marjan van den Belt. "The Value of the World's Ecosystem Services and Natural Capital." *Nature* 387, no. 6230 (May 15, 1997): 253–60.

Council on Environmental Quality. "Ecology and Living Resources—Biological Diversity." In *Environmental Quality: 11th Annual Environmental Quality Report.* Washington, D.C.: U.S. Government Printing Office, 1980.

———. *Environmental Quality: 21st Annual Report.* Washington, D.C.: U.S. Government Printing Office, 1991.

Council on Environmental Quality and the U.S. Department of State. *The Global 2000 Report to the President of the U.S.* New York: Pergamon Press, 1980.

Cowardin, Lewis, Virginia Carter, Francis Golet, and Edward LaRoe. *Classification of Wetlands and Deepwater Habits of the United States.* Washington, D.C.: U.S. Department of the Interior, U.S. Fish and Wildlife Service, 1979.

Cracraft, Joel. "Species Diversity, Biogeography, and the Evolution of Biotas." *American Zoologist* 34 (1994): 33–47.

Craighead, Frank, Jr. *Track of the Grizzly.* San Francisco: Sierra Club Books, 1979.

Cramer, Jerome, and J. Madeleine Nash. "War over the Wetlands." *Time,* August 36, 1991, 53.

Cronin, William. "The Trouble with Wilderness; or, Getting Back to the Wrong Nature." In *Uncommon Ground: Toward Reinventing Nature,* edited by William Cronin. New York: W. W. Norton, 1995.

Crossley, James W. "Managing Ecosystems for Integrity: Theoretical Considerations for Resource and Environmental Managers." *Society & Natural Resources* 9, no. 5 (1996): 465–81.

Crumpacker, David W. "Prospects for Sustainability of Biodiversity Based on Conservation Biology and US Forest Service Approaches to Ecosystem Management." *Landscape and Urban Planning* 40 (1998): 47–71.

Cuarón, Alfredo D. "Extinction Rate Estimates." *Nature* 366 (November 11, 1993): 118.

Culotta, Elizabeth. "Biological Immigrants Under Fire." *Science* 254, no. 5037 (December 6, 1991): 1444–46.

Cutler, Alan. "What Is a Species?" *Washington Post,* August 9, 1995, H1.

Dahl, Robert A. "On Removing Certain Impediments to Democracy in the United States." In *The Moral Foundations of the American Republic,* edited by Robert H. Horowitz. Charlottesville: University of Virginia Press, 1979.

Dahl, T. E. *Wetland Losses in the United States 1780s–1980s.* Washington, D.C.: U.S. Department of the Interior, U.S. Fish and Wildlife Service, 1990.

Dahl, T. E., and C. E. Johnson. *Status and Trends of Wetlands in the Conterminous United States, Mid-1970s to Mid-1980s.* Washington, D.C.: U.S. Department of the Interior, Fish and Wildlife Service, 1991.

Darling, F. Fraser, and Noel D. Eichhorn. *Man and Nature in the National Parks*. 2nd ed. Washington, D.C.: Conservation Foundation, 1969.

De Blij, H. J. *Nature on the Rampage*. Washington, D.C.: Smithsonian Institution, 1994.

De Blij, H. J., Michael H. Glantz, Stephen L. Harris, Patrick Hughes, Richard Lipkin, Jeff Rosenfeld, and Richard S. Williams Jr. *Restless Earth: Disasters of Nature*. Washington, D.C.: National Geographic Society, 1997.

de Haes, Helias A. Udo, and Frans Klijn. "Environmental Policy and Ecosystem Classification." In *Ecosystem Classification for Environmental Management,* edited by Frans Klijn. Dordrecht: Kluwer Academic Publishers, 1994.

Degner, Richard. "Five of 21 'Endangered Ecosystems' Found in N.J.," *Press* (Atlantic City), February 10, 1996.

Denevan, William M. "The Pristine Myth: The Landscape of the Americas in 1492." *Annals of the Association of American Geographers* 82, no. 3 (September 1992): 369–85.

Derr, Thomas S. "The Challenge of Biocentrism." In *Creation at Risk,* edited by Michael Cromarie. Grand Rapids, Mich.: William B. Eerdmans, 1995.

Devall, Bill, and George Sessions. *Deep Ecology: Living as if Nature Mattered.* Salt Lake City: Peregrine Smith Books, 1985.

Dewar, Heather. "Biologists Say Florida, California, Southeast Face Greatest Risk of Losing Their Wild Lands." Knight-Ridder/Tribune News Service, December 20, 1995.

Dial, Roman. "Extinction or Miscalculation?" *Nature* 370 (July 14, 1994): 104.

Dietrich, Bill. "Natural Balance," *Seattle Times,* March 30, 1995, A13.

Dombeck, Mike. "A Gradual Unfolding of a National Purpose: A Natural Resource Agenda for the 21st Century." Speech delivered to Forest Service employees in Washington, D.C., March 2, 1998.

Doolittle, William E. "Agriculture in North America on the Eve of Contact: A Reassessment." *Annals of the Association of American Geographers* 82, no. 3 (September 1992): 386–401.

Downes, David R. "Global Trade, Local Economies, and the Biodiversity Convention." In *Biological Diversity and the Law,* edited by William Snape III. Washington, D.C.: Island Press, 1996.

Easterbrook, Greg. *A Moment on Earth: The Coming Age of Environmental Optimism.* New York: Penguin, 1995.

Easter-Pilcher, Andrea. "Implementing the Endangered Species Act." *BioScience* 46 (May 1996): 355–63.

Echeverria, John D., and Raymond Booth Eby, eds. *Let the People Judge: Wise Use and Property Rights Movement.* Washington, D.C.: Island Press, 1995.

Egerton, Frank N. "Changing Concepts in the Balance of Nature." *Quarterly Review of Biology* 48 (1973): 322–50.

———. "The History and Present Entanglements of Some General Ecological Perspectives." In *Humans as Components of Ecosystems,* edited by Mark J. McDonnell and Steward T. A. Pickett. New York: Springer-Verlag, 1993.

Ehrenfeld, David. "Ecosystem Health." *Orion* (winter 1993): 12–15.

———. "Why Put a Value on Biodiversity." In *Biodiversity,* edited by E. O. Wilson. Washington, D.C.: National Academy Press, 1988.

Ehrlich, Anne H., and Paul R. Ehrlich. "Needed: An Endangered Humanity Act." In *Balancing on the Brink of Extinction,* edited by Kathryn A. Kohm. Washington, D.C.: Island Press, 1991.

Ehrlich, Paul R., and Anne H. Ehrlich. *Betrayal of Science and Reason: How Anti-Environmental Rhetoric Threatens Our Future.* Washington, D.C.: Island Press, 1996.

———. *Healing the Planet.* Reading, Mass.: Addison-Wesley, 1991.

Elder, Jane. "At Home on the Planet." *Sierra* (March/April 1994): 53–56.

Eldredge, Niles. "Introduction." In *Systematics, Ecology, and the Biodiversity Crisis,* edited by Niles Eldridge. New York: Columbia University Press, 1992.

———. "Where the Twain Meet: Causal Intersections between Genealogical and Ecological Realms." In *Systematics, Ecology, and the Biodiversity Crisis,* edited by Niles Eldridge. New York: Columbia University Press, 1992.

Epstein, Richard. *Takings: Private Property and the Power of Eminent Domain.* Cambridge: Harvard University Press, 1985.

Espenshade, Edward B., Jr., and Joel L. Morrison, eds. *Goode's World Atlas.* 18th ed. Chicago: Rand McNally, 1991.

Executive Office of the President. *Budget of the United States for Fiscal Year 1995.* Washington, D.C.: U.S. Government Printing Office, 1994.

Faith, Daniel P. "Phylogenetic Pattern and the Quantification of Organismal Biodiversity." In *Biodiversity Measurement and Estimation,* edited by D. L. Hawksworth. Oxford: The Royal Society, 1995.

Fanning, Delvin S., and Mary C. Fanning. *Soil: Morphology, Genesis, and Classification.* New York: John Wiley & Sons, 1989.

Farber, Daniel A. "Environmental Protection as a Learning Experience." *Loyola of Los Angeles Law Review* 27, no. 3 (April 1994): 791–807.

Fautin, Dahpne Gail. "Preface." In *Annual Review of Ecology and Systematics,* vol. 26, edited by Dahpne Gail Fautin. Palo Alto, Calif.: Annual Reviews, 1995.

Federal Geographic Data Committee Vegetation Subcommittee. *FGDC Vegetation Classification and Information Standards.* Reston, Va.: U.S. Department of the Interior, U.S. Geological Survey, Federal Geographical Data Committee Secretariat, June 3, 1996.

Ferkiss, Victor. *Nature, Technology and Society: Cultural Roots of the Current Environmental Crisis.* New York: New York University Press, 1993.

Fiedler, Peggy L., Peter S. White, and Robert A. Leidy. "The Paradigm Shift in Ecology and Its Implications for Conservation." In *The Ecological Basis of Conservation,* edited by S. T. A. Pickett, R. S. Ostfeld, M. Shachak, and G. E. Likens. New York: Chapman & Hall, 1997.

Fishbein, Seymour. *Yellowstone Country.* Washington D.C.: National Geographic Society, 1989.

Fitzsimmons, Allan K. "Clearing Ecosystem Misunderstandings." *Regulation* 40, no. 4 (fall 1997): 3–4.

———. "Sound Policy or Smoke and Mirrors: Does Ecosystem Management Make Sense?" *Water Resources Bulletin* 32, no. 2 (April 1996): 217–27.

———. "Why a Policy of Federal Management and Protection of Ecosystems is a Bad Idea." *Landscape and Urban Planning* 40 (1998): 195–202.

Flader, Susan L. "Evolution of a Land Ethic." In *Aldo Leopold: The Man and His Legacy,* edited by Thomas Tanner. Ankeny, Iowa: Soil Conservation Society of America, 1987.

———. *Thinking Like a Mountain.* Madison: University of Wisconsin Press, 1994.

Flournoy, Alyson C. "Coping with Complexity." *Loyola of Los Angeles Law Review* 27, no. 3 (April 1994): 809–24.

Foreman, Dave. *Confessions of an Eco-Warrior.* New York: Harmony Books, 1991.

———. "Missing Links." *Sierra* (September/October 1995): 52–57, 96–98.

Foreman, Dave, and Howie Wolke. *The Big Outside.* New York: Harmony Books, 1992.

Fowler, Robert Booth. *The Greening of Protestant Thought.* Chapel Hill: University of North Carolina Press, 1995.

Fox, Matthew. *The Coming of the Cosmic Christ: The Healing of Mother Earth and the Birth of a Global Renaissance.* San Francisco: HarperSanFrancisco, 1988.

Fox, Matthew, and Rupert Sheldrake. *Natural Grace.* New York: Doubleday, 1996.

Fox, Stephen. *John Muir and His Legacy: The American Conservation Movement.* Boston: Little, Brown, 1981.

Franklin, Jerry F. "The Fundamentals of Ecosystem Management with Applications in the Pacific Northwest." In *Defining Sustainable Forestry,* edited by Gregory H. Aplet, Nels Johnson, Jeffrey T. Olson, and V. Alaric Sample. Washington, D.C.: Island Press, 1993.

Freemuth, John. "Ecosystem Management and Its Place in the National Park System." *Denver University Law Review* 74 (1997): 697–727.

Freyfogle, Eric. "Ownership and Ecology." *Case Western Reserve Law Review* 43 (1993): 1269–97.

———. "The Owning and Taking of Sensitive Lands." In *Land Use & Environmental Law Review 1996,* edited by Stuart L. Deutsch and A. Dan Tarlock. Deerfield, Ill.: Clark, Boardman, Callaghan, 1996.

Freyfogle, Eric T. *Justice and the Earth.* Urbana: University of Illinois Press, 1993.

Frissell Christopher A., and David Bayles. "Ecosystem Management and the Conservation of Biodiversity and Ecological Integrity." *Water Resources Bulletin* 32, no. 2 (April 1996): 229–40.

Fritsch, Albert, and Angela Ladavaia-Cox. *Eco-Church: An Action Manual.* San Jose, Calif.: Resource Publications, 1992.

Garver, John C. Jr. "New Perspective on the World." *National Geographic,* 174, no. 6 (December 1988): 910–13.

German Cartographic Society. "The So-Called Peters Projection." *The Cartographic Journal* 22, no. 2 (December 1985): 108–10.

Gigliotti, Larry. "Environmental Education: What Went Wrong? What Can Be Done?" *Environmental Educator* 22, no. 1 (fall 1990): 9–12.

Glacken, Clarence. *Traces on the Rhodian Shore.* Berkeley: University of California Press, 1967.

Glacken, Clarence J. "Changing Ideas of the Habitable World." In *Man's Role in Changing the Face of the Earth,* edited by William L. Thomas Jr. Chicago: University of Chicago Press, 1956.

Glick, Dennis, Mary Carr, and Bert Harting, eds. *An Environmental Profile of the*

Greater Yellowstone Ecosystem. Bozeman, Mont.: The Greater Yellowstone Coalition, 1991.

Goklany, Indur M. "Factors Affecting Environmental Impacts: The Effect of Technology on Long-Term Trends in Cropland, Air Pollution and Water-Related Diseases." *Ambio* 25, no. 8 (December 1996): 497–503.

Goldstein, Bruce. "The Struggle of Ecosystem Management at Yellowstone." *BioScience* 42, no. 3 (March 1992): 183–87.

Golley, Frank. *A History of the Ecosystem Concept in Ecology*. New Haven, Conn.: Yale University Press, 1993.

Goodchild, Michael, and Sucharita Gupta. "Preface." In *The Accuracy of Spatial Databases,* edited by Michael Goodchild and Sucharita Gupta. London: Taylor and Francis, 1989.

Goodland, Robert, and Herman Daly. "Environmental Sustainability: Universal and Non-Negotiable." *Ecological Applications* 6, no. 4 (November 1966): 1002–17.

Gordon, Robert. "When the 'Best Available Data' Is B.A.D.: The Data Error Plague." *NWI Resource* 4, nos. 2–3 (summer 1993): 3–7.

Gordon, Robert E., James K. Lacy, and James R. Streeter. "Conservation under the Endangered Species Act." *Environment International* 23, no. 3 (November 1997): 359–419.

Gore, Al. *Creating a Government That Works Better and Costs Less: Reinventing Environmental Management*. Washington, D.C.: Office of the Vice President, September 1993.

———. *Earth in the Balance: Ecology and the Human Spirit*. New York: Penguin, 1993.

Gottlieb, Roger S. "Introduction: Religion in an Age of Environmental Crisis." In *This Sacred Earth, Religion, Nature, Environment,* edited by Roger S. Gottlieb. New York: Routledge, 1996.

———, ed. *This Sacred Earth:: Religion, Nature, Environment*. New York: Routledge, 1996.

Goulden, Clyde E. "Ecological Comprehensiveness." *Science* 264 (April 29, 1994): 726–27.

Graber, Linda. *Wilderness as Sacred Space*. Washington, D.C.: Association of American Geographers, 1976.

Graf, William L. *Wilderness Preservation and the Sagebrush Rebellions*. Lanham, Md.: Rowman & Littlefield, 1990.

Grande, Lance, and Olivier Rieppel, eds. *Interpreting the Hierarchy of Nature*. San Diego: Academic Press, 1994.

Greater Yellowstone Coordinating Committee. *A Framework for Coordination of National Parks and National Forests in the Greater Yellowstone Area*. N.p.: U.S. Department of the Interior, National Park Service and U.S. Department of Agriculture, National Forest Service, September 1991.

———. *The Greater Yellowstone Area: An Aggregation on National Park and National Forest Plans*. N.p.: U.S. Department of the Interior, National Park Service and U.S. Department of Agriculture, National Forest Service, September 1987.

———. "Vision for the Future: A Framework for Coordination in the Greater Yellowstone Area." Draft. Billings, Mont.: Greater Yellowstone Coordinating Committee, August 1990.

Greve, Michael S. *Demise of Environmentalism in American Law*. Washington, D.C.: AEI Press, 1996.

Grumbine, Ed. "Protecting Biological Diversity through the Greater Ecosystem Concept." *Natural Areas Journal* 10, no. 3 (1990): 114–20.

Grumbine, R. Edward. "Introduction." In *Environmental Policy and Biodiversity,* edited by R. Edward Grumbine. Washington, D.C.: Island Press, 1994.

———. "What Is Ecosystem Management?" *Conservation Biology* 8, no. 1 (March 1994): 27–38.

Haeuber, Richard. "Setting the Environmental Policy Agenda: The Case of Ecosystem Management." *Natural Resources Journal* 36, no. 1 (winter 1996): 1–27.

Hagen, Joel B. *An Entangled Bank: The Origins of Ecosystem Ecology*. New Brunswick, N.J.: Rutgers University Press, 1992.

Hahn, Robert W. "Toward a New Environmental Paradigm." *Yale Law Journal* 102, no. 7 (May 1993): 1719–61.

Hannon, Bruce. "Accounting in Ecological Systems." In *Ecological Economics: The Science of Management of Sustainability,* edited by Robert Costanza. New York: Columbia University Press, 1991.

Harden, Garrett. "Tragedy of the Commons." *Science* 162 (December 13, 1968): 1243–48.

Hardt, Scott W. "Federal Land Management in the Twenty-First Century: From Wise Use to Wise Stewardship." *Harvard Environmental Law Review* 18, no. 2 (1994): 345–403.

Harper, John L., and David L. Hawksworth. "Preface." In *Biodiversity Measurement and Estimation,* edited by D. L. Hawksworth. Oxford: The Royal Society, 1995.

Hart, John Fraser. "Presidential Address, the Highest Form of the Geographer's Art." *Annals of the Association of American Geographers* 72, no. 1 (March 1982): 1–29.

Hartmann, William K., and Ron Miller. *The History of Earth*. New York: Workman Publishing, 1991.

Hartshorne, Richard. *Perspectives on the Nature of Geography*. Chicago: Rand McNally, 1959.

Haskell, Benjamin D., Bryan D. Norton, and Robert Costanza. "Introduction: What Is Ecosystem Health and Why Should We Worry about It?" In *Ecosystem Health: New Goals for Environmental Management,* edited by Benjamin D. Haskell, Bryan D. Norton, and Robert Costanza. Washington, D.C. : Island Press, 1992.

Hayes, Michael T. *Incrementalism and Public Policy*. New York: Longman, 1992.

Hazilla, Michael, and Raymond J. Kopp. "Social Costs of Environmental Quality Regulations: A General Equilibrium Analysis." *Journal of Political Economy* 98, no. 4 (1990): 853–73.

Heilbroner, Robert. "Socialism." In *The Fortune Encyclopedia of Economics,* edited by David R. Henderson. New York: Warner Books, 1993.

Heimlich, Ralph, and Jeanne Melanson. "Wetlands Lost, Wetlands Gained." *National Wetlands Newsletter* 17, no. 3 (May–June, 1995): 1–23, 25.

Helvarg, David. *The War against the Greens*. San Francisco: Sierra Club Books, 1994.

Henderson, Rick. "The Swamp Thing." *Reason,* April 1991, 30–35.

Heywood, Vernon H., Georgina M. Mace, Robert M. May, and S. N. Stuart. "Uncertainties in Extinction Rates." *Nature* 368 (March 10, 1994): 105.

Hidore, John J., and John E. Oliver. *Climatology: An Atmospheric Science*. New York: Macmillan, 1993.

Hilborn, R., C. J. Walters, and D. Ludwig. "Sustainable Exploitation of Renewable Resources. " In *Annual Review of Ecology and Systematics*. Palo Alto, Calif.: Annual Reviews, 1995.

Hofstadter, Richard. "The Founding Fathers: An Age of Realism." In *The Moral Foundations of the American Republic,* edited by Robert H. Horowitz. Charlottesville: University of Virginia Press, 1979.

Holling, C. S. "What Barriers? What Bridges?" In *Barriers and Bridges to the Renewal of Ecosystems and Institutions,* edited by Lance H. Gunderson, C. S. Holling, and Stephen S. Light. New York: Columbia University Press, 1995.

Holsinger, Kent E. "Population Biology for Policy Makers." *BioScience* (Special Supplement on Science and Biodiversity Policy, June 1995): S10–S20.

Hook, Donal. "Wetlands: History, Current Status, and Future." *Environmental Toxicology and Chemistry* 12 (1993): 2157–166.

Hopkins, Thomas D. *Regulatory Costs in Profile*. Policy Study 132. St. Louis: Center for the Study of American Business, Washington University, 1996.

Houck, Oliver A. "On the Law of Biodiversity and Ecosystem Management." *Minnesota Law Review* 81, no. 4 (April 1997): 869–979.

———. "Reflections on the Endangered Species Act." In *Land Use & Environmental Law Review 1996*. Edited by Stuart L. Deutsch and A. Dan Tarlock. Deerfield, Ill.: Clark, Boardman, Callaghan, 1996.

Hudson, John C. "Scale in Space and Time." In *Geography's Inner Worlds: Pervasive Themes in Contemporary American Geography,* edited by Ronald F. Abler, Melvin G. Marcus, and Judy M. Olson. Washington, D.C.: Association of American Geographers, 1992.

Huffman, James L. "Avoiding the Takings Clause through the Myth of Public Rights: The Public Trust and Reserved Rights Doctrines at Work." *Journal of Land Use and Environmental Law* 3, no. 2 (fall 1987): 171–211.

———. "A Fish Out of Water: The Public Trust Doctrine in a Constitutional Democracy." *Environmental Law* 19, no. 3 (spring 1989): 527–72.

Hughes, J. Donald, and Jim Swan. "How Much of the Earth Is Sacred Space?" *Environmental Review* 10, no. 4 (winter 1986): 247–59.

Hunter, David. "An Ecological Perspective on Property: A Call for Judicial Protection of the Public's Interest in Environmentally Critical Resources." *Harvard Environmental Law Review* 12, no. 2 (1988): 311–83.

Hunter, Malcom, Jr. "Coping with Ignorance: The Coarse-Filter Strategy for Maintaining Biodiversity." In *Balancing on the Brink of Extinction,* edited by Kathryn A. Kohm, Washington, D.C.: Island Press, 1991.

Independent Commission on Environmental Education. *Are We Building Environmental Literacy?* Washington, D.C.: ICEE, 1997.

Ingram, Helen H., Brinton Milward, and Wendy Laird. "Scientists and Agenda Setting: Advocacy and Global Warming." In *Risk and Society: The Interaction of Science, Technology and Public Policy,* edited by Marvin Waterstone. Dordrecht: Kluwer Academic Press, 1992.

Ise, John. *Our National Park History*. Baltimore: The Johns Hopkins University Press, 1961.

IUCN (International Union for the Conservation of Nature). *1997 IUCN Red List of Threatened Plants*. Gland, Switzerland: IUCN, 1998.

IUCN (International Union for the Conservation of Nature). *1994 IUCN Red List of Threatened Animals*. Gland, Switzerland: IUCN, 1993.

James, Preston. *All Possible Worlds: A History of Geographical Ideas*. Indianapolis: Bobbs-Merrill, 1972.

Jaunich, R. Prescott. "The Environment, the Free Market, and Property Rights: Post-Lucas Privatization of the Public Trust." *Public Land Law Review* 15 (1994): 167–97.

Jensen, Mark E., Patrick Bourgeron, Richard Everitt, and Iris Goodman. "Ecosystem Management: A Landscape Ecology Perspective." *Water Resources Bulletin* 32, no. 2 (April 1996): 203–16.

Johnson, Lionel. "The Far-from-Equilibrium Ecology Hinterlands." In *Complex Ecology: The Part-Whole Relation in Ecosystems,* edited by Bernard C. Patten and Sven E. Jorgensen. Englewood Cliffs, N.J.: Prentice Hall, 1995.

Johnson, Nels. "Introduction." In *Defining Sustainable Forestry,* edited by Gregory H. Aplet, Nels Johnson, Jeffrey T. Olson, and V. Alaric Sample. Washington, D.C.: Island Press, 1993.

Johnson, Ronald. "Ecosystem Management and Reinventing Government." In *Breaking the Policy Gridlock,* edited by Terry L. Anderson. Stanford, Calif.: Stanford University Press, 1997.

Kalen, Sam. "Commerce to Conservation: The Call for a National Water Policy and the Evolution of Federal Jurisdiction over Wetlands." *North Dakota Law Review* 69, no. 4 (1993): 873–913.

Kaplan, Ruth. *Our Earth, Ourselves*. New York: Bantam, 1990.

Karp, James. "A Private Property Duty of Stewardship: Changing Our Land Ethic." *Environmental Law* 23, no. 3 (1993): 735–62.

Kaufman, Wallace. *No Turning Back, Dismantling the Fantasies of Environmental Thinking*. New York: Basic Books, 1994.

Kay, James J. "On the Nature of Ecological Integrity: Some Closing Comments." In *Ecological Integrity and the Management of Ecosystems,* edited by Stephen Woodley, James Kay, and George Francis. Delray Beach, Fla.: St. Lucie Press, 1993.

Kay, James J., and Eric Schneider. "Embracing Complexity the Challenge of the Ecosystem Approach." In *Prospective on Ecological Integrity,* edited by Laura Westra and John Lemons. Dordrecht: Kluwer, 1995.

Keiter, Robert B. "Beyond the Boundary Line: Constructing a Law of Ecosystem Management." *University of Colorado Law Review* 65, no. 2 (June 1994): 293–333.

———. "Conservation Biology and the Law: Assessing the Challenge Ahead." *Chicago-Kent Law Review* 69, no. 4 (1994): 911–33.

———. "An Introduction to the Ecosystem Management Debate." In *The Greater Yellowstone Ecosystem,* edited by Robert B. Keiter and Mark S. Boyce. New Haven, Conn.: Yale University Press, 1991.

———. "NEPA and the Emerging Concept of Ecosystem Management on the Public Lands." *Land and Water Law Review* 25, no. 1 (1990): 43–60.

———. "Preserving Nature in the National Parks: Law, Policy, and Science in a Dynamic Environment." *Denver University Law Review* 74, no. 3 (1997): 649–95.

———. "Toward Legitimizing Ecosystem Management on the Public Domain." *Ecological Applications* 6, no. 3 (1996): 727–30.

Keiter, Robert B., and Mark S. Boyce, eds. *The Greater Yellowstone Ecosystem*. New Haven, Conn.: Yale University Press, 1991.

Keller, Edward A. *Environmental Geology*. 7th ed. Upper Saddle River, N.J.: Prentice Hall, 1996.

Kellert, Stephen R. *The Value of Life*. Washington, D.C.: Island Press, 1996.

Kennedy, James J., and Thomas M. Quigley. "Evolution of USDA Forest Service Organizational Culture and Adaption Issues in Embracing an Ecosystem Management Paradigm," *Landscape and Urban Planning* 40 (1998): 113–22.

Kenworthy, Tom. "In the Desert Southwest, a Vigorous Species Act Endangers a Way of Life." *Washington Post,* February 1, 1998, A3.

———. "Park, Forest Officials allege GOP Pressure: Two Caught in Crossfire over Public Lands." *Washington Post,* September 23, 1991, A23.

Kessler, Winifred B., Hal Salwasser, Charles W. Cartwright Jr., and James A Caplan. "New Perspectives for Sustainable Natural Resources Management." *Ecological Applications* 2, no. 3 (1992): 221–25.

Keystone Center. *The Keystone National Policy Dialogue on Ecosystem Management—Final Report*. Keystone, Colo.: The Keystone Center, October 1996.

———. *Report of a Keystone Policy Dialogue—Biological Diversity on Federal Land*. Keystone, Colo.: The Keystone Center, 1991.

Kimball, Peter. "Shawnee Forest Fight Cuts Both Ways." *Chicago Tribune,* September 23, 1996.

King, Anthony W. "Considerations of Scale and Hierarchy." In *Ecological Integrity and the Management of Ecosystems,* edited by Stephen Woodley, James Kay, and George Francis. Delray Beach, Fla.: St. Lucie Press, 1993.

Kinsley, David. *Ecology and Religion: Ecological Spirituality in a Cross-Cultural Perspective*. Englewood Cliffs, N.J.: Prentice Hall, 1995.

Knudsen, Tom. "Sierra's Problem: Too Many People." *Sacramento Bee,* June 11, 1996, B1.

Küchler, A. W. "The Classification of Vegetation." In *Vegetation Mapping,* edited by A. W. Küchler and I. S. Zonneveld. Dordrecht: Kluwer Academic Publishers, 1988.

———. "Potential Natural Vegetation." Map. In *The National Atlas of the United States*. Washington, D.C.: U.S. Department of the Interior, U.S. Geological Survey (map revision of 1985).

Küchler, A. W., and I. S. Zonneveld, eds. *Vegetation Mapping*. Dordrecht: Kluwer Academic Publishers, 1988.

Kuhlman, Walter. "Wildlife's Burden." In *Biological Diversity and the Law,* edited by William Snape III. Washington, D.C.: Island Press, 1996.

Kuhn, Thomas. *The Structure of Scientific Revolutions*. Chicago: University of Chicago Press, 1962.

Kunich, John Charles. "The Fallacy of Deathbed Conservation under the Endangered Species Act." *Environmental Law* 24, no. 2 (1994): 501–80.

Kusler, Jon. "Wetlands Delineation: An Issue of Science or Politics?" *Environment* 34, no. 2 (March 1992): 7–11.

Lackey Robert. "Ecosystem Management: Implications for Fisheries Management." *Renewable Resources Journal* 13, no. 4 (winter 1995–96): 11–13.

———. "Seven Pillars of Ecosystem Management." *Landscape and Urban Planning* 40 (1998): 21–30.

Lackey, Robert T. "Ecosystem Management: Paradigms and Prattle, People and Prizes." *Renewable Resources Journal* 16, no. 1 (1998): 8–13.

———. "Pacific Salmon, Ecological Health, and Public Policy." *Ecosystem Health* 2, no. 1 (March 1996): 61–68.

Ladd, Everett, and Karlyn Bowman. *Attitudes toward the Environment: Twenty-Five Years after Earth Day*. Washington, D.C.: American Enterprise Institute, 1995.

Lawton, John H., and Robert M. May, eds. *Extinction Rates*. Oxford: Oxford University Press, 1995.

Lazarus, Richard J. "Shifting Paradigms of Tort and Property in the Transformation of Natural Resources Law." In *Natural Resources Policy and Law,* edited by Lawrence J. MacDonnell and Sarah F. Bates. Washington, D.C.: Island Press, 1993.

Leder, Lawrence H. *Liberty and Authority: Early American Political Ideology 1689–1763*. Chicago: Quadrangle Books, 1968.

Leighly, John. "Some Comments on Contemporary Geographic Method." *Annals of the Association of American Geographers* 27, no. 3 (September 1937): 125–41.

Lemons, John. "The Conservation of Biology: Scientific Uncertainty and the Burden of Proof." In *Scientific Uncertainly and Environmental Problem Solving,* edited by John Lemons. Cambridge: Blackwell Science, 1996.

Leopold, Aldo. *A Sand County Almanac: With Essays on Conservation from Round River*. New York: Ballantine, 1966.

Levin, Simon. "The Problem of Pattern and Scale in Ecology." *Ecology* 73, no. 6 (December 1992): 1943–67.

Lewis, Ed. "The View from Greater Yellowstone." *Greater Yellowstone Report* (fall 1990): 3.

Lewis, Martin. *Green Delusions: An Environmental Critique of Radical Environmentalism*. Durham, N.C.: Duke University Press, 1992.

Lichtman, Pamela, and Tim W. Clark. "Rethinking the 'Vision' Exercise in the Greater Yellowstone Ecosystem." *Society and Natural Resources* 7, no. 5 (1994): 459–78.

Likens, Gene. *The Ecosystem Approach: Its Use and Abuse*. Oldendorf/Luhe, Germany: Ecology Institute, 1992.

Lloyd, Julian. "The Prairie-Dog Divide: To Reduce or Protect." *Christian Science Monitor,* 8 May 1996.

Lovejoy, Thomas E. "Will Expectedly the Top Blow Off?" *BioScience* (Special Supplement on Science and Biodiversity Policy, June 1995): S3–7.

Loveland, Thomas R., James W. Merchant, Jesslyn F. Brown, Donald O. Ohlen, Bradley C. Reed, Paul Olson, and John Hutchinson. "Seasonal Land-Cover Regions of the United States." *Annals of the Association of American Geographers* 85, no. 2 (June 1995): 339–55.

Ludwig, Donald, Ray Hilborn, and Carl Walters. "Uncertainty, Resource Exploitation, and Conservation: Lessons from History." *Science* 260 (April 2, 1993): 17, 36.

Lutz, Donald. *A Preface to American Political Theory*. Lawrence: University of Kansas Press, 1992.

Lydolf, Paul E. *Weather and Climate,* Totowa, N.J.: Rowman & Allanheld, 1985.

MacArthur, Robert H. *Geographical Ecology*. Princeton, N.J.: Princeton University Press, 1972.

MacArthur, Robert H., and Edward O. Wilson. *The Theory of Island Biogeography*. Princeton, N.J.: Princeton University Press, 1967.

MacCleery, Douglas. *American Forests: A History of Resiliency and Recovery*. Durham, N.C.: Forest History Society, 1994.

Madison, James. *The Federalist Papers: Number Ten*. New York: Mentor, 1961.

Mann, Charles C. "Extinction: Are Ecologists Crying Wolf?" *Science* 253 (August 16, 1991): 736–37.

Mann, Charles C., and Mark L. Plummer. "The High Cost of Biodiversity." *Science* 260 (June 25, 1993): 1868–71.

———. *Noah's Choice*. New York: Alfred Knopf, 1996.

Margulis, Lynn, and Karlene V. Schwartz. *Five Kingdoms: An Illustrated Guide to the Phyla of Life on Earth*. New York: W. H. Freeman, 1988.

Marsh, George Perkins. *Man and Nature; or Physical Geography as Modified by Human Action*. New York: Charles Scribner, 1864. Republished in 1965 by Belknap Press, Cambridge, and edited by David Lowenthal.

Maser, Chris. *Sustainable Forestry: Philosophy, Science, and Economics*. Delray Beach, Fla.: St. Lucie Press, 1994.

Masterman, Margaret. "The Nature of a Paradigm." In *Criticism and the Growth of Knowledge,* edited by Imre Lakatos and Alan Musgrave. Cambridge: Cambridge University Press, 1970.

May, Robert M. "Conceptual Aspects of the Quantification of the Extent of Biological Diversity." In *Biodiversity Measurement and Estimation,* edited by D. L. Hawksworth. Oxford: The Royal Society, 1995.

———. "Foreword." In *Ecological Understanding: The Nature of Theory and the Theory of Nature,* edited by Steward T. A. Pickett, Jurek Kolasa, and Clive G. Jones. San Diego: Academic Press, 1994.

Mayr, Ernst ed. *The Species Problem*. 1974. Arno Press reprint edition of Publication No. 50 of the American Association for the Advancement of Science, Washington, D.C.: 1957.

McAllister, Bill. "Yellowstone Report Controversy Prompts House Inquiry." *Washington Post,* September 12, 1991, A21.

McElfish, James. "Back to the Future." *The Environmental Forum* (September/October 1995): 14–23.

McFague, Sallie. "The Scope of the Body: The Cosmic Christ." In *This Sacred Earth: Religion, Nature, Environment,* edited by Roger S. Gottlieb. New York: Routledge, 1996.

McIntosh, Robert P. *The Background of Ecology: Concept and Theory*. Cambridge: Cambridge University Press, 1985.

———. "Pluralism in Ecology." In *Annual Review of Ecology and Systematics*. Palo Alto, Calif.: Annual Reviews, 1987.

McKnight, Tom L. *Physical Geography: A Landscape Appreciation*. 5th ed. Upper Saddle River, N.J.: Prentice Hall, 1996.

Meiners, Roger E. "Elements of Property Rights: The Common Law Alternative." In *Land Rights,* edited by Bruce Yandle. Lanham, Md.: Rowman & Littlefield, 1995.

Merchant, Carolyn. *Radical Ecology*. New York: Routledge, 1992.

Metrick, Andrew, and Martin L. Weitzman. "Patterns of Behavior in Endangered Species Preservation." *Land Economics* 72, no. 1 (February 1996): 1–16.

Meyer, Judy L. "The Dance of Nature: New Concepts in Ecology." *Chicago-Kent Law Review* 69, no. 4 (1994): 875–86.

Meyer, William B. "When Dismal Swamps Became Priceless Wetlands." *American Heritage* 45, no. 3 (May/June 1994): 108–16.

Meyers, Gary D. "Diving Common Law Standards for Environmental Protection: Application of the Public Trust Doctrine in the Context of Reforming NEPA and the Commonwealth Environmental Protection Act." *Environment and Planning Law Journal* (August 1994): 289–306.

———. "Old Growth Forests, the Owl, and Yew: Environmental Ethics versus Traditional Dispute Resolution under the Endangered Species Act and Other Public Lands and Resources Laws." *Boston College Environmental Affairs Law Review* 18, no. 4 (summer 1991): 623–68.

———. "Variation on a Theme: Expanding the Public Trust Doctrine to Include Protection of Wildlife." *Environmental Law* 19, no. 3 (spring 1989): 723–35.

Miller, Julie Ann. "Biosciences and Ecological Integrity." *BioScience* 41, no. 4 (April 1991): 206–10.

Mills, L. Scott, Michael E. Soulé, and Daniel F. Doak. "The Keystone-Species Concept in Ecology and Conservation." *BioScience* 43, no. 4 (April 1993): 219–24.

Minelli, Alessandro. *Biological Systematics; The State of the Art*. London: Chapman & Hall, 1993.

Miniter, Richard. "Muddy Waters." *Policy Review* no. 56 (spring 1991): 70–77.

Mitsch, William, and James Gosselink, *Wetlands*. New York: Van Nostrand, 1993.

Moghissi, A. Alan. "Editorial—Management of the Endangered Species Act by the U.S. Government: Urgent Need for Corrective Action." *Environment International* 23, no. 3 (November 1997): 269–72.

Monmonier, Mark. *Drawing the Line*. New York: Henry Holt, 1995.

———. *How to Lie with Maps*. 2nd ed. Chicago: University of Chicago Press, 1996.

Montgomery, David R., Gordon E. Grant, and Kathleen Sullivan. "Watershed Analysis as a Framework for Implementing Ecosystem Management." *Water Resources Bulletin* 31, no. 3 (June 1995): 1–18.

Mooney, Hal, and Clifford J. Gabriel. "Preface." *BioScience* (Special Supplement on Science and Biodiversity Policy, June 1995).

Moote Margaret, Sabrina Burke, Hanna J. Cortner, and Mary G. Wallace. *Principles of Ecosystem Management*. Tucson: Water Resources Research Center, University of Arizona, January 1994.

Muir, John. *Nature Writings,* edited by William Cronin. New York: Penguin, 1997.

———. *Our National Parks*. Boston: Houghton Mifflin, 1909.

———. *The Yosemite*. 1912. Reprinted in *John Muir: Nature Writings*. New York: Library of America, 1997.

Mulder, N. J. "Digital Image Processing, Computer-Aided Classification and Mapping." In *Vegetation Mapping,* edited by A. W. Küchler and I. S. Zonneveld. Dordrecht: Kluwer Academic Publishers, 1988.

Naar, Jon. "The Green Cathedral." *The Amicus Journal* 14, no. 4 (winter 1993): 22–28.

Nash, Roderick. "Aldo Leopold and the Limits of American Liberalism." In *Aldo Leopold: The Man and His Legacy,* edited by Thomas Tanner. Ankeny, Iowa: Soil Conservation Society of America, 1987.

———. *The American Environment: Readings in the History of Conservation*. 2nd ed. Reading, Mass.: Addison-Wesley, 1976.

———. "Historical and Philosophical Considerations of Ecosystem Management." In *Ecosystem Management: Status and Potential*. Report of a workshop convened by the Congressional Research Service, March 24–25, Senate Committee on Environment and Public Works, 103rd Cong., 2d sess., December 1994, S. Prt. 103–98.

———. *The Rights of Nature*. Madison: University of Wisconsin Press, 1989.

National Audubon Society. *Almanac of the Environment,* edited by Valerie Harm. New York: G. P. Putnam, 1994.

National Council of the Churches of Christ. "A Service of Worship: The Earth Is the Lord's—a Liturgy of Celebration, Confession, Thanksgiving, and Commitment." In *This Sacred Earth:: Religion, Nature, Environment,* edited by Roger S. Gottlieb. New York: Routledge, 1996.

National Oceanic and Atmospheric Administration. *Climatic Atlas of the United States*. Washington, D.C.: U.S. Department of Commerce, 1983.

National Research Council. *A Biological Survey for the Nation*. Washington, D.C.: National Academy Press, 1993.

———. *Oil in the Sea: Inputs, Fates and Effects*. Washington, D.C.: National Academy Press, 1985.

———. *Our Seabed Frontier*. Washington, D.C.: National Academy Press, 1989.

———. *Science and the Endangered Species Act*. Washington, D.C.: National Academy Press, 1995.

———. *Wetlands: Characteristics and Boundaries*. Washington, D.C.: National Academy Press, 1995.

Nelson, Robert H. "Environmental Calvinism: The Judeo-Christian Roots of Ecotheology." In *Taking the Environment Seriously,* edited by Roger E. Meiners and Bruce Yandle. Lanham, Md.: Rowman & Littlefield, 1993.

———. "Federal Zoning: The New Era in Environmental Policy." In *Land Rights,* edited by Bruce Yandle. Lanham, Md.: Rowman & Littlefield, 1995.

———. "How Much Is Enough? An Overview of the Benefits and Costs of Environmental Protection." In *Taking the Environment Seriously,* edited by Roger E. Meiners and Bruce Yandle. Lanham, Md.: Rowman & Littlefield, 1993.

———, ed. *The Failure of Scientific Management: Rethinking the Public Lands*. Lanham, Md.: Rowman & Littlefield, 1995.

Nesmith, Jeff. "Texas 'Ecosystems' Ranked among Endangered in Report." *Austin American-Statesmen,* December 21, 1995, A1.

Norton, Bryan G. "A New Paradigm for Environmental Management." In *Ecosystem Health: New Goals for Environmental Management.* edited by Robert Costanza, Bryan G. Norton and Benjamin D. Haskell. Washington, D.C.: Island Press, 1992.

Norton, Bryan G., and Robert E. Ulanowicz. "Scale and Biodiversity Policy: A Hierarchical Approach." In *Ecosystem Management,* edited by Fred B. Samson and Fritz L. Knopf. New York: Springer, 1996.

Norton, Seth. "Property Rights, the Environment, and Economic Well-Being." In *Who Owns the Environment?,* edited by Peter J. Hill and Roger E. Meiners. Lanham, Md.: Rowman & Littlefield, 1998.

Noss, Reed F. "From Endangered Species to Biodiversity." In *Balancing on the Brink of Extinction,* edited by Kathryn A. Kohm. Washington, D.C.: Island Press, 1991.

———. "Some Principles of Conservation Biology, as They Apply to Environmental Law." *Chicago-Kent Law Review* 69, no. 4 (1994): 893–909.

———. "Sustainable Forestry or Sustainable Forests?" In *Defining Sustainable Forestry,* edited by Gregory H. Aplet, Nels Johnson, Jeffrey T. Olson, and V. Alaric Sample. Washington, D.C.: Island Press, 1993.

———. "The Wildlands Project: Land Conservation Strategy." *Wild Earth* (special issue, 1992): 13–24.

Noss, Reed F., and Allen Y. Cooperrider. *Saving Nature's Legacy.* Washington, D.C.: Island Press, 1994.

Noss, Reed F., Edward T. LaRoe III, and J. Michael Scott. *Endangered Ecosystems of the United States: A Preliminary Inquiry.* Biological Report No. 28. Washington, D.C.: U.S. Department of the Interior, National Biological Service, February 1995.

Noss, Reed F., and Robert L. Peters. *Endangered Ecosystems: A Status Report on America's Vanishing Habitat and Wildlife.* Washington, D.C.: Defenders of Wildlife, 1995.

O'Donnell, Anthony, G. Michael Goodfellow, and David L. Hawksworth. "Theoretical and Practical Aspects of the Quantification of Biodiversity among Microorganisms." In *Biodiversity Measurement and Estimation,* edited by D. L. Hawksworth. Oxford: The Royal Society, 1995.

Odum, Eugene. *Basic Ecology.* Philadelphia: Saunders College, 1983.

———. "The Strategy of Ecosystem Development." *Science* 164, no. 3877 (April 18, 1969): 262–70.

Oelschlaeger, Max. *Caring for Creation.* New Haven, Conn.: Yale University Press, 1994.

Office of Technology Assessment. U.S. Congress. *Harmful Non-Indigenous Species in the United States.* OTA-F-565. Washington, D.C.: U.S. Government Printing Office, 1993.

Olinger, David. "Florida Tops List of Endangered Habitats." *St. Petersburg Times,* December 21, 1995, B1, B6.

Oliver, John E., and L. Wilson. "Climatic Classification." In *The Encyclopedia of Climatology,* edited by John E. Oliver and Rhodes W. Fairbridge. New Work: Van Nostrand Reinhold, 1987.

Olivier, Michael J. "The Wetlands Dilemma." *Economic Development Review* 10 (summer 1992): 56–57.

Omernik, James M. "Ecoregions of the Conterminous United States." *Annals of the Association of American Geographers* 77, no. 1 (March 1987): 118–25.

Omernik, James M., and Robert G. Bailey. "Distinguishing between Watersheds and Ecoregions." *Journal of the American Water Resources Association* (formerly *Water Resources Bulletin*) 33, no. 5 (October 1997): 935–49.

Omernik, James, and G. E. Griffith. "Ecological Regions versus Hydrologic Units: Frameworks for Managing Water Quality." *Journal of Soil and Water Conservation* 46, (1991): 224–340.

O'Neill, R. V. "Get Ready to Rewrite Your Lecture Notes." *Ecology* 77, no. 2 (March 1996): 660.

O'Neill R. V., D. L. DeAngelis, J. B. Waide, and T. F. H. Allen. *A Hierarchical Concept of Ecosystems*. Princeton, N.J.: Princeton University Press, 1986.

O'Neill, Robert V., Carolyn T. Hunsaker, K. Bruce Jones, Kurt H. Riitters, James D. Wickham, Parul M. Schwartz, Iris A. Goodman, Barbara L. Jackson, and William S. Biallargeon. "Monitoring Environmental Quality at the Landscape Scale." *BioScience* 47, no. 8 (September 1997): 513–19.

Pahl-Wostl, Claudia. *The Dynamic Nature of Ecosystems: Chaos and Order Intertwined*. Chichester: John Wiley & Sons, 1995.

Patlis, Jason, "Biodiversity, Ecosystems, and Endangered Species." In *Biological Diversity and the Law,* edited by William Snape III. Washington, D.C.: Island Press, 1996.

Patton, Duncan T. "Defining the Greater Yellowstone Ecosystem." In *The Greater Yellowstone Ecosystem,* edited by Robert B. Keiter and Mark S. Boyce. New Haven, Conn.: Yale University Press, 1991.

Payne, Daniel G. *Voices in the Wilderness: American Nature Writing and Environmental Politics*. Hanover, N.H.: University Press of New England, 1996.

Pennisi, Elizabeth. "Conservation's Ecocentrics." *Science News* 144 (September 11, 1993): 168–70.

Peters, Robert H. *A Critique for Ecology*. Cambridge: Cambridge University Press, 1991.

Pickett, Steward T. A., Jurek Kolasa, and Clive G. Jones. *Ecological Understanding: The Nature of Theory and the Theory of Nature*. San Diego: Academic Press, 1994.

Pickett, Steward T., A. V. Thomas Parker, and Peggy L. Fiedler. "The New Paradigm in Ecology: Implications for Conservation Biology above the Species Level." In *Conservation Biology,* edited by Peggy L. Fiedler and Subodh K. Jain. New York: Chapman & Hall, 1992.

Pimentel, David, Ulrich Stachow, David A. Takacs, Hans W. Brubaker, Amy R. Dumas, John J. Meaney, John A. S. O'Neil, Douglas E. Onsi, and David B. Corzilius. "Conserving Biodiversity in Agricultural/Forestry Systems." *BioScience* 42, no. 5 (May 1992): 354–62.

Pimentel, David, Christa Wilson, Christine McCullum, Rachel Huang, Paulette Dwen, Jessica Flack, Quynh Tran, Tamara Saltman, and Barbara Cliff. "Economic and Environmental Benefits of Biodiversity." *BioScience* 47, no. 11 (December 1997): 747–57.

Pimm, Stuart. "An American Tale." *Nature* 379 (July 21, 1994): 188–89.

Pimm, Stuart L., Gareth J. Russell, John L. Gittleman, and Thomas M. Brooks. "The Future of Biodiversity." *Science* 269 (July 21, 1995): 347–50.

Pinchot, Gifford. *The Fight for Conservation.* Seattle: University of Washington Press, 1967.

Plater, Zygmunt J. B. "From the Beginning, a Fundamental Shift of Paradigms: A Theory and Short History of Environmental Law." *Loyola of Los Angeles Law Review* 27, no. 3 (April 1994): 981–1008.

Platt, Rutherford H. *Land Use and Society: Geography, Law, and Public Policy.* Washington, D.C.: Island Press, 1996.

Platt, Rutherford H., and James M. Kendra. "The Sears Island Sage: Law in Search of Geography." *Economic Geography* (extra issue, 1998): 46–61.

Pollot, Mark. "The Unconventional Convention." In *Technical Review of the Convention on Biological Diversity.* Washington, D.C.: National Wilderness Institute and the Alexis de Tocqueville Institution, 1994.

Pope John Paul II. "The Ecological Crisis: A Common Responsibility." A Message of His Holiness Pope John Paul II for the Celebration of the World Day of Peace, 1 January 1990. Vatican City, December 8, 1989.

Pope John Paul II. *The Gospel of Life.* New York: Times Books, 1995.

Portney, Paul R., and Wallace E. Oates. "On Prophecies of Environmental Doom." *Resources,* no. 131 (spring 1998): 17.

Pounder, Bruce M. "Reforming Livestock Grazing on the Public Domain: Ecosystem Management–Based Standards and Guidelines Blaze a New Path for Range Management." *Environmental Law* 27, no. 2 (summer 1997): 513–610.

Power, Mary E., David Tilman, James A. Estes, Bruce A. Menge, William J. Bond, L. Scott Milles, Gretchen Daily, Juan Carlos Castilla, Jane Lubchenco, and Robert T. Paine. "Challenges in the Quest for Keystones." *BioScience* 46, no. 8 (September 1996): 609–20.

Pramaggiore, Anne. "The Supreme Court's Trilogy of Takings: Keystone, Glendale, and Nolan." *DePaul Law Review* 38 (1989): 441–86.

Prance, G. T. "Biodiversity." In *Encyclopedia of Environmental Biology,* vol. 1, editor-in-chief William Nierenberg. San Diego: Academic Press, 1995.

Press, Daniel. *Democratic Dilemmas in the Age of Ecology.* Durham, N.C.: Duke University Press, 1994.

Rabkin, Jeremy. "The Yellowstone Affair: Environmental Protection, International Treaties, and National Sovereignty." Washington, D.C.: Competitive Enterprise Institute, May 1997.

Rapport, David J. "Ecosystem Health: An Emerging Integrative Science" In *Evaluating and Monitoring the Health of Large Scale Ecosystems,* edited by David J. Rapport, Connie L. Gaudet, and Peter Calow. New York: Springer, 1995.

Rapport, David J. "What Is Clinical Ecology?" In *Ecosystem Health: New Goals for Environmental Management,* edited by Robert Costanza, Bryan G. Norton, and Benjamin D. Haskell. Washington, D.C.: Island Press, 1992.

Rapport, David, Robert Costanza, Paul R. Epstein, Connie Gaudet, and Richard Levins, eds. *Ecosystem Health.* London: Blackwell Science, 1998.

Ray, Dixie Lee, and Lou Guzzo. *Environmental Overkill: Whatever Happened to Common Sense?* Washington, D.C.: Gateway, 1993.

Ray, Dixie Lee, with Lou Guzzo. *Trashing the Planet.* Washington, D.C.: Regnery Gateway, 1990.

Reese, Rick. *Greater Yellowstone: The National Park and Adjacent Wildlands.* 2nd ed. Montana Geographic Series, no. 6. Helena, Mont.: American and World Geographic Publishing, 1991.

Regier, Henry A. "The Notion of Natural and Cultural Integrity." In *Ecological Integrity and the Management of Ecosystems,* edited by Stephen Woodley, James Kay, and George Francis. Delray Beach, Fla.: St. Lucie Press, 1993.

Reidel, Carl, and Jean Richardson. "Strategic Environmental Leadership in a Time of Change." Inaugural Donlon Lecture. Syracuse: State University of New York, College of Environmental Science and Forestry, spring 1995.

Reiser, Alison. "Ecological Preservation as a Public Property Right: An Emerging Doctrine in Search of a Theory." *The Harvard Environmental Law Review* 15, no. 2 (1991): 393–433.

Richardson, Curtis J. "Ecological Functions and Human Values in Wetlands: A Framework for Assessing Forestry Impacts." *Wetlands* 14, no. 1 (1993): 1–9.

Ricklefs, Robert. "Structure in Ecology." *Science* 236 (April 10, 1987): 206–7.

Ricklefs, Robert E. *Ecology.* 3rd ed. New York: W. H. Freeman, 1990.

Ridley, Mark. *Evolution and Classification.* London: Longman, 1986.

Risser, Paul. "The Status of the Science Examining Ecotones." *BioScience* 45, no. 5 (May 1995): 318–25.

Robinson, Arthur H. *Early Thematic Mapping in the History of Cartography.* Chicago, University of Chicago Press, 1982.

Robinson, Arthur H., Joel L. Morrison, Philip C. Muehrcke, A. Jon Kimerling, and Stephen C. Guptill. *Elements of Cartography.* 6th ed. New York: John Wiley & Sons, 1995.

Rodgers, William, Jr. *Environmental Law.* St. Paul: West Publishing, 1986.

Rose, Carol M. "Property as the Keystone Right?" In *Land Use and Environmental Law Review 1997,* edited by Stuart L. Deutsch and A. Dan Tarlock. Deerfield, Ill.: Clark, Boardman, Callaghan, 1997.

Rosenzweig, Michael L. *Species Diversity in Space and Time.* Cambridge: Cambridge University Press, 1995.

Roush, Wade. "Putting a Price Tag on Nature's Bounty." *Science* 276 (May 16, 1997): 1029.

Russell, Colin A. *The Earth, Humanity and God.* London: UCL Press, 1994.

Rykeil, Ed. "Ecosystem Science for the Twenty-First Century." *BioScience* 47, no. 10 (November 1997): 705–7.

Sagoff, Mark. "Muddle or Muddle Through? Takings Jurisprudence Meets the Endangered Species Act." *William and Mary Law Review* 38, no. 3 (March 1997): 825–993.

———. "Where Ickes Went Right, or Reason and Rationality in Environmental Law." *Ecology Law Quarterly* 14 (1987): 265–82.

Salwasser, Hal, Douglas W. MacCleery, and Thomas A. Snellgrove. "The Pollyannas vs. the Chicken Littles—Enough Already." *Conservation Biology* 11, no. 1 (February 1997): 283–86.

Samson, Fred B., and Fritz L. Knopf. "Preface." In *Ecosystem Management: Selected Readings,* edited by Fred B. Samson and Fritz L. Knopf. New York: Springer, 1996.
Sanera, Michael, and Jane S. Shaw. *Facts Not Fear*. Washington, D.C.: Regnery, 1996.
Sauer, Carl O. "The Agency of Man on Earth." In *Man's Role in Changing the Face of the Earth,* edited by William L. Thomas Jr. Chicago: University of Chicago Press, 1956.
Savage, Melissa. "Structural Dynamics of a Southwestern Pine Forest under Chronic Human Influence." *Annals of the Association of American Geographers* 81, no. 2 (June 1991): 271–89.
Sax, Joseph L. "Nature and Habitat Conservation and Protection in the United States." *Ecology Law Quarterly* 20 (1993): 47–56.
———. "Property Rights and the Economy of Nature: Understanding Lucas v. South Carolina Coastal Council." *Stanford Law Review* 45 (1993): 1433–55.
———. "The Public Trust Doctrine in Natural Resources Law: Effective Judicial Intervention." *Michigan Law Review* 68 (1970): 471–526.
———. "The Search for Environmental Rights." *Journal of Land Use and Environmental Law* 6 (1990): 93–105.
Scheffer, Victor. *The Shaping of Environmentalism in America*. Seattle: University of Washington Press, 1990.
Schlickeisen, Rodger. "The Argument for a Constitutional Amendment to Protect Living Nature." In *Biodiversity and the Law,* edited by William J. Snape III. Washington, D.C.: Island Press, 1996.
Schmidt, Karen F. "Green Education under Fire." *Science* 274, no. 5294 (December 13, 1996): 1828–30.
Seaber, Paul R., F. Paul Kapinos, and George L. Knapp. *Hydrologic Unit Maps*. Water Supply Paper No. 2294. Washington, D.C.: U.S. Department of the Interior, U.S. Geologic Survey, 1987.
Sedjo, Roger A. "Ecosystem Management: An Uncharted Path for Public Forests." *Resources* 10 (fall 1995): 10, 18–20.
———. "Forest Resources: Resilient and Serviceable." In *America's Renewable Resources: Historical Trends and Current Challenges*, edited by Kenneth D. Frederick and Roger A. Sedjo. Washington, D.C.: Resources for the Future, 1991.
Seligmann, Jean. "What on Earth Is a Wetland?" *Newsweek,* August 26, 1991: 48–49.
Shrader-Frechette, K. S., and E. D. McCoy. *Method in Ecology*. Cambridge: Cambridge University Press, 1993.
Simberloff, Daniel. "Biogeographic Approaches and the New Conservation Biology." In *The Ecological Basis for Conservation,* edited by S. T. A. Pickett, R. S. Ostfeld, M. Shachak, and G. E. Likens. New York: Chapman & Hall, 1997.
Simmons, Randy T. "Fixing the Endangered Species Act." In *Breaking the Environmental Policy Gridlock,* edited by Terry L. Anderson. Stanford, Calif.: Hoover Institution Press, 1997.
Simon, Julian L. *The Ultimate Resource 2*. Princeton, N.J.: Princeton University Press, 1996.
Simon Julian L., and Aaron Wildavsky. "Species Loss Revisited." *Society* 30 (November/December 1992): 41–46.

Sirico, Robert. "Despoiler or Problem-Solver." *National Catholic Register,* October 23, 1994, 5.

Slobodkin, Lawrence B. "Islands of Peril and Pleasure." *Nature* 381, no. 6579 (May 16, 1996): 205–6.

Slocombe, D. Scott. "Environmental Planning, Ecosystem Science, and Ecosystem Approaches for Integrating Environment and Development." *Environmental Management* 17, no. 3 (1993): 289–303.

———. "Implementing Ecosystem-Based Management." *BioScience* 43, no. 9 (October 1993): 612–22.

Smith, Fraser D. M., Robert M. May, Robin Pellew, Timothy H. Johnson, and Kerry S. Walter. "Estimating Extinction Rates." *Nature* 364 (August 5, 1993): 494.

Smith, V. Kerry. "Mispriced Planet." *Regulation* 40 (summer 1997): 16–17.

Snape, William J. III. "International Protection: Beyond Human Boundaries." In *Biological Diversity and the Law,* edited by William Snape III. Washington, D.C.: Island Press, 1996.

Soil Survey Staff. *Soil Taxonomy: A Basic System of Soil Classification for Making and Interpreting Soil Surveys*. Handbook No. 436. Washington, D.C.: U.S. Department of Agriculture, 1975.

Solow, Andrew, Stephen Polasky, and James Broadus. "On the Measurement of Biological Diversity." *Journal of Environmental Economics and Management* 24, no. 1 (1993): 60–68.

Somerville, Richard C. J., and Diane Manuel. "Weather and Climate." In *Encyclopedia of Climate and Weather,* edited by Stephen H. Schneider. New York: Oxford University Press, 1996.

Soulé, Michael E. "The Social Siege of Nature." In *Reinventing Nature? Responses to Postmodern Deconstruction,* edited by Michael E. Soulé and Gary Lease. Washington, D.C.: Island Press, 1995.

Stanley, Thomas R., Jr. "Ecosystem Management and the Arrogance of Humanism." *Conservation Biology* 9, no. 2 (April 1995): 255–62.

Steel, Brent ed. *Public Land Management in the West: Citizens, Interest Groups, Values*. Westport, Conn.: Praeger, 1997.

Stoms, David. "Scale Dependence of Species Richness Maps." *Professional Geographer* 46, no. 3 (August 1994): 346–58.

Stone, Richard. "Wetlands Reform Bill Is All Wet, Say Scientists." *Science* 268 (May 19, 1995): 970.

Stork, Nigel E., and Christopher H. C. Lyal. "Extinction or Co-Extinction Rates?" *Nature* 366 (November 25, 1993): 307.

Strahler, Alan H., and Arthur N. Strahler. *Modern Physical Geography*. 4th ed. New York: John Wiley & Sons, 1992.

Suplee, Curt. "Earth's Biotic Wealth Faces Unprecedented Threat." *Washington Post,* 20 November, 1995, 3 (A).

Tansley, A. G. "The Use and Abuse of Vegetational Concepts and Terms." *Ecology* 16, no. 3 (1935): 284–307.

Tarlock, A. Dan. "The Nonequilibrium Paradigm in Ecology and the Partial Unraveling of Environmental Law." *Loyola of Los Angeles Law Review* 27, no. 3 (April 1994): 1121–44.

Tattersall, Ian. “Systematic Versus Ecological Diversity: The Example of Malagasy Primates.” In *Systematics, Ecology, and the Biodiversity Crisis,* edited by Niles Eldridge. New York: Columbia University Press, 1992.

Taylor, Jonathon, and Nina Burkhardt. “Introduction: The Greater Yellowstone Ecosystem—Biosphere Reserves and Economics.” *Society and Natural Resources* 6 (1993): 105–8.

Taylor, Paul W. *Respect for Nature: A Theory of Environmental Ethics*. Princeton, N.J.: Princeton University Press, 1986.

“Environmental Scares, Plenty of Doom.” *The Economist* 345, no. 8048 (December 20, 1997): 19–21.

“Godliness and Greenness: Thou Shalt Not Covet the Earth.” *The Economist* 341, no. 7997 (December 21, 1996): 108–10.

Thomas, Jack Ward. “Ecosystem Management.” A speech delivered to U.S. Forest Service public affairs personnel, Washington, D.C., April 11, 1993.

———. “Forest Service Perspective on Ecosystem Management.” *Ecological Applications* 6, no. 3 (1996): 703–5.

———. “Foreword.” In *Biodiversity in Managed Landscapes,* edited by Robert C. Szaro and David W. Johnston. New York: Oxford University Press, 1996.

Thomas, William L. “Introductory.” In *Man’s Role in Changing the Face of the Earth,* edited by William L. Thomas Jr. Chicago: University of Chicago Press, 1956.

Thompson, Douglas A., and Thomas G. Yocum. “Uncertain Ground.” *Technology Review* 96, no. 6 (August/September 1993): 22–29.

Thompson, Rebecca W. “Ecosystem Management: Great Idea, But What Is It, Will It Work, and Who Will Pay?” *Natural Resources and the Environment* 9, no. 3 (winter 1995): 42–45, 70–72.

Thuermer, Angus M., Jr. “Mintzmyer Calls ‘Vision’ Paper a Political Fraud.” *Jackson Hole News,* September 25, 1991.

Tolman, Jonathon. “Achieving No Net Loss.” *National Wetlands Newsletter* 17, no. 3 (May-June, 1995): 5–7.

Trewartha, Glenn T., and Lyle H. Horn. *An Introduction to Climatology*. 5th ed. New York: McGraw-Hill, 1980.

Tsonis, Anastasios A. “Climate.” In *Encyclopedia of Climate and Weather,* edited by Stephen H. Schneider. New York: Oxford University Press, 1996.

Turner, Monica G., Robert H. Gardner, and Robert V. O’Neill. “Ecological Dynamics at Broad Scales.” *The Science and Biodiversity Policy Supplement to BioScience* (June 1995): S29–37.

Turner, Monica, Robert O’Neill, Robert Gardner, and Bruce Milne. “Effects of Changing Spatial Scale on the Analysis of Landscape Pattern.” *Landscape Ecology* 3 (1989), 153–62.

United Nations. *Convention on Biological Diversity*. New York: United Nations, 1992.

———. *Report of the United Nations Conference on Environment and Development, Rio Declaration on Environment and Development*. Rio de Janeiro, June 3–14, 1992, A/Conf.151/26 (Vol. I), August 12, 1992.

———. *Report of the United Nations Conference on Environment and Development, Agenda 21*. Rio de Janeiro, June 3–14, 1992, A/Conf.151/26 (Vol. II), August 12, 1992.

U.S. Catholic Conference. "Renewing the Earth." A Pastoral Statement of the United States Catholic Conference. Washington, D.C.: U.S. Catholic Conference, November 14, 1991.

———. *Renewing the Face of the Earth.* Washington, D.C.: U.S. Catholic Conference, 1994.

U.S. Department of Agriculture, Economic Research Service. *Major Uses of Land in the United States, 1992,* Agriculture Economic Report No. 723, by Arthur B. Daugherty. Washington, D.C.: U.S. Department of Agriculture, 1995.

U.S. Department of Agriculture, Forest Service. *A Framework for Ecosystem Management In the Interior Columbia Basin and Portions of the Klamath and Great Basins,* Pacific Northwest Research Station General Technical Report No. PNW-GTR-374, edited by Richard W. Haynes, Russell T. Graham, and Thomas M. Quigley. June 1996.

U.S. Department of Agriculture, Forest Service, and Department of the Interior, Bureau of Land Management. *Eastside Draft Environmental Impact Statement.* Vol. 1. Walla Walla, Wash.: Interior Columbia Basin Ecosystem Management Project, May 1997.

———. *Upper Columbia River Basin Draft Environmental Impact Statement.* Vol. 1. Boise, Idaho: Interior Columbia Basin Ecosystem Management Project, May 1997.

U.S. Department of Agriculture, Soil Conservation Service (now National Resource and Conservation Service) and Iowa State University Statistical Laboratory. *Summary Report 1987 National Resources Inventory.* Statistical Bulletin No. 790. Washington, D.C.: U.S. Government Printing Office, 1990.

U.S. Department of Energy. *National Energy Strategy.* Washington, D.C.: U.S. Government Printing Office, 1991.

U.S. Department of the Interior, Bureau of Land Management. "Ecosystem Management from Concept to Commitment." Washington, D.C.: Bureau of Land Management, n.d.

U.S. Department of the Interior, Fish and Wildlife Service. "Directorate Decision on the U.S. Fish and Wildlife Service Approach to Ecosystem Conservation: An Assessment by Ohio State University." February 1998.

U.S. Department of the Interior, National Park Service. *Management Policies.* Washington, D.C.: National Park Service, 1988.

U. S. Environmental Protection Agency. *EPA Strategic Plan.* EPA/190-R-97–002. September 1997.

———. *National Air Quality and Emissions Trends Report, 1995.* Washington, D.C.: U.S. Government Printing Office, 1997.

U.S. Fish and Wildlife Service. *An Ecosystem Approach to Fish and Wildlife Conservation.* Washington, D.C.: Fish and Wildlife Service, February 1995.

U.S. Fish and Wildlife Service, Division of Endangered Species. "Listed Species and Recovery Plans as of May 31, 1998."

U.S. Forest Service. "National Forest System Land Planning and Resource Management Planning Proposed Rule." *Federal Register* 60 (April 13, 1995): 18886–932.

———. *National Hierarchical Framework of Ecological Units.* Washington, D.C.: ECOMAP, October 29, 1993.

U.S. General Accounting Office. *Ecosystem Management: Additional Actions Needed to Adequately Test a Promising Approach*. GAO/RCED-94–111. Washington, D.C.: U.S. General Accounting Office, August 1994.

———. *Land Ownership: Information on the Acreage, Management, and Use of Federal and Other Lands*. GAO/RCED-96–40. Washington, D.C.: U.S. General Accounting Office, March 1996.

———. *Wetlands Protection: The Scope of the Section 404 Program*. GAO-RCED 93–26. April 1993.

Van Dyne, George M. "Ecosystems, Systems Ecology, and Systems Ecologists." In *Complex Ecology: The Part-Whole Relation in Ecosystems,* edited by Bernard C. Pattern and Sven Jorgensen. Englewood Cliffs, N.J.: Prentice Hall, 1995.

———, ed. *The Ecosystem Concept in Natural Resource Management*. New York: Academic Press, 1969.

Van Matre, Lynn. "Fox River Groups Join to Help Ecosystem Partnership Qualifies for State Program to Safeguard Resources." *Chicago Tribune,* September 25, 1996.

von Hayek, Friedrich. *The Road to Serfdom*. 50th anniversary ed. Chicago: University of Chicago Press, 1994.

Watson, Paul. "On the Precedence of Natural Law." *Journal of Environmental Law and Litigation* 3 (1988): 79–90.

Wayne, Robert K., and John L. Gittleman. "The Problematic Red Wolf." *Scientific American* 273, no. 1 (July 1995): 36–39.

Weis, Julie. "Eliminating the National Forest Management Act's Diversity Requirement as a Substantive Standard." *Environmental Law* 27, no. 2 (summer 1997): 641–62.

Welch, Lee Ann. "Property Rights Conflicts under the Endangered Species Act: Protection of the Red-Cockaded Woodpecker." In *Land Rights,* edited by Bruce Yandle. Lanham, Md.: Rowman & Littlefield, 1995.

Welner, Jon. "Natural Communities Conservation Planning: An Ecosystem Approach to Protecting Endangered Species." *Stanford Law Review* 47 (January 1995): 319–61.

Westra, Laura. *An Environmental Proposal for Ethics: The Principle of Integrity*. Lanham, Md.: Rowman & Littlefield, 1994.

Whelan, Robert, Joseph Kirwan, and Paul Haffner. *The Cross and the Rainforest: A Critique of Radical Green Spirituality*. Grand Rapids, Mich.: William B. Eerdmans, 1996.

White, C. Langdon, Edwin J. Foscue, and Tom L. McKnight. *The Regional Geography of Anglo-American*. 5th ed. Englewood Cliffs, N.J.: Prentice Hall, 1979.

White, Lynn. "The Historical Roots of Our Ecological Crisis." *Science* 155, no 3767 (March 10, 1967): 1203–7.

White House Interagency Ecosystem Management Task Force. *The Ecosystem Approach: Healthy Ecosystems and Sustainable Economies. Vol. I: Overview*. Washington, D.C.: The White House, June 1995.

———. *The Ecosystem Approach: Healthy Ecosystems and Sustainable Economies. Vol. II: Implementation Issues*. Washington, D.C.: The White House, November 1995.

Whittaker, Robert H., ed. *Classification of Plant Communities*. The Hague: Dr. W. Junk, 1978.

Whittlesey, Derwent. "The Regional Concept and the Regional Method." In *American*

Geography Inventory and Prospect, edited by Preston E. James and Clarence F. Jones. Syracuse, N.Y.: Association of American Geographers and Syracuse University Press, 1954.

Wilbanks, Thomas J. "Sustainable Development in Geographical Perspective." *Annals of the Association of American Geographers* 84, no. 4 (December 1994): 541–56.

Wilcox, Louisa. "The Yellowstone Experience." *BioScience* (Special Supplement on Science and Biodiversity Policy, June 1995): S79–83.

Wilkins, Thurman. *John Muir: Apostle of Nature*. Norman: University of Oklahoma Press, 1995.

Williams, James D., and Ronald M. Nowak. "Vanishing Species in Our Own Backyard: Extinct Fish and Wildlife of the United States and Canada." In *The Last Extinction,* edited by Les Kaufman and Kenneth Mallory. Cambridge: MIT Press, 1993.

Williams, Michael. *Americans and Their Forests: An Historical Geography*. Cambridge: Cambridge University Press, 1989.

Wilson, Edward O. "Biodiversity: Challenge, Science, and Opportunity." *American Zoologist* 34, (1994): 5–11.

———. "Toward Renewed Reverence for Life." *Technology Review* 95, no. 2 (November/December 1992): 72–73.

Wilson, E. O. *Diversity of Life*. Cambridge, Mass.: Harvard University Press, 1992.

———, ed. *Biodiversity*. Washington, D.C.: National Academy Press, 1988.

Wood, Christopher A. "Ecosystem Management: Achieving the New Land Ethic." *Renewable Resources Journal* 12, no. 1 (spring 1994): 6–12.

Wood, Denis. "The Power of Maps." *Scientific American* (May 1993): 89–93.

World Commission on Environment and Development. *Our Common Future*. New York: Oxford University Press, 1987.

Worster, Donald. *Nature's Economy—a History of Ecological Ideas*. 2nd ed. Cambridge: Cambridge University Press, 1994.

Wright, Robert R., and Morton Gitelman. *Case and Materials on Land Use*. 4th ed. St. Paul: West Publishing, 1991.

Wuichert, John W. "Toward an Ecosystem Management Policy Grounded in Hierarchy Theory." *Ecosystem Health* 1, no. 3 (September 1995): 161–69.

Yaffee, Stephen et al. *Ecosystem Management in the United States*. Washington, D.C.: Island Press, 1996.

Yandle, Bruce. *Common Sense and Common Law for the Environment*. Lanham, Md.: Rowman & Littlefield, 1997.

———, ed. *Land Rights: The 1990s Property Rights Rebellion*. Lanham, Md.: Rowman & Littlefield, 1995.

Yoon, Carol. "Counting Creatures Great and Small." *Science* 260 (April 30, 1993): 620–22.

Zaslowsky, Dyan, and the Wilderness Society. *These American Lands*. New York: Henry Holt, 1986.

Zeide, Boris. "Another Look at Leopold's Land Ethic." *Journal of Forestry* 96, no. 1 (January 1998): 13–19.

———. "Assessing Biodiversity." *Environmental Monitoring and Assessment* 48 (1997): 249–60.

Zimmerer, Karl S. "Human Geography and the 'New Ecology': The Prospect and Promise of Integration." *Annals of the Association of American Geographers* 84, no. 1 (March 1994): 108–25.

Zonneveld, Isaak S. "The Land Unit—a Fundamental Concept in Landscape Ecology, and Its Applications." *Landscape Ecology* 3, no. 2 (1989): 67–86.

Index

About the Author

Allan K. Fitzsimmons heads Balanced Resource Solutions, an environmental consultancy in Woodbridge, Virginia, and is an adjunct scholar with the Political Economy Research Center in Bozeman, Montana. His work on natural resources, land management, energy, and environmental issues spans twenty-five years. He served as a senior staffer to policymakers in the federal Departments of Interior and Energy during the Reagan and Bush administrations. Prior to that he was an assistant, then associate, professor of geography and chair of the environmental studies program at George Washington University for five years and he also taught at the Universities of Kentucky and Alberta. He has provided expert testimony to Congress on ecosystem management as well as conducted seminars on the subject for business leaders, congressional staff, and land users. He participated in the Keystone Center's national dialogue on ecosystem management. He is widely published, with work appearing in journals such as *Science, Natural Resources Journal, Bioscience, Geographical Review, Regulation, Policy Analysis,* and *Landscape and Urban Planning*. He holds a Ph.D. in geography from the University of California at Los Angeles, and an M.A. in geography and a B.A. in mathematics from California State University—Northridge. An avid rower, he coaches crew for Woodbridge Senior High School.